FOUNDATIONS OF BIOENERGETICS

Foundations of Bioenergetics

HAROLD J. MOROWITZ
Department of Molecular Biophysics and Biochemistry
Yale University
New Haven, Connecticut

ACADEMIC PRESS New York San Francisco London 1978
A Subsidiary of Harcourt Brace Jovanovich, Publishers

ACADEMIC PRESS, INC.
111 Fifth Avenue, New York, New York 10003

United Kingdom Edition published by
ACADEMIC PRESS, INC. (LONDON) LTD.
24/28 Oval Road, London NW1 7DX

Library of Congress Cataloging in Publication Data

Morowitz, Harold J.
Foundations of bioenergetics.

Includes bibliographies.
1. Bioenergetics. I. Title.
QH510.M67 574.1'9121 77-11217
ISBN 0-12-507250-3

PRINTED IN THE UNITED STATES OF AMERICA

This book is dedicated to a group of exceptional scholar-teachers who edified an often very confused and uncertain seeker of knowledge: *Brand Blanshard, Laurence Heinemann, G. Evelyn Hutchinson, Henry Margenau, F. S. C. Northrop, Lars Onsager, Leigh Page, Ernest C. Pollard,* and the ghost of *Josiah Willard Gibbs.*

Contents

Preface

The subject of bioenergetics often seems unnecessarily confusing because it is acquired in bits and pieces from many different courses of study. Basic thermodynamics and energetics come from physics and chemistry programs. Pathways and molecular apparatus of energy conversion are treated in biochemistry courses. The applications of these principles to organelles, cells, and organisms are then spread among botany, physiology, ecology, nutrition, and a dozen other specialties. These applications pay lip service to the fundamentals, often without firmly establishing the connection between the basic principles and the experimental examples. As a result, a subject that is already difficult takes on awesome proportions and some may respond by simply ignoring it.

However, bioenergetics is too important to be ignored. It provides many keys to a broader understanding of biology and may well take on major engineering significance in the societal quest for energy sources and efficient methods of energy utilization. Therefore, it seems prudent to attempt a more unified presentation of the physical foundations of bioenergetics and the methods of applying these constructs to biological problems. In short, we are combining parts of thermal physics, biochemistry, ecology, and cellular and organismic biology into a single coherent work. In such a synthesis something must go. What goes is a certain amount of depth and a vast amount of detail. This is an unavoidable sacrifice in an effort to establish the unity and generality of energetic concepts in biology.

Much of the material in this presentation is from "Entropy for Biologists,"† an introductory thermodynamics book aimed particularly at life scientists. Some of the topics originally appeared in the monograph "Energy

† Academic Press, New York, 1970.

Flow in Biology."[†] The current volume expands on that material with respect to biological applications and attempts to bridge the gulf between physics and biology. Also included are an introduction to irreversible thermodynamics and some of the newer methods of viewing nonequilibrium systems.

The motivation for this work is the desire to make the backgrounds of bioenergetics as clear and unambiguous as possible. The subject is inevitably fuzzy at the growing edges, but an effort to clarify the fundamentals should eventually lead to a better awareness of the unsolved problems that are presented to us by our study of energetic processes in living systems.

What was initially envisioned as a book embracing all aspects of bioenergetics began to grow in scope until it became clear that the more macroscopic parts could best be served by a volume of their own. This book fulfills that function with the hope that we may soon codify the material on molecular mechanisms of energy transduction so as to form a companion volume.

To make the material from divergent disciplines as coherent as possible, we have recast all of the measures into the international system of units, SI (Système International d'Unités), based on the meter, kilogram, second, ampere, kelvin, and candela. Traces of earlier units may persist to accord with conventional usages. However, the gain in consistency and clarity that results from using SI units is so great that every effort should be made to orient our formulas as well as our thinking in these terms. Appendix I reviews the system and presents the major physical constants in these units.

This book is intended to serve readers with a wide variety of backgrounds. Someone with a thorough knowledge of classical thermodynamics could begin with Chapter 6. For those persons who have studied statistical mechanics Chapter 7 will review familiar material. All chapters are, however, written with biology in mind so that the approach is somewhat nontraditional for physicists and chemists. Sections 2.C, 3.E, 4.G, 5.H, 13.D, 13.E, as well as Chapters 9, 12, 14, and 18 contain material not likely to be found elsewhere except in journal articles. Much of the background work for these sections has been supported by the National Aeronautics and Space Administration and the National Science Foundation, and I would like to take this opportunity to acknowledge that support.

It is the author's hope that the form of presentation in this book will help biologists, biochemists, chemists, and physicists to speak the same language in an effort to deal with the questions of bioenergetics and the applications of that knowledge to the solutions of human problems.

[†] Academic Press, New York, 1968.

The manuscript was read and commented on by three very dear friends: Alan Bearden, Robert Macnab, and Alexander Mauro. I thank them for the most precious gift of all, the gift of time. To another friend, John Wyatt, I want to express my graditude both for Appendix V and for numerous discussions that have left their mark throughout the text. The manuscript was transformed from a highly entropic pile of scraps of paper to its final ordered form by my wife, Lucille. Words are but poor instruments to acknowledge my feelings, but once again, thank you love.

List of Symbols

(*Italic*)

A	Helmholtz free energy
B	number of molecules in container
C	concentration
C_x	specific heat
D	diffusion coefficient
E	total energy of the system
F	force
G	Gibbs free energy
H	enthalpy
I	information content
J	flux
K	equilibrium constant
L	dimensions of a box
L_{ij}	Onsager coefficients
M	mass per mole or M.W.
N	number of elements
P	pressure
Q	heat
R	gas constant
S	entropy
T	temperature
U	total energy of the system or internal energy
V	volume
W	weight
W	work
X	generalized fluctuating force
Z	partition function
a	slope
b	intercept of line
c	velocity of light
d	differential
e	exponential base
f	function
f_i	probability
g	gravitational acceleration
g	function
h	Planck's constant
h	function
h_c	heat of compression
i	square root of minus one
k	Boltzmann constant
k_x	rate constant
m	mass
n	mole number
p	momentum
q	generalized coordinate
r	radius
s	entropy per unit mass
s_V	entropy per unit volume
t	time
t_{ij}	transition probability
u	energy density
v	velocity
x	spatial coordinate
y	spatial coordinate
z	spatial coordinate

(Greek)

Γ surface excess
Δ difference
Θ function
Λ interaction energy coefficient
Ξ vibration period
Π osmotic pressure
$\sum$ summation sign
Υ generalized vector
Φ dissipation
Ψ function—describes the behavior of the system
Ω function

α constant of proportionality
α_i Onsager variable
β undetermined multiplier
γ absorption coefficient
γ ratio of C_P to C_v
δ angle
δ width of well
δ_{ij} Kronecker delta function
ε molecular energy
ζ scale factor
ζ momentum coefficient
η efficiency of heat engine
η coefficient of viscosity
θ empirical temperature
θ_c critical point temperature
κ spring constant
λ numerical factor
λ wavelength
μ chemical potential
ν stoichiometric coefficient
ν vibrational & photon frequency
ξ degree of advancement of a chemical reaction
ρ density
σ surface energy or surface tension
σ rate of production
τ time
ϕ kinetic theory temperature
ψ psi function—quantum mechanics
ω angular velocity

(Fraktur–Gothic)

$\mathfrak{A}$ affinity
$\mathfrak{B}$ constant of integration
$\mathfrak{C}$ Stefan–Boltzmann coefficient
$\mathfrak{D}$ Stefan–Boltzmann coefficient
$\mathfrak{E}$ emissive power
$\mathfrak{G}$ collision number
$\mathfrak{H}$ surface absorbtivity

$\mathfrak{a}$ polarizability
$\mathfrak{b}_{ij}$ $= \partial^2 \Delta S/\partial\alpha_i\, \partial\alpha_j$
$\mathfrak{e}$ charge
$\mathfrak{f}$ frictional coefficient
$\mathfrak{g}$ conductance
$\mathfrak{l}$ elongation
$\mathfrak{n}$ quantum number

$\mathfrak{K}$	constant of integration
$\mathfrak{L}$	size of a step
$\mathfrak{M}$	macroscopic property
$\mathfrak{N}$	mole fraction
$\mathfrak{N}_i$	mass fraction
$\mathfrak{O}$	property of fluid
$\mathfrak{R}$	magnetic field
$\mathfrak{T}$	transfer coefficient
$\mathfrak{U}$	autocorrelation function
$\mathfrak{V}$	average volume per molecule
$\mathfrak{V}_i$	average volume per molecule of the ith species
$\mathfrak{W}$	property being transported
$\mathfrak{p}$	Peltier coefficient
$\mathfrak{q}$	coefficient of heat conduction
$\mathfrak{r}$	rotational quantum number
$\mathfrak{s}$	Seebeck coefficient
$\mathfrak{t}$	Tompson coefficient
$\mathfrak{v}$	specific volume
$\mathfrak{v}$	vibrational quantum number
$\mathfrak{z}$	molecular partition function

(*Script*)

$\mathscr{A}$	area
$\mathscr{C}$	capacitance
$\mathscr{E}$	electromotive force
$\mathscr{F}$	Faraday
$\mathscr{G}$	gravitational potential
$\mathscr{G}$	function
$\mathscr{H}$	Hamiltonian
$\mathscr{I}$	moment of inertia
$\mathscr{J}$	integrating factor
$\mathscr{J}$	function
$\mathscr{K}$	function
$\mathscr{L}$	function
$\mathscr{M}$	degree of magnetization
$\mathscr{M}$	function
$\mathscr{N}$	Avogadro's number
$\mathscr{O}$	extensive variable
$\mathscr{P}$	probability
$\mathscr{R}$	electrical resistance
$\mathscr{S}$	solubility
$\mathscr{T}$	tension
$\mathscr{U}$	steps per second
$\mathscr{V}$	potential
$\mathscr{W}$	Boltzmann probability
$\mathscr{W}$	number of ways
$\mathscr{X}$	number of living states
$\mathscr{Z}$	the square of the position coordinate

1 Energy

A. INTRODUCTION

The modern concept of "energy" developed in the 19th century as a result of many convergent lines of study. Both Julius Mayer and Hermann von Helmholtz, two of the earliest formulators of the first law of thermodynamics, based their heuristic ideas of energy on its role in physiological systems. That early association between biology and thermodynamics has now matured into a full appreciation of the profoundly energetic foundations of life processes. Ecologists, cellular biologists, and physiologists alike utilize notions of energy flow and storage in formulating the first principles of their subjects. For biologists and biochemists, the understanding of the energy construct has become a *sine qua non* for the comprehension and unification of their disciplines.

We now know that, given the nature of our understanding of the physical world, the profoundly energetic character of biological theory follows by necessity. The paradigm of physics, as applied at the level of biology and chemistry, presents us with space, matter, energy, and time as the subjects of discourse. Any reductionist discipline must return to these constructs for its underlying ideas. It is not our purpose to pause here to move back into the foundations of physics, but it is clear that the science which lies at the roots of biophysics is profoundly shaped by the concept of energy. All attempts to understand the living world in terms of the constructs of physical science will inevitably involve us in a study of bioenergetics.

One of the central themes of biochemistry is that energy processing occurs through the release of chemical potential stored in molecules of adenosine triphosphate (ATP) or related nucleotide triphosphates. The formation of these compounds from their precursors and the utilization of the energy for macromolecular synthesis, transport, motility, and other cell functions are core parts of the operation of prokaryotic cells. When we move on to consider the eukaryotes, the dominant energy-producing organelles are the mitochondria, transducers of oxidation energy into the chemical potential of the ubiquitous ATP. We also encounter cilia and microtubules associated with protoplasmic streaming and chromosomal motion. A special case of interest is muscle tissue, the primary site of conversion from chemical to mechanical energy in the animal world. At the physiological level, energy balance and temperature regulation are important features in stabilizing organisms. Examining populations and biomes we note the profound impact on ecology of Lindeman's trophic dynamic view (see Bibliography) and subsequent developments along those lines. Finally, underlying all of biological activity is photosynthesis, the transduction of electromagnetic into chemical bond energy. This conversion is the primary process in the operation of the biosphere.

The subject matter of bioenergetics has been fraught with ambiguities and it is well to make them explicit, at the outset, to alert the reader to areas where one's own confusion is shared by the scientific community. Firstly, thermodynamics is a conceptually deep subject which requires a good deal of thought and study. There seems to be no shortcut to the effort necessary to understand this discipline. Secondly, thermodynamics has been formulated rigorously only for equilibrium systems. For near-to-equilibrium cases a formalism is emerging which has had considerable biological application. Far from equilibrium, where much of biological interest takes place, there is at present very little in the way of thermodynamic guidance. One must deal with kinetic models that are incomplete and mathematically very tedious. Thus the physics necessary for a total understanding of bioenergetics does not exist as a finished structure. Indeed, biological questions are, at the moment, one of the incentives for the development of this branch of science. Thirdly, many processes of energy transformation have not been satisfactorily elaborated. For example, there are at present several competing theories to explain oxidative phosphorylation in mitochondria. Consequently, much of our discussion of bioenergetics is in the nature of a progress report rather than a set of paradigms of wide acceptance.

Another area of confusion arises from the frequent failure to make explicit the epistemological nature of statements that are introduced in the foundations of bioenergetics. Therefore, in initiating our inquiry into thermal physics, it is instructive to give some thought to the logical character of the statements we shall employ. The primary assertions are generally of four categories: definitions, empirical generalizations, postulates, and conclusions that can be logically inferred from the other three categories.

Definitions are arbitrary but serve to structure the subject and place emphasis on certain concepts. When we say that kinetic energy is equal to $\frac{1}{2}mv^2$ we are defining "kinetic energy" and at the same time indicating that the quantity is of sufficient importance to deserve a name. Thus, although we could take the quantity $3m^2v$ and assign it a name "momentamass," such a definition would serve no purpose, as the quantity is without significance in the development of dynamics.

Empirical generalizations occur in all branches of science but appear to be of special significance in thermodynamics. They are inductive statements which should be based on a large number of observations and the absence of contradictory observations. An example is the generalization that no engine acting in a cycle can transform heat into work without causing other changes. The particular power of thermodynamic reasoning is that such inductions can form the basis of an elaborate and impressive intellectual superstructure.

Postulates are assertions whose validity rests entirely on comparison between experiments and the predictions based on the postulates. Thus when we assume that gases are made of very small molecules which interact on collision like hard elastic spheres, we are making statements that ultimately can be checked by deducing what the macroscopic properties of such a system would be. The validity of assumptions rests on the agreement between the observed behavior of gases and predicted behavior of a collection of small spheres.

Conclusions depend on the acceptance of the rules of logic and mathematics and the application of these rules to statements of the other three types.

In fully understanding the subtleties of thermodynamic reasoning, it is important to preserve the above distinction in logical character among various types of primary statements. In the following pages this will be done by printing them in italics and following them by either (definition), (empirical generalization), (postulate), or (conclusion).

B. CONSERVATION OF MECHANICAL ENERGY

From a historical point of view the science of mechanics preceded thermodynamics by over one hundred years. This order has imposed a certain character on the usual development of the subject so that we shall first introduce "energy" from the point of view of classical mechanics, the study of motion of material bodies under forces. We assume a familiarity with the concepts of position, velocity, acceleration, mass, and force. We then develop our introductory concepts in terms of the motion of a single particle. From this introduction it is a relatively straightforward matter to generalize.

For simplicity in the mathematical development we will consider a particle that is free to move in one dimension only, along the x axis. The argument can easily be extended to three dimensions and, for those who desire a more rigorous treatment, we develop the full argument in vector notation in Appendix II.

We then introduce the concept of work, which is given as follows: *The work done in moving a mass from one point to another is defined as the distance moved times the force directed along the line of motion* (definition). Since the force may change along the path, we must use differentials and then sum up to get the total work. In one dimension we can state the previous definition in mathematical notation as

$$dW = F\,dx \tag{1-1}$$

where F represents the magnitude of the force directed along the axis. The amount of work done in moving an object from point x_1 to point x_2 is then given by

$$W_{1,2} = \int_{x_1}^{x_2} F\,dx \tag{1-2}$$

For purely mechanical systems with no friction $W_{1,2}$ depends just on the end points x_1 and x_2 and is independent of the detailed way of getting from x_1 to x_2 (empirical generalization). Such systems are called conservative and they constitute the subject material of pure mechanics.

The idea of a frictionless mechanical system may at first sight seem like too much of an abstraction to be useful; nevertheless in celestial mechanics we actually encounter systems where to a very high order of approximation friction can be ignored. The theory of planetary orbits has been a very strong confirmation of the basic mechanics of conservative systems.

The empirical generalization just stated imposes strong mathematical limitations on the function F and in fact requires the first equality in the following equation:

$$F = \frac{dW}{dx} = -\frac{d\mathscr{V}}{dx} \tag{1-3}$$

For mathematical convenience we often do not use the function W, but use the negative of W, which we designate as $\mathscr{V}$. *$\mathscr{V}$ is called the potential, and forces which conform to eq.* (1-3) *are said to be derivable from a potential* (definition). *All of the forces of mechanics and electrostatics are derivable from a potential* (empirical generalization).

We can now extend our analysis by introducing Newton's second law of motion, which states that the force on a body is equal to the mass of the body times its acceleration, or

$$F = m\frac{d^2x}{dt^2} \tag{1-4}$$

It has been pointed out that Newton's second law is a definition of force as that which accelerates a material body. The law assumes definite content only when we are able to specify the force or potential as a function of the coordinate. For readers interested in these problems of the foundations of mechanics, we suggest referring to Chapter 3 of "Foundations of Physics" by R. B. Lindsay and H. Margenau (see Bibliography).

We next proceed to calculate the change of potential in going from x_1 to x_2. To carry this out, we start with eq. (1-1) and remember that $dW = -d\mathscr{V}$; we then substitute Newton's law [eq. (1-4)] for the force and we have

$$-d\mathscr{V} = m\frac{d^2x}{dt^2}\,dx \tag{1-5}$$

The change of potential may be obtained by integrating eq. (1-5) from the point on its path x_1 to the point x_2. If we remember that the acceleration is the derivative of the velocity, we get

$$\int_{x_1}^{x_2} -d\mathscr{V} = m\int_{x_1}^{x_2}\frac{dv}{dt}\,dx = m\int_{x_1}^{x_2} dv\,\frac{dx}{dt} = m\int_{x_1}^{x_2} v\,dv \tag{1-6}$$

The last two equalities in eq. (1-6) represent formal rearrangement of the symbols to get the expression in more convenient form. When we integrate eq. (1-6) we get

$$\mathscr{V}(x_1) - \mathscr{V}(x_2) = \tfrac{1}{2}mv^2(x_2) - \tfrac{1}{2}mv^2(x_1) \tag{1-7}$$

This may be rewritten in a form that makes its meaning clearer,

$$\mathscr{V}(x_1) + \tfrac{1}{2}mv^2(x_1) = \mathscr{V}(x_2) + \tfrac{1}{2}mv^2(x_2) \tag{1-8}$$

The quantity $\mathscr{V} + \frac{1}{2}mv^2$ thus has the same value at the point x_2 as at the point x_1. Since these are arbitrary points along the path, the quantity $\mathscr{V} + \frac{1}{2}mv^2$ is thus an invariant or constant of motion. *The term $\frac{1}{2}mv^2$ is known as the kinetic energy* (definition) and $\mathscr{V}$ has already been designated as the potential or potential energy. *The quantity $\mathscr{V} + \frac{1}{2}mv^2$ is designated the total energy of a particle* (definition). We have thus shown that the total energy of a particle is invariant and consists of the sum of two terms, the potential and the kinetic energy. The potential energy is a function of the coordinates only. It is considered to be the stored energy that a particle possesses by virtue of the work which had to be done to bring it to its present position. The kinetic energy is a function of the velocity only. It is thus associated with motion, the amount of energy possessed by a particle by virtue of its moving at velocity v. Equation (1-8) states that, as the particle moves along its trajectory, potential may be converted to kinetic energy and vice versa, but the sum of the two quantities remains constant. Equation (1-8) is the law of conservation of mechanical energy and is the cornerstone for many developments in physics.

The dimensions of kinetic energy are mass times velocity squared or mass times length squared divided by time squared. The dimensions of work or potential energy are force times distance. Force is given dimensionally as mass times acceleration or mass times distance divided by time squared. Potential energy thus has the dimensions of mass times distance squared divided by time squared and is dimensionally the same as kinetic energy, as we would expect from eq. (1-8). In SI units mass is measured in kilograms and force in newtons, which is the force that will accelerate a one-kilogram mass by one meter per second squared. Energy is measured in joules, the work done by a one-newton force acting through one meter.

We may sum up the foregoing as follows. Energy is a measure of a system's capacity for doing work. For conservative mechanical systems, this is the sum of two components: the kinetic energy, which the system possesses by virtue of its motion, and the potential energy, which results from its configuration. Work in the mechanical sense is the displacement of any body against an opposing force, or the product of the force times the distance displaced.

The previous development of the law of conservation of energy is based on a macroscopic approach. There is no reason to suggest that it would not hold at a microscopic (molecular) level, but a difficulty enters when we try to account for energy at this level. An ordinary piece of matter contains an enormous number of atoms, so that any detailed accounting of the dynamical variables becomes clearly impossible. We also know that in any real situation there are frictional processes in which macroscopic mechanical energy is converted into energy in microscopic modes. Therefore, the methods of mechanics are inadequate to a wide range of physical problems in which we need to be concerned with the energy distribution among a large number of molecules. Three closely related methods are used by physicists and chemists to deal with these situations where detailed mechanical treatments are not possible. These are thermodynamics, statistical mechanics, and kinetic theory. Thermodynamics uses only macroscopic parameters and derives certain equations that must obtain between experimentally determined quantities and which are called phenomenological relations. Statistical mechanics considers average properties of systems that we can deal with in spite of our ignorance of the details of individual molecules. To date, this type of averaging has been worked out almost exclusively for equilibrium systems, so that statistical mechanics has not been very useful in the study of transport properties such as diffusion and heat flow. Kinetic theory deals with mechanical model systems at a molecular level and the appropriate averaging over the molecules to predict measurable properties. While it can thus be used for transport properties, it suffers from the difficulty that for all real cases the mathematics becomes exceedingly complex and often impossible.

Each of the three branches of thermal physics thus has advantages and disadvantages and they can best be studied together since the various approaches interact and supplement each other. Our understanding of various concepts is therefore improved by examining their development in the three disciplines. In solving problems we are often called upon to decide the best method of approach and it is well to realize that we may have a choice so that we can opt for the best available technique.

C. THERMAL ENERGY

With the previous considerations of mechanical energy in mind we can generalize our point of view and turn to the concept of thermal energy. *It is frequently observed that for real systems the conservation of mechanical energy [eq. (1-8)] fails to hold. Such situations are almost invariably associated with a rise in temperature somewhere in the system* (empirical generalization). For example, if a car is traveling along a road at 50 km/h, it has a certain kinetic energy ($\frac{1}{2}mv^2$). If on a level surface we apply the brakes and the car comes to a stop, the kinetic energy goes to zero and no potential energy appears. The brakes heat up in this process. Equation (1-7) makes no mention of temperature, so that mechanics by itself provides no method of dealing with the preceding situation. Temperature does not enter into the formulation of mechanics, so new methods must be devised for systems involving friction or dissipative forces.

The loss of mechanical energy, with the subsequent rise in temperature, was a most important observation in the historical development of thermodynamics. In 1798 Benjamin Thompson, Count Rumford, carried out his classical experiments on the boring of cannon. He showed in a rough way that the heat produced was proportional to the total work done. He wrote[†]

> It is hardly necessary to add, that any thing which any insulated body, or system of bodies, can continue to furnish without limitation, cannot possibly be a material substance: and it appears to me to be extremely difficult, if not quite impossible, to form any distinct idea of any thing capable of being excited and communicated, in the manner the heat was excited and communicated in these experiments, except it be Motion.

This is usually cited as an example of the importance of this type of process to the formulation of thermodynamics. We are now aware, of course, that the mechanical energy "lost" in dissipative processes appears as kinetic energy of molecules and that the temperature and average kinetic energy of atoms are very closely related. This result goes beyond thermodynamics and results in a more detailed view of the nature of matter.

TRANSITION

The science of bioenergetics is based upon the concept of energy which emerged from the study of the laws of motion in mechanics. One of the integrals of the equation of motion is energy consisting of two terms, the velocity-dependent kinetic energy and the potential which is often a function of the coordinates only. Energy is a constant of motion in mechanics. Under certain conditions, the conservation law fails and parts of the systems warm up. To proceed with our understanding of such circumstances, we move on to the concept of temperature that is introduced in the next chapter.

BIBLIOGRAPHY

Bridgman, P. W., "The Nature of Thermodynamics," Harper and Row, New York, 1961.

A witty and profound essay on some conceptual difficulties in thermodynamics.

Callen, H. B., "Thermodynamics." Wiley, New York, 1965.

A presentation of thermodynamics from a postulational approach. Provides understanding of the basic structure of thermodynamic theory.

Katchalsky, A., and Curran, P. F., "Nonequilibrium Thermodynamics in Biophysics." Harvard Univ. Press, Cambridge, 1965.

A presentation of near-to-equilibrium thermodynamics including a number of biological applications.

[†] B. Thompson, "Philosophical Transactions of the Royal Society," London, 1798.

Lindeman, R. L., *Ecology* **23**, 399 (1942).

The original paper setting forth the analysis of ecology in terms of trophic levels and energy transfers.

Lindsay, R. B., and Margenau, H., "Foundations of Physics." Dover, New York, 1963.

A detailed discussion of the concepts involved in the formulation of physics. Particularly insightful on probability and the statistical point of view.

Margenau, H., "The Nature of Physical Reality." Ox Bow Press, Woodbridge, Connecticut, 1977.

An analysis of the philosophical foundations of physics. This work is especially useful in answering epistemological questions that arise in formulating thermodynamics.

2 Temperature

A. EMPIRICAL THERMOMETERS

Before we can proceed with a further study of energy, we must look more thoroughly into the concept of temperature, a notion that does not occur in mechanics per se, but is the cornerstone of thermodynamic analysis. Our primitive ideas of temperature are rooted in the physiological sensations of hot and cold, which seems most appropriate to our biologically oriented presentation of thermodynamics.

Indeed hotness and coldness allow us to establish a crude qualitative temperature scale since, *given two objects A and B within the range of hotness such that we can touch them, we can always say that A is hotter than B, A is colder than B, or A and B are indistinguishable with respect to hotness* (empirical generalization).

The preceding can be extended by noting that *it has been observed that a number of mechanical and electrical properties vary in a regular way with the degree of hotness as determined by physiological temperature* (empirical generalization). Among these properties are the electrical resistance of wires, the volume of liquids, the pressure of gases at constant volume, the chirping rate of crickets, and the rates of chemical reactions. Those properties which vary in a monotonic way with the degree of hotness (always increase with increasing temperature or always decrease with increasing temperature) can be used to construct an empirical thermometer; that is, we can use the measure of the property as the measure of the temperature. The most familiar example is a mercury-in-glass thermometer, where the length of the mercury column is a monotonic increasing function of physiological temperature, such that we can use the measure of length as some kind of numerical measure of the temperature. Such a device can then be placed in contact with an object and the empirical temperature determined. This use of empirical thermometers requires one further statement. *When two objects are kept in physical contact for a sufficient period of time they come to the same temperature; that is, they become indistinguishable with respect to degree of hotness* (empirical generalization). The existence of an empirical thermometer allows us to measure temperature, although the measure is on some completely arbitrary scale.

Using the instruments just described allows the discovery of temperature fixed points. We observe that *pure ice/water mixtures at one atmosphere pressure always give the same reading on a given empirical thermometer* (empirical generalization). Similarly, we observe that *all pure boiling-water samples at one atmosphere pressure give the same reading on a given empirical thermometer* (empirical generalization). We may then use boiling water and melting ice as temperature fixed points and adjust all empirical thermometers to agree at these fixed points. If we call the ice temperature θ_f and the boiling-water temperature $\theta_f + 100$, we can divide the intermediate measure into 100 equal divisions. Setting $\theta_f = 0$ yields the conventional centigrade scale. Note that different temperature measuring instruments which are set to give the same readings at the boiling and freezing points will not necessarily be in agreement at any intermediate temperatures. Indeed each empirical thermometer establishes its own temperature scale.

Among the great variety of possible empirical thermometers, one class has been of particular utility in developing more theoretically founded notions of temperature. This class can be designated "gas-law thermometers." *It is found for a number of gases under a wide variety of pressures that the pressure of the gas at constant volume and the volume at constant pressure both increase in a regular way with physiological temperature* (empirical generalization). Either pressure or volume is then a suitable parameter for the design of an empirical temperature scale. We may then construct a thermometer which consists of a constant-volume vessel filled with gas attached to a pressure gage.

B. *GAS-LAW THERMOMETERS*

Given the existence of empirical thermometers, we can now carry out experiments on temperature-dependent properties of systems. One such set of experiments shows that *there is a group of gases (helium, neon, argon, etc.) for which the pressure at constant empirical temperature varies as a linear function of the concentration of gas in the vessel* (empirical generalization). *Such gases are called ideal* (definition).

We can now begin to analyze a gas-law thermometer in the following way. The temperature measure θ of a constant-volume thermometer is by definition proportional to the pressure P, so that we may write

$$\theta = \alpha P \tag{2-1}$$

where α is some constant of proportionality. If the gas-law thermometer is filled with one of the ideal gases, we then know that the constant of proportionality is inversely proportional to the concentration of gas in the thermometer. This may be formally written

$$\alpha = \left(\frac{1}{C}\right)\left(\frac{1}{R}\right) \tag{2-2}$$

where $1/R$ is a new constant of proportionality and the concentration C is the number of moles of gas n divided by the volume of the thermometer V:

$$C = \frac{n}{V} \tag{2-3}$$

If we substitute eqs. (2-2) and (2-3) into eq. (2-1) we get

$$PV = nR\theta \tag{2-4}$$

This looks like the familiar form of the ideal gas law of Boyle and Charles. It is, however, logically something quite different; it is a definition of the empirical temperature θ for a constant-volume, ideal-gas thermometer. Indeed we could have arrived at eq. (2-4) more abruptly by asserting that the quantity PV is a monotonic increasing function of the physiological temperature; hence, an empirical temperature may be established as a constant of proportionality times this quantity. The constant R may now be determined by use of the temperature fixed points. If we build an empirical thermometer of known volume V and amount of gas n, then we may measure the pressure at the boiling and freezing points of water, P_b and P_f. Th equations at these two temperatures may be written

$$P_b V = nR\theta_b \tag{2-5}$$

$$P_f V = nR\theta_f \tag{2-6}$$

If we subtract eq. (2-6) from (2-5) we get

$$\frac{(P_b - P_f)V}{(\theta_b - \theta_f)n} = R \tag{2-7}$$

All quantities on the left-hand side of the equation can be measured except $\theta_b - \theta_f$. However, if we define our temperature scale by calling this difference 100, then R may be directly computed. Once R has been established this empirical thermometer provides a complete temperature scale.

The constant R as discussed above applies only to the specific thermometer under consideration. To show that it is a universal constant, we need to invoke Avogadro's law, which states that *equal volumes of different (ideal) gases at the same temperature and pressure contain equal numbers of molecules* (postulate based on certain empirical generalizations). Applying this result to eq. (2-4) for several ideal gases shows that R must be a universal constant. It is called the gas constant and has the dimensions of energy divided by temperature. Its value has been determined as 8.31434 J $mole^{-1}$ K^{-1} (joules per mole per degree kelvin, units which will later be developed).

Starting with a good working empirical thermometer, we can establish a further property of water. *There is only one temperature and pressure at which pure water can coexist in the solid, liquid, and vapor phases* (empirical generalization). This is known as the triple point of water and is a convenient fixed point for very precise calibration in thermometry. If we call the temperature at the critical point θ_c, then we can write eq. (2-5) for ideal gases as

$$R = \frac{P_c V}{\theta_c n} \tag{2-8}$$

Thus a single fixed point suffices to determine R and establish an unambiguous gas-law temperature scale. This is done by immersing an ideal gas thermometer into a water triple point cell and allowing it to come to equilibrium. The internationally agreed upon value of the triple point of water is 273.16. The choice of this value is in principle completely arbitrary and establishes the size of a degree in the final temperature scale. In fact the choice was made to preserve agreement with the centigrade scale. When we refer to empirical temperature, unless otherwise stated we will mean the temperature of an ideal-gas thermometer calibrated at the θ_c value given above. The *kelvin*, a unit of thermodynamic temperature that we will later elaborate upon, is the fraction 1/273.16 of the thermodynamic temperature of the triple point of water.

C. *TEMPERATURE IN BIOLOGY*

Before investigating a more sophisticated measuring scale, we can briefly review some of the phenomenological aspects associated with the temperature dependence of biological phenomena. We consider an empirical temperature scale with the freezing point of pure water at zero degrees, and the boiling point of pure water at one hundred degrees. For practical purposes we employ a working thermometer such as mercury in glass to establish the intermediate points.

It is first noted that growth temperatures of almost all organisms are between the freezing and the boiling points of water (empirical generalization). The few exceptions are marine organisms that survive in sea water slightly below zero degrees on the centigrade scale. *Living forms may roughly be divided into two groups: homeotherms and poikilotherms* (empirical generalization). Homeotherms maintain an almost constant body temperature independent of the environmental temperature. This group, which includes the mammals and birds, thus attains a relative freedom from one of the thermodynamically most pervasive external parameters and is able to function year round. Homeothermic animals are referred to as "warm blooded" and universally maintain internal temperatures in the rather narrow range of 35–45 degrees centigrade. The mammals show temperature homeostasis in a particularly close range as is indicated in table 2-1.

Birds span a somewhat greater range, the widest variation being reported for the swifts and hummingbirds (apodiformes) with reported values between 35.6 and 44.6 degrees centigrade.

Poikilotherms show much less temperature regulation, and exhibit changes in internal temperature which tend to follow environmental temperatures. The group includes all plants and microorganisms and those animals designated "cold blooded," which is something of a misnomer since their blood can indeed get quite warm. The widest range of growth temperatures is reported among the bacteria. Some thermophiles are reported to grow in

TABLE 2-1
Rectal Temperatures[a] of Various Mammals

Species	Mean temperature	Normal range
Shrew	35.7	———
Rat	38.1	37.5–38.6
Rabbit	39.4	38.6–40.1
Cat	38.6	38.1–39.2
Dog	38.6	38.1–39.2
Sheep	39.1	38.3–39.9
Horse	37.7	37.2–38.2
Cattle	38.6	38.0–39.2
Camel	36.4	34.9–37.8
Chimpanzee	37.0	36.3–37.8
Man	37.0	36.2–37.8
Elephant	36.2	35.7–36.7
Seal	38.3	———
Whale	36.5	36.0–37.0

[a] Temperatures are reported here in the conventional centrigrade scale. In subsequent sections we will use the Kelvin absolute temperature scale which will be developed in the following chapters.

the 80–90 centigrade range whereas some cryophiles function at temperatures as low as minus four degrees centigrade.

The poikilotherms are not entirely passive about temperature regulation. Reptiles may alternatively seek sun or shade in order to control internal temperature to within a few degrees. Even plants may control transpiration or amount of leaf area available to the sun as devices for regulating temperature.

In addition to the growth range of organisms, temperature is a significant parameter in a number of other ways. All biochemical and physiological processes have reaction rates which depend sharply on empirical temperature. In ecological systems, ambient temperature constitutes a prime factor in the distribution of species and the overall character of the habitats.

D. KINETIC THEORY

Before we consider temperature in biology from a deeper perspective, we must further relate the concept to the rest of thermal physics. This can best be done from the point of view of the kinetic theory of gases which roots the temperature concept in the framework of mechanics. We shall outline kinetic theory in an elementary way since our purpose is not to develop the details of gas behavior but to deepen our understanding of the physical basis that underlies our sensations of hot and cold.

We proceed by making a series of postulates about the nature of matter, deducing the consequences of these postulates, and checking these results against experiment. In this context we introduce the following postulates:

P(1) *Ideal monatomic gases are made up of very small spherical molecules.*

P(2) *At normal pressures the molecules occupy only a small fraction of the total volume of the system.*

P(3) *The molecules move rapidly and randomly in all directions. There is a state of the system (the equilibrium state) in which there is no net motion of the entire system, so that if we look at a small volume of gas, as many molecules have a velocity* $\mathbf{v}$ *as have a velocity* $-\mathbf{v}$.

P(4) *Molecules interact with each other and with the walls only by elastic collisions involving short-range repulsive forces.*

P(5) *The laws of Newtonian mechanics hold for the molecular system.*

The postulates just presented introduce a number of new ideas. The first of these is the notion of equilibrium, a concept that has a number of subtleties that require discussion. *First consider an infinite isothermal reservoir, which is an object very large in comparison to an experimental system we wish to study and has the property that an empirical thermometer gives the same reading regardless of where or when it is placed in the reservoir* (definition). An infinite isothermal reservoir is clearly an idealization in the sense that point masses and point charges are idealizations. Nonetheless, we can approach such a reservoir arbitrarily closely, a point stressed with great vigor by the manufacturers of incubators and refrigerators and other constant-temperature devices.

A second idealization is *an adiabatic wall, which is a structure through which no energy may flow* (definition). Since we have not completed our discussion of energy, the adiabatic wall must stand as a somewhat vague concept. We have, however, noted the existence of potential, kinetic, and thermal energy, so that an adiabatic wall must, as a minimum, not permit the passage of these forms. Since matter flow in general is accompanied by energy flow, it follows that an adiabatic wall must be impermeable to the flow of matter. As we shall later see, an adiabatic wall is also an idealization which may be experimentally approached although never perfectly attained. We may for most purposes consider a rigid wall, which has no holes or pores and is a very good insulator with respect to heat flow. An adiabatic wall which is movable allows us to do work on a system even though there is no flow of energy through the wall itself.

If a system is placed in contact with an infinite isothermal reservoir for a very long time, it approaches the following conditions:

(1) *The empirical temperature is uniform throughout* (empirical generalization).

(2) *All measurable properties of the system become independent of time* (empirical generalization).

(3) *The system can be completely described (macroscopically) in terms of a small number of parameters. If a system contains N chemically distinct types of molecules, $N + 2$ parameters suffice to completely describe the system* (empirical generalization).

A system of this type is designated an equilibrium system, or alternatively we designate it as being in an equilibrium state. The concept of equilibrium, while very awkward to talk about, is extremely fundamental to our understanding of thermal physics. Equilibrium states, because of their particular simplicity, are at the present time amenable to much more exact analysis than are nonequilibrium states.

The three conditions used to describe the equilibrium state require some further discussion. The uniformity of empirical temperature is subject to temperature fluctuations which become more and more significant as we look at smaller and smaller regions of the system. In taking a more microscopic point of view, we will later examine these fluctuations, which are themselves fundamental to the description of the system. The independence of parameters with time depends on the time scale that we use; that is, we may achieve equilibrium with respect to relatively rapid processes but not be in equilibrium with respect to slower processes. For example, consider a container with two moles of hydrogen gas and one mole of oxygen gas in contact with a reservoir at zero degrees centigrade. After a short time the system will be in apparent equilibrium. If, however, we continue the study of the system over a very long time, there will be a slow production of water. The final equilibrium state will contain virtually no hydrogen and oxygen but will be almost entirely water. The initial equilibrium takes place in minutes or hours, while the final equilibrium may require years or centuries. For many purposes we can regard the initial equilibrium as a true equilibrium and treat the system accordingly.

Equilibrium may also be achieved by surrounding a system with adiabatic walls and allowing it to age in that condition for a long time. The previous three statements apply to such a system, as do our other comments on the concept of equilibrium.

We next undertake a detailed account of the mechanical features involved in an ideal-gas molecule colliding with a wall. When this process is understood we can use the postulates of kinetic theory to study the properties of ideal gases and ultimately to extend our understanding of the temperature concept. Consider then a fixed, rigid wall being approached by a gas molecule at velocity v as shown in fig. 2-1. For simplicity we assume that the velocity vector is in the x direction and is perpendicular to the plane of the wall. Ideally, such gas molecule would act as a completely hard sphere; that is, no forces would exist between the molecule and the wall until the distance between the center of the sphere and the wall was r, the radius of the sphere. The force would then be repulsive and would assume a very high value, effectively infinity. The potential energy curve corresponding to this is shown

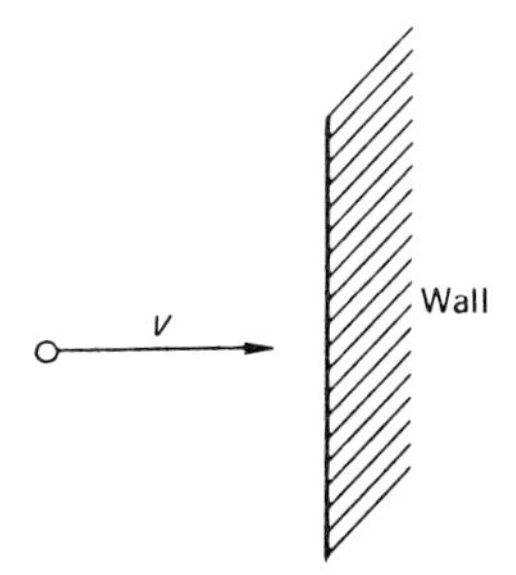

FIGURE 2-1. Ideal gas molecule moving with velocity v toward a rigid wall.

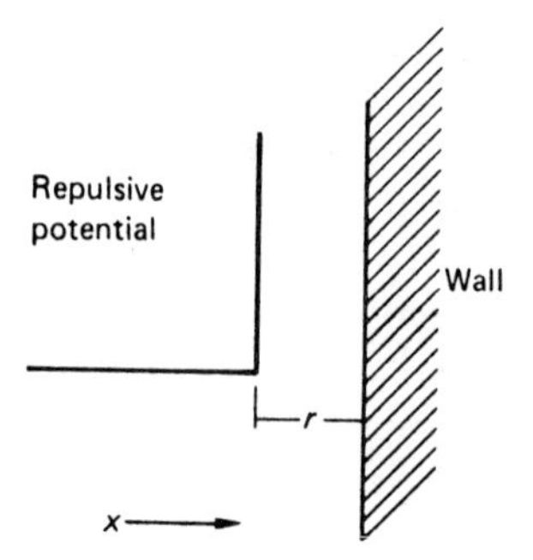

FIGURE 2-2. Interaction potential between molecule and wall as a function of the distance between them, x. Hard sphere approximation is shown with r equal to the molecular radius.

in fig. 2-2. In fact, we need not invoke complete rigidity but can assume a force field like that shown in fig. 2-3, which more closely represents what would happen at an actual wall. This problem can now be examined in terms of classical mechanics, starting with the conservation of mechanical energy. For the moving gas molecule we can write the following equation:

$$\tfrac{1}{2}mv_0^{\,2} = \tfrac{1}{2}mv(x)^2 + \mathscr{V}(x) \tag{2-9}$$

As the particle approaches the wall it has a kinetic energy $\frac{1}{2}mv_0^{\,2}$ and no potential energy. As it gets into the force field work is done, the kinetic energy decreases, and the potential energy increases. Eventually a point x is reached where $\mathscr{V}(x) = \frac{1}{2}mv_0^{\,2}$; the energy is now all potential, the velocity is zero, and the particle has reached its point of closest approach to the wall. The repulsive force is still operative and the molecule now begins to

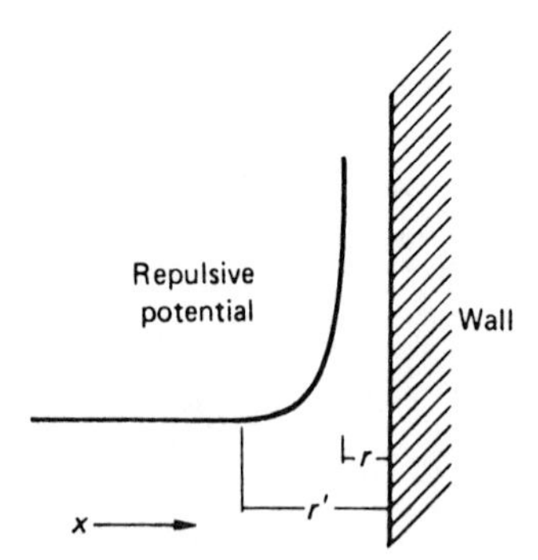

FIGURE 2-3. Interaction potential between molecule and wall as a function of the distance between them, x. Actual case is shown where the repulsive force sets in at a distance r', and the distance of closest approach is r.

move away from the wall. When it reaches the point where it originally entered the force field the velocity is now $-v_0$ and the energy is entirely kinetic. Assume that the particle enters the force field at time t_1 and leaves at time t_2. The average force on the wall during this time is, by Newton's third law, the negative of the force on the particle, F,

$$\bar{F} = \frac{1}{t_2 - t_1} \int_{t_1}^{t_2} - F\, dt \tag{2-10}$$

Equation (2-10) represents the general method of determining the time average of any quantity. If we now apply Newton's second law, we get

$$\bar{F} = \frac{-1}{t_2 - t_1} \int_{t_1}^{t_2} m \frac{d^2x}{dt^2}\, dt = \frac{-1}{t_2 - t_1} \int_{t_1}^{t_2} m \frac{dv}{dt}\, dt$$

$$= -\int_{t_1}^{t_2} \frac{m\, dv}{t_2 - t_1} = \frac{2mv_0}{t_2 - t_1} \tag{2-11}$$

We have again used the fact that the acceleration d^2x/dt^2 is the time derivative of the velocity dv/dt. The final equality in the above equation comes about since $v(t_1)$ is v_0, and $v(t_2)$ is $-v_0$, as already indicated. The calculated value then represents the average force that one gas molecule exerts on the wall during the period of collision.

Return now to the postulates of kinetic theory and consider an ideal gas in a container having plane walls. Assume there are B molecules in a container of volume V, or B/V molecules per unit volume. The molecules are small, hard spheres moving randomly in all directions with an average speed $\bar{v}$. Consider next a plane wall of the container. At a distance r from the wall construct an imaginary plane as in fig. 2-4. The number of molecules passing the plane from left to right in a unit time is $\frac{1}{6}(B/V)\bar{v}$. This quantity is arrived at by assuming that $\frac{1}{3}$ of the molecules are moving in the x direction, $\frac{1}{3}$ in the y direction, and $\frac{1}{3}$ in the z direction. Of those moving in the x direction, $\frac{1}{2}$ are moving from right to left and $\frac{1}{2}$ are moving from left to right. Therefore $\frac{1}{2}$ of $\frac{1}{3}$, or $\frac{1}{6}$ of the molecules are moving toward the plane we wish to consider. Since the concentration of molecules is B/V, the quantity $\frac{1}{6}(B/V)$ is the concentration of molecules moving toward the plane we are considering.

The molecules reaching the plane in one second are those heading in the right direction which are within a distance of $\bar{v}$ times one second, which is just $\bar{v}$. If we look at unit area of surface, one square meter, the molecules arriving in one second are contained in a rectangular parallelepiped of base 1×1 and height $\bar{v}$. The volume of this figure is $\bar{v}$, and the concentration of appropriately directed molecules is $\frac{1}{6}(B/V)$; therefore, the number of molecules passing a unit area of the plane in a second is $\frac{1}{6}(B/V)\bar{v}$.

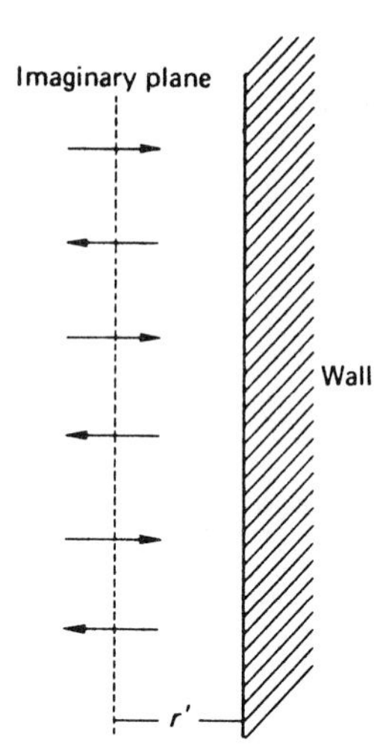

FIGURE 2-4. Molecules passing an imaginary plane a distance r' from a rigid wall.

Each molecule is in contact with the wall from the time it passes the imaginary plane until the time it passes again in the opposite direction, having been reflected by the contact. This time is $t_2 - t_1$, and we have already calculated in eq. (2-11) the average force F exerted on the wall during the contact. Per unit area of wall the average number of molecules in contact at any time is therefore

$$\frac{1}{6}\left(\frac{B}{V}\right)\bar{v}(t_2 - t_1)$$

and the force per unit area is

$$\frac{F}{\mathscr{A}} = \frac{1}{6}\left(\frac{B}{V}\right)\bar{v}(t_2 - t_1)\bar{F} = \frac{1}{6}\left(\frac{B}{V}\right)\bar{v}(t_2 - t_1)\frac{2m\bar{v}}{t_2 - t_1} = \frac{1}{3}\frac{B}{V}m\bar{v}^2 \qquad \text{(2-12)}$$

Note that the quantity $t_2 - t_1$ drops out of the final expression. This quantity would be very difficult to calculate since it depends on the detailed nature of the force field in the neighborhood of the wall. Equation (2-12) shows that the general result is independent of this detail. The force per unit area of wall is, however, simply the pressure exerted on the wall by the gas, which we have already designated as P, so that eq. (2-12) can be written as

$$P = \frac{1}{3}\left(\frac{B}{V}\right)m\bar{v}^2 \qquad \text{(2-13)}$$

We can now compare eq. (2-13) with eq. (2-4) since they both relate to gas pressure,

$$PV = nR\theta = \tfrac{1}{3}Bm\bar{v}^2 \qquad \text{(2-14)}$$

The average kinetic energy of a gas molecule is equal to $\frac{1}{2}m\bar{v}^2$, and B molecules are equal to $B/\mathcal{N}$ or n moles, where $\mathcal{N}$ is Avogadro's number. Thus

$$\left(\frac{R}{\mathcal{N}}\right)\theta = \frac{2}{3} \text{ kinetic energy} \tag{2-15}$$

The empirical temperature of an ideal-gas thermometer is thus directly proportional to the average kinetic energy of a molecule.

This result turns out to be more general than for ideal gases, so that under a wide variety of equilibrium conditions temperature is a measure of the average kinetic energy of the molecules making up the system. The ratio of the gas constant to Avogadro's number $R/\mathcal{N}$ is itself a constant k, usually called the Boltzmann constant. Equation (2-15) thus indicates that the average kinetic energy per molecule is equal to $\frac{3}{2}k\theta$; that is, the kinetic energy is three halves times the Boltzmann constant times the empirical temperature. For an ideal gas, temperature is therefore a straightforward measure of the average kinetic energy of translation per molecule.

The previous result can be generalized beyond ideal gases and beyond equilibrium situations. We will define the kinetic theory temperature ϕ as follows (definition):

$$Bk\phi = \frac{2}{3}\sum_i \frac{1}{2} m_i v_i^2 \tag{2-16}$$

where the summation is over all molecules in the system. For an ideal gas at equilibrium, ϕ can simply be related to θ.

We are now able to assume a much more intuitive idea of temperature at the molecular level. All matter is made of atoms and molecules. These building blocks are very small and are in constant motion as well as constant collision with each other. As a result of these processes there is a certain amount of kinetic energy which is rather uniformally distributed over the particles. The macroscopic notion of temperature is simply and directly related to the amount of kinetic energy associated with a given sample. The existence of an absolute zero of temperature in classical physics is then easy to understand, since it is that state where the particles have zero kinetic energy. Clearly, no lower temperature will be possible.

The idea of a temperature range for biological phenomena can also be looked at from this perspective. At low temperatures the molecules possess little energy, and processes which require an activation energy are slowed way down (see Chapter 13). Above the normal temperature range the biological structures are subject to many high-energy collisions from neighboring molecules and this tends to break down the macromolecular configurations and leads to inactivation.

Inherent in this view is the notion that matter is not at all static, but, when examined at the molecular level, is in violent agitation. It is well to incorporate this thought into our ideas of biology since in the end it will give us a truer picture of life at the molecular level.

The concept of temperature as measuring kinetic energy of molecules or, as we shall later see, of atoms within molecules opens a view of cellular systems which is both important and seldom considered. One of the most profound insights of the last 150 years of science is the realization that all matter consists of atoms and molecules that are in ceaseless thermal motion. The more microscopic a view we take, the more we are impressed with the magnitude of this "thermal noise." One of the deep problems of theoretical biophysics is to explain in detail how the very precise order of molecular biology takes place in spite of the constant disordering due to thermal energy. Indeed, many people date the modern era in biophysics to 1944 when Erwin Schrödinger published his thoughts on the problems raised by considering the effects of thermal noise on the molecular structures of biological systems (see Bibliography).

In thinking about molecular biology, we often disregard thermal motion and emerge with a perception that is much too static. We do not always realize that the exquisite molecular models that result from x-ray diffraction studies are time averages. To appreciate molecular function, we must consider the continuing motion of all the parts of the molecule, which is as much a component of the description as is the structure itself. By way of analogy, if we inspect a micrograph of a cell we get a rather static impression, while if we look at a time lapse movie we get a feeling of a seething, pulsating, squirming, highly active living cell. Similarly, a molecular model is a static view of a structure which at the atomic level is showing a fantastic amount of vibrating, turning, moving, rotating, and oscillating, displaying the large quantity of kinetic energy that is distributed among molecular modes of motion. We will later turn to a more detailed description of the problem of "thermal noise" in biology (Chapter 16).

TRANSITION

Our primitive ideas of temperature are derived from the senses of hot and cold. These ordering relations allow us to develop empirical thermometers, instruments for measuring the degree of hotness on some arbitrary scale. Of special utility among these instruments is the ideal gas thermometer. Elementary kinetic theory analysis of ideal gases demonstrates that the temperature defined by such substances is proportional to the average kinetic energy of the gas molecules. This provides a deeper insight into the construct of temperature. Once temperature is measurable we can move on to ideas about equilibrium and equilibrium states, whose existence makes possible a well-defined measure of the energy of a system. This will be the main topic of the next chapter.

BIBLIOGRAPHY

Chapman, S., and Cowling, T. G., "The Mathematical Theory of Non-Uniform Gases." Cambridge Univ. Press, London and New York, 1960.

A very sophisticated treatment which deals in part with the kinetic approach to nonequilibrium problems.

Jeans, J. H., "An Introduction to the Kinetic Theory of Gases," Cambridge Univ. Press, London and New York, 1946.

A classical introductory volume on gas theory.

Kauzmann, W., "Kinetic Theory of Gases." Benjamin, New York, 1966.

A clear treatment of the elementary properties of gases.

Plumb, H. H. (ed.), "Temperature—Its Measurement and Control in Science and Industry" Volume 4. Instrument Society of America, Pittsburgh, Pennsylvania, 1972.

An encyclopedic work covering all aspects of temperature. Volumes 1, 2, and 3 are earlier versions. Not all of the earlier material is found in Volume 4.

Precht, H., Christophersen, J., Hensel, H., and Larcher, W., "Temperature and Life." Springer-Verlag, Berlin and New York, 1973.

A compendium covering all aspects relating living processes to environmental temperature.

Rose, A. H. (ed.), "Thermobiology." Academic Press, New York, 1967.

A series of papers on the effects of heat and cold on a wide variety of organisms.

Schrödinger, E., "What is Life?" Cambridge Univ. Press, London and New York, 1967.

A somewhat philosophically oriented forerunner of modern biophysics.

Winslow, C. E. A., and Herrington, L. P., "Temperature and Human Life." Princeton Univ. Press, Princeton, New Jersey, 1949.

This short work relates temperature and human biology.

3 The Measurability of Energy

A. ENERGY CHANGES IN A SYSTEM

Having deepened our understanding of temperature, we can now move on to energy in a more general context than envisioned in our previous discussion of mechanics. A central idea is that of the measurability of energy in an isolated physical system, and we now turn our attention to that problem.

In the preceding chapters we have distinguished potential energy from kinetic energy and we have further distinguished the energy in macroscopic modes, which is accessible to study and measurement by the methods of mechanics, from the energy in microscopic modes, which is ultimately associated with the notion of temperature. In this chapter we will begin by a consideration of the total energy of a system which is isolated from the rest of the universe by a set of rigid adiabatic walls.

Since the adiabatic conditions we have established preclude the entry or exit of energy from the system and since energy is a conserved quantity, the total energy of the system must remain constant and we will designate this quantity by the symbol U. We now introduce a postulate which underlies the first law of thermodynamics. *For an adiabatically isolated system of given total energy U, given total volume V, and given compositional variables (mole numbers of constituents), there exists a unique equilibrium state which fixes the values of all other macroscopic parameters* (postulate based on empirical generalization). Thus for such a state the value of any other macroscopic parameter of interest, the temperature for instance, can be represented by

$$\theta = \theta(U, V, n_1, n_2, \ldots, n_N) \tag{3-1}$$

Alternatively, we might note that the total energy of an equilibrium system is uniquely determined by the volume compositional variables and any one additional independent macroscopic variable, which we will for the moment choose to be temperature. It then follows that

$$U = U(V, \theta, n_1, n_2, \ldots, n_N) \tag{3-2}$$

The right-hand side now consists of macroscopically measurable quantities and we turn our attention to the measurability of U.

Start with a system which is adiabatically isolated and has a certain set of values $V, \theta, n_1, n_2, \ldots, n_N$. Next permit a relaxation of the rigid-wall condition and allow external work to be done on the system by some arrangement of pistons, levers, pulleys, wheels, and other macroscopic devices. Now re-isolate the system and allow it to equilibrate. A measured amount of mechanical work has been done on the system. For historical reasons we represent the work done *by* the system as ΔW so that the work done *on* the system is $-\Delta W$. Since energy is conserved, we must argue that

$$\Delta U = -\Delta W \tag{3-3}$$

Thus the energy difference between these two particular states of the system can be measured since ΔW can be determined from the methods of mechanics. We next consider any two arbitrary equilibrium states I and II of the system. If it were always possible to go from one state to another by external macroscopic mechanical work alone, then we could always measure the energy difference $\Delta W_{I,II}$, as is done above. In the previous statements we must envision the possibility of the system doing work on the surroundings as well as the surroundings doing work on the system.

In an elegant series of experiments carried out largely by James Prescott Joule in the mid 1800s, it was in fact found *that given two equilibrium states I and II of an isolated system, it is always possible by external mechanical work alone to go either from state I to state II or from state II to state I* (empirical generalization). It is not always possible to go both ways by external mechanical work alone. This is related to the concept of reversibility and will be considered later. Joule's empirical generalization coupled with the previous postulate on total energy means that the total energy difference between any two equilibrium states is always measurable in terms of mechanical work or that the total energy U is always measurable relative to some arbitrary zero level of energy. This arbitrary zero is a property of energy in mechanics and need not be of special concern to us in our thermal studies; we need only adopt some standard state and can then measure relative to that state.

The quantity U is often designated internal energy rather than total energy of the system and this designation will be used in subsequent pages. We may now sum up the preceding ideas. For isolated systems at equilibrium there exists an internal energy which is a function of the macroscopic variables of the system. A corollary of being a function of the values of the equilibrium macroscopic variables only is that the value of U is independent of how the system arrived at its present state. The internal energy is a measurable quantity which ultimately relates thermodynamics to mechanics.

B. *THE FIRST LAW OF THERMODYNAMICS*

The next problem is to investigate energy exchanges between the system and the environment by processes other than just the performance of external mechanical work. To allow for such exchange, we must relax the condition of adiabatic walls and allow energy exchanges by placing the system in contact with hotter or colder systems in the environment. To treat this process formally, we must reconsider the concept of equilibrium as introduced in Chapter 2. Systems may achieve equilibrium either by adiabatic isolation or by being in contact with a very large isothermal reservoir. Consider a constant-volume system which has come to equilibrium with a large thermal reservoir. Such a system can be characterized by its macroscopic parameters $V, \theta, n_1, n_2, \ldots, n_N$. *If we now surround the system by adiabatic walls, the values of the macroscopic parameters will remain unchanged* (empirical generalization). The isolated systems will have an internal energy U and some relation $U = U(V, \theta, n_1, n_2, \ldots, n_N)$ will obtain. Surrounding the system by adiabatic walls does not change its total energy, so that the value of U and the relation between U and $V, \theta, n_1, n_2, \ldots, n_N$ must be the same for the system before and after adiabatic isolation. The same equilibrium states exist under the two conditions of achieving equilibrium.

Next we consider a system in equilibrium state I which interacts with the environment both by external mechanical work and by the flow of thermal energy. These interactions take the system to a new equilibrium state II. We can now write the following generalized statement of the conservation of energy:

$$U_{II} - U_I = \text{energy exchanges of the system by external mechanical work} + \text{energy exchanges of the system by all other processes}$$

$$\Delta U = -\Delta W + \Delta Q \tag{3-4}$$

We have already noted that ΔU is measurable and ΔW is available from a consideration of mechanics. Equation (3-4), which is called the first law of thermodynamics, serves to define ΔQ, the heat flow or energy exchange of a system by all processes not subject to analysis by consideration of macroscopic mechanics. The functions Q and W have a different status from the function U in that they depend not only on the state of the system but the path undergone in transitions of the system. Consider a simple example to clarify this point; an initial state consisting of x grams of ice and y grams of water at zero centigrade and a final state of $x + y$ grams of water at 25 centigrade. Consider two possible paths from the initial to the final state:

(a) The system is adiabatically isolated, connected to the outside by a paddle wheel, which is turned by an external torque, the turning continuing until the system is at 25 centigrade.
(b) The system is placed in contact with an isothermal reservoir at 25 centigrade and is kept there until its temperature comes to equilibrium.

In both cases ΔU is the same since the same initial and final states are employed. In the first case, $-\Delta W = \Delta U$ and $\Delta Q = 0$, while in the second case, $\Delta Q = \Delta U$ and $\Delta W = 0$. It is clear that ΔQ and ΔW are not functions of the state of the system, but depend on the path.

The first law of thermodynamics is thus seen as a generalized statement of the conservation of energy for systems involving thermal exchanges. The generalized conservation of energy is postulated and the heat term ΔQ is introduced as a bookkeeping device to maintain the balance of eq. (3-4). The unique aspect of the internal energy U is that for equilibrium conditions it is a function of state only and is independent of how the equilibrium state was reached.

Heat is not a form of energy, it is a method of energy flow. The only system property corresponding to heat is internal kinetic energy, but the exact distribution between kinetic modes and potential energy within the system is a function of the detailed structure and does not, at equilibrium, depend upon how energy enters. It is of course true that a system which has not reached equilibrium will be in a state which depends on its past history. The study of these cases, which may be very important for biology, is the subject matter of nonequilibrium thermodynamics, a field which is in its formative stages (see Chapters 19–21).

C. ENERGY UNITS

The equivalence of heat and mechanical work is indicated in eq. (3-4) and the preceding discussion requires a consideration of the units of measurement. The existence of separate heat and work units reflects in part the history of the subject insofar as there existed separate sciences of calorimetry and mechanics before the experiments of Joule and others established the first law of thermodynamics and the concept of the mechanical equivalent of heat. The heat unit chosen, the 15-degree calorie (the name itself is a carry-over from the old caloric theory), is defined as the quantity of heat necessary to change the temperature of 1 gram of water from 15.0 to 16.0 centigrade. The large calorie or kilogram calorie is 1000 times the size of the calorie. The mean calorie is 1/100th of the quantity of heat necessary to raise the temperature of one gram of water from ice temperature to boiling water temperature. The heat flow in warming one gram of water one degree at 15.0 centigrade is then one calorie. We could carry out the same change of state by performing work under adiabatic conditions and find that $\Delta W =$ 4.18580 joules. A wide variety of experiments of this type have been completed and, as is implied from the first law, the relation between ΔW and ΔQ is a universal one, so that the quantity 4.18580 joules/calories is a universal constant designated as the mechanical equivalent of heat.

In current Standard International usage based on the meter, kilogram, and second, the fundamental energy unit is the joule, which is equal to the newton-meter. The joule is equal to 0.23890 calories. More commonly used in chemical applications is the kilojoule which is related to the kilocalorie by the same factor of 0.23890. The calorie can in principle be abandoned as a unit of measurement since we now have sufficient information to formulate all of thermodynamics in internationally agreed upon energy units. However, tradition dies hard.

In addition to the mechanical and thermal measures of energy, a third set of units has been introduced from electricity. This turns out to be of great practical importance in calorimetry because the high precision possible with electrical measurements has made this the unit of choice in introducing a known amount of energy into a closed system. The advent of potentiometry and the discovery of reversible electrodes has made it possible to equate the changes accompanying certain reactions with the potential difference between the electrodes. We will later have the opportunity to develop this relation in some detail.

The practical unit of electrical potential is the volt, which is the potential difference between two points (electrodes in this case), when one joule of work is involved in the transfer of one coulomb of charge from one point to the other (definition). The energy in electrical measurements is usually given in one of three sets of units:

(a) electron-volt
(b) volt-coulomb (volt-ampere-second)
(c) volt-faraday

The electron-volt is the amount of energy available when the charge of one electron is transferred through a potential difference of one volt. Since the charge of the electron is 1.5921×10^{-19} coulomb, an electron-volt is

1.5921×10^{-19} joules. The volt-coulomb is equal to a joule and the volt-faraday is the amount of energy developed in the transfer of a mole of electrons and is 96,501 joules. The units of energy that we will most frequently use are the electron-volt and the joule or kilojoule. Much of the available literature data is still in kilocalories per mole.

D. GENERALIZATION OF THE CONSERVATION OF ENERGY

The preceding allows us to extend eq. (3-4). Previously ΔW was the energy exchange due to mechanical work and for the simple case of a piston moving against the pressure of a gas ΔW could be equated to $\int P\,dV$. The term ΔW can now be reformulated allowing for electrochemical processes:

$$\Delta W = \Delta W_{\text{mechanical}} + \Delta W_{\text{electrical}} \tag{3-5}$$

For a two-electrode system with an external current flow, the $\Delta W_{\text{electrical}}$ term can be formulated as $\int \mathscr{E}\,de$ where $\mathscr{E}$ is the electrical potential and e the charge. We are introducing another method of doing work by the system which has the same formal status as mechanical work. Equation (3-4) can now be written

$$\Delta U = \Delta Q - \Delta W_{\text{mechanical}} - \Delta W_{\text{electrical}} \tag{3-6}$$

This is a statement of the conservation of energy appropriate to systems which are closed to the flow of matter but open to heat flow and the performance of electrical and mechanical work. The equation is a bookkeeping relation, which says that the change of energy of a system can be measured by keeping track of all the energy flows in and out of the system (deposits and withdrawals). For this relation to be true, *energy can be neither created nor destroyed within the system; that is to say, energy is conserved* (empirical generalization).

The results embodied in eq. (3-6) largely follow as an empirical generalization from a series of experiments which we noted was carried out by James Prescott Joule from 1840 to 1850. These experiments used heat as a common measure of energy and, although somewhat more restricted than the given equation, serve to establish the principle.

The next extension of energy conservation consists of combining two systems, or relaxing the restriction closing the system to the flow of matter. We proceed by placing two separate systems inside a single adiabatic container and thus combining them. Then by the defined properties of the container,

$$U_{\text{total}} = U_1 + U_2 \tag{3-7}$$

If system 1 is much larger than system 2 we can regard the procedure as the addition of material to system 1 and note that

$$U_{\text{total}} - U_1 = \Delta U_1 = \Delta U_{\text{mass}} \tag{3-8}$$

The extended conservation equation can now be written

$$\Delta U = \Delta Q - \Delta W_{\text{mechanical}} - \Delta W_{\text{electrical}} + \Delta U_{\text{mass}} \tag{3-9}$$

E. ENERGY CONSERVATION IN LIVING ORGANISMS

The question of the applicability of the conservation of energy to living systems is the theme that appeared frequently in the formulation of thermodynamics in the mid 1800s. The experimental basis of these considerations goes back to two French savants, Lavoisier and Laplace, who constructed an ice calorimeter in 1781 and measured the heat given off by hot objects, by exothermic chemical reactions, and by animals. The measure of heat was the amount of water produced by melting ice. They confined a guinea pig in their calorimeter for several hours and determined the amount of heat produced. They compared the amount of animal heat and the production of carbon dioxide by guinea pigs with the same parameters in the combustion of carbon. In experiments carried out by Lavoisier and Armand Sequin in 1789–1791, the general notion was finally established that the combustion of food in the animals led to CO_2 and H_2O production and a parallel release of heat.

During the one hundred years following the original calorimetry experiments, a number of other more precise measurements were made, and, in 1894, M. Rubner reported on a series of experiments with dogs where he showed that the heat produced equals the heat of combustion of nutrient catabolized minus the heat of combustion of the excreted urine. In 1894, F. Laulanie reported on similar experiments for guinea pigs, rabbits, ducks, and dogs. This type of experiment was refined and extended by W. A. Atwater and F. G. Benedict (see Bibliography). They were able to conclude

> For practical purposes we are therefore warranted in assuming that the law (conservation of energy) obtains in general in the living organism as indeed there is every reason *a priori* to believe that it must.

The concept of the energy U as a measurable quantity as introduced in eqs. (3-2)–(3-4), implies the restriction of the measure to equilibrium systems.

That limitation can be lifted in the following way. Assume we have a nonequilibrium system whose internal energy we wish to measure. Isolate the system by enclosing it in adiabatic walls and allow it to equilibrate by aging. After it comes to equilibrium, the quantity U can be measured relative to any arbitrary state of the system that we wish to define. (Remember that even in mechanics all energy measures are relative to an undetermined constant and only energy differences have significance.) Since no energy entered or left the system after adiabatic isolation, the total energy of the system at equilibrium, U, must be equal to its value in the nonequilibrium state. We have therefore reduced the total energy of the system to a measurable quantity, independent of whether or not the system is at equilibrium.

The measurability of total energy turns out to be a useful concept. For in far-from-equilibrium systems, such as those characteristic of living organisms, most of the classical variables of equilibrium thermodynamics can no longer be measured or, indeed, assigned a well-defined meaning. This applies to entropy, free energy functions, chemical potentials, and in some cases temperature and pressure. Thus it is well to have at least one parameter to make contact between equilibrium and far-from-equilibrium constructs. However, before dwelling on the difficulties of nonequilibrium theory, we must develop the cornerstone of classical theory, the notion of entropy.

TRANSITION

The ability to transform a system between two equilibrium states by mechanical work alone provides the necessary criteria for establishing the energy of that system as a measureable parameter. This parameter, along with the measure of work, allows us to define the heat change and formulate the first law of thermodynamics as a bookkeeping device for the energy entering and leaving a domain of interest. All energy units may thus be reduced to work units, joules, and a unified system of accounting is possible. This equivalence leads to a much more general statement of the conversation of energy. The consequences of the conservation law have been explored for living organisms which were found to be in accord with these generalizations. Much of thermodynamics deals with the efficiency of conversion from one form of energy to another. We will next discuss the formalism that has been developed to deal with problems of this type.

BIBLIOGRAPHY

Atwater, W. A., and Benedict, F. G., Experiments on the metabolism of matter and energy in the human body, *U.S. Dept. Agr. Exp. Sta. Bull. 136* (1903).

The basic work establishing energy conservation for Homo sapiens.

Callen, H. B., "Thermodynamics." Wiley, New York, 1965.

The subject of thermodynamics is presented from a postulational approach. This is an elegant presentation with an in-depth discussion of many of the basic issues of thermodynamics.

Spanner, D. C., "Introduction to Thermodynamics." Academic Press, New York, 1964.

A discussion of introductory thermodynamics that is oriented toward biological problems.

4 The Second Law of Thermodynamics

A. HEAT ENGINES AND REFRIGERATORS

There are few ideas in science that have been as difficult to comprehend as the second law of thermodynamics and the accompanying introduction of the entropy function. The concepts were first introduced in an engineering context; indeed Sadi Carnot's original 1824 monograph was entitled "Reflections on the Motive Power of Fire and on Machines Fitted to Develop that Power." From these considerations of steam engines a series of concepts has arisen which have profoundly influenced physics, chemistry, and a number of related sciences, including biology. The recently discovered relationships between entropy, work, and information and the profoundly informational character of biology have served to stress the deep connections between biology and thermal physics.

In the following chapter we introduce this subject in a traditional way. In order to understand thermodynamics, it is necessary to comprehend how certain verbal statements about what cannot be done have been elaborated into some of our profoundest thoughts about the nature of the universe. Subsequently we will develop informational ideas which more closely relate entropic concepts to biology and allow us to sense the relation between the observer and the phenomena being observed.

The classical approach to thermodynamics starts with a consideration of heat engines and refrigerators. In the final analysis the discipline is designed to deal with real engines and real refrigerators; however, in establishing the theory, certain limiting idealizations are used. If we are careful to define our concepts, this difference between real and idealized engines and refrigerators will not be a cause of misunderstanding. The entities that we wish to deal with are then:

1. Heat Engine. A heat engine is a device which takes heat from a hot source and performs mechanical work (definition). The hot source can be combusting gasoline, the steam from a boiler, the sun, or any other high-temperature generator. In order to continuously produce work, it is necessary for an engine to operate in a cycle; that is, it must take in heat, produce work, and return to its original state so that it can repeat the process and continuously develop output. *It is found that if an engine does not operate in a cycle, its internal state changes until it reaches a point where it can no longer function* (empirical generalization). Our concern will then be with heat engines which operate in a cyclic fashion.

2. Refrigerator. A refrigerator is a device which transfers heat from a cold body to a hotter body (definition). Refrigerators which operate continuously also operate in cycles; that is, they must transfer heat and return to their original state so as to repeat the process. *Work (generally mechanical) must be performed in order to accomplish this kind of heat transfer* (empirical generalization).

B. STATEMENTS OF THE SECOND LAW

The second law of thermodynamics was originally formulated in two different ways, one about heat engines and the second about refrigerators. In subsequent considerations it is necessary to use both statements interchangeably, so that we must prove their equivalence. These inductions are the foundation upon which all of classical thermodynamics is based. Our procedure will be to set down the two forms of the second law and then to prove their logical equivalence:

(1) Engine statement of the second law: *It is impossible to construct an engine that, operating in a cycle, will produce no effect other than the extraction of heat from a reservoir and the performance of an equivalent amount of work* (empirical generalization).

(2) Refrigerator statement of the second law: *It is impossible to construct a device that, operating in a cycle, will produce no effect other than the transfer of heat from a cooler to a hotter body* (empirical generalization).

To prove the logical equivalence of the two statements we will use formal symbolic logic, utilizing the following symbols:

(a) Equivalence $\equiv$; $A \equiv B$; A is logically equivalent to B.

(b) Implication $\supset$; $A \supset B$; A implies B, the truth of A implies the truth of B.

(c) Negation $-$; $-A$, which is read "not A"; either A is true or $-A$ is true.

We now quote a result of formal logic which seems intuitively clear without proof. $A \equiv B$ if $A \supset B$ and $B \supset A$; that is, A is equivalent to B if the truth of A implies the truth of B and the truth of B implies the truth of A. An alternative formulation of equivalence is $A \equiv B$ if $-A \supset -B$ and $-B \supset -A$; that is, A is equivalent to B if the falsity of A implies the falsity of B and the falsity of B implies the falsity of A. We shall use the second form of proving equivalence; that is, we may assume that the engine statement of the second law is logically equivalent to the refrigerator statement of the second law if the falsity of the engine statement implies the falsity of the refrigerator statement and the falsity of the refrigerator statement implies the falsity of the engine statement.

The engine statement specifically rules out an engine that can extract energy from a heat source and convert it to work with no other effects. The classical example of such an engine would be one that could extract heat energy from the ocean and drive an ocean liner without the necessity of buying fuel. In fact, engines not only take heat from a hot reservoir but give up heat to a cold reservoir.

We now assume the falsity of the refrigerator statement and consider the device shown in fig. 4-1. The engine operates by taking up an amount of

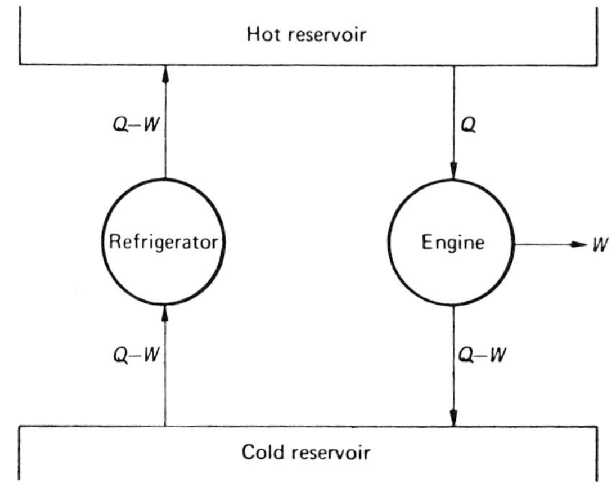

FIGURE 4-1. An engine and a refrigerator working between two reservoirs at different temperatures. The engine takes in an amount of heat Q, does an amount of work W, and rejects an amount of heat $Q - W$. The refrigerator (acting contrary to the refrigerator statement of the second law) transfers an amount of heat $Q - W$ from the cold reservoir to the hot reservoir.

heat Q from the hot reservoir, doing an amount of work W, and giving up an amount of heat $Q - W$ to the cold reservoir. If the refrigerator could violate the refrigerator statement of the second law, it could transfer an amount of heat $Q - W$ from the cold to the hot reservoir. The engine and refrigerator working together would then take up an amount of heat W from the hot reservoir and do an amount of work W and produce no other effect. The whole device would then violate the engine statement of the second law. Therefore the assumed falsity of the refrigerator statement implies the falsity of the engine statement.

Next assume the falsity of the engine statement of the second law. Referring to fig. 4-2, the engine would now take up an amount of energy Q and do an amount of work $W = Q$. This work could be used to drive a refrigerator which would take up an amount of heat Q_2 and transfer $Q_2 + Q$ to the hot reservoir. The entire device of engine and refrigerator now acts cyclically to cause no other effect than the transfer of Q_2 units from the cold reservoir to the hot reservoir, which violates the refrigerator statement of the second law.

If we designate the engine statement as *Eng* and the refrigerator statement as *Ref*, we have now demonstrated that

$$-Eng \supset -Ref \qquad \text{and} \qquad -Ref \supset -Eng$$

which implies that $Ref \equiv Eng$. The two statements of the second law are logically equivalent and can be used interchangeably.

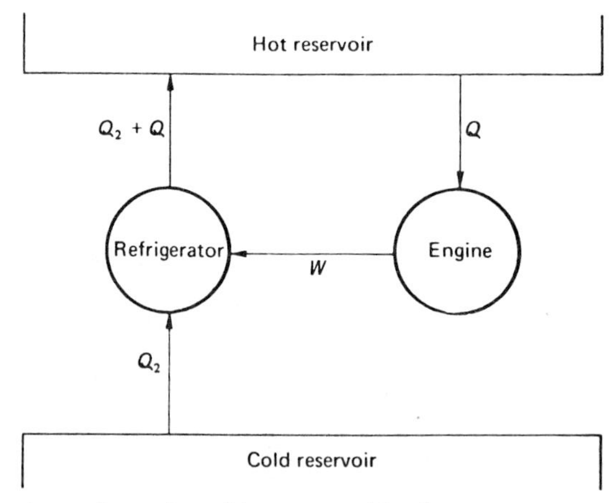

FIGURE 4-2. An engine and a refrigerator working between two reservoirs at different temperatures. The engine (in violation of the engine statement of the second law) takes in an amount of heat Q and does an equivalent amount of work W. The work is done on a refrigerator which then takes an amount of heat Q_2 from the cold reservoir and transfers an amount of heat $Q_2 + W$, which is $Q_2 + Q$, to the hot reservoir.

C. CYCLES

The second law can now be used to derive two important results, known as Carnot's theorems, about the efficiency of engines. These theorems can in turn be used to develop an absolute thermodynamic temperature scale. We shall then be able to show the equivalence of this scale with the ideal-gas temperature scale. This should then extend our formalization of the concept of temperature.

Three notions are required prior to our statement and proof of Carnot's theorems. These are the definition of efficiency, the detailing of the concept of reversibility, and the introduction of the ideal reversible heat engine.

1. Efficiency. The efficiency of a heat engine is the ratio of the mechanical work produced to the heat taken in from the hot source (definition). This may be formally stated as

$$\eta = \frac{W}{Q_1} \tag{4-1}$$

where W is work and Q_1 is the heat taken in. As we have already noted, an engine also gives up heat to a cold reservoir in completing its cycle. If we designate this amount of heat as Q_2, we can then utilize the first law of thermodynamics to write

$$\Delta U = Q_1 - W - Q_2 \tag{4-2}$$

For a complete cycle the engine returns to its original state and $\Delta U = 0$ since we have already noted that U is a function of the state of the system only. Equation (4-2) can then be rewritten as

$$W = Q_1 - Q_2 \tag{4-3}$$

and the efficiency can be formulated as

$$\eta = \frac{W}{Q_1} = \frac{Q_1 - Q_2}{Q_1} = 1 - \frac{Q_2}{Q_1} \tag{4-4}$$

The engine form of the statement of the second law is equivalent to saying that η must always be less than one, all heat engines must have efficiencies less than unity.

2. Reversibility. Consider a simple one-component system which is originally at an equilibrium state characterized by P_1 and θ_1 and which under the influence of external changes moves to another equilibrium state characterized by P_2, θ_2. For simplicity allow the system to be in a cylinder whose pressure is maintained by an external piston and whose temperature is maintained by contact with an external heat reservoir as in fig. 4-3. Consider

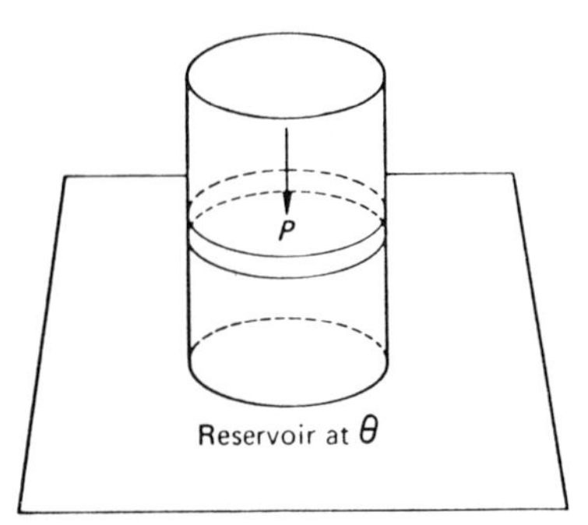

FIGURE 4-3. Cylinder containing a one-component system which is in contact with a reservoir at temperature θ. An external piston allows the pressure to be brought to any value P. The sides of the cylinder are adiabatic and an adiabatic barrier may be introduced between the cylinder and the reservoir.

the following pathway from P_1, θ_1 to P_2, θ_2. First change the pressure by a tiny amount ΔP and allow the system to reequilibrate. After it has equilibrated change the pressure by another tiny increment ΔP and again wait for equilibrium. Repeat this process until the system is at P_2. Next place the cylinder in contact with a reservoir at temperature $\theta + \Delta\theta$ and allow it to equilibrate. After equilibration place the cylinder in contact with a reservoir at temperature $\theta + 2\Delta\theta$ and again let the system come to an equilibrium state. Continue this process until the system comes to temperature θ_2.

The process we have just described takes the system from state 1 to state 2 through an intermediate series of states all of which are equilibrium states. In fact, this is not quite true since every time we change the external conditions we must introduce transients and until these transients die out the system is not in equilibrium. However, if we make each ΔP and $\Delta\theta$ sufficiently small and if we carry out the process sufficiently slowly, then the system can be maintained arbitrarily close to equilibrium. Such slow small transformations are called quasistatic changes in the system. The changes are also called reversible, since at any point reversing the succession of ΔP's and $\Delta\theta$'s with $-\Delta P$'s and $-\Delta\theta$'s will reverse the succession of equilibrium states.

Returning to our previous example, we could have gone from state 1 to state 2 by changing the piston pressure suddenly from P_1 to P_2 and simultaneously placing the cylinder in contact with a reservoir at temperature θ_2. The system will eventually equilibrate at the new boundary conditions P_2, θ_2, but the intermediate states will involve pressure transients, thermal gradients, and other nonhomogeneous phenomena, so that it will be impossible to describe the intervening states by a small number of parameters such as P and θ. In such a case it will be impossible to retrace the pathway and the transformation is designated an irreversible process. All real processes are irreversible, reversible changes being a limiting case. The idealization of reversibility is nevertheless an extremely important abstraction for equilibrium thermodynamics and it is critically important to the continued development of the subject. Consider fig. 4-4, which represents a P–θ diagram. Any path that can be represented on this diagram must be reversible, since points of well-defined P and θ required to be represented on the graph are of necessity equilibrium points. The quantities P and θ are only meaningful for the whole system when it is uniform throughout, which implies equilibrium.

In general, any processes which involve friction or the local generation of heat will be irreversible since the reverse pathway would involve converting all this heat back to work, which violates the engine statement of the second law. The conditions for reversibility are that a process takes place quasi-statically so as to always be arbitrarily near to equilibrium and that the process is not accompanied by any frictional effects.

3. Heat Engines. The concept of a reversible engine arose out of an attempt to ask the question as to what is the maximum efficiency that can be obtained from a device working in a cycle between a heat source at temperature θ_1, and a heat sink at temperature θ_2. The characteristics of such an engine are that it must take up a certain amount of heat Q_1 from the hot reservoir, perform a certain amount of work W, give up a certain amount of heat Q_2 to the cold reservoir, and return to the initial state. Heat engines are usually given efficiency ratings in terms of the work output divided by the heat input or W/Q_1. This ratio has been designated η [see eq. (4-1)].

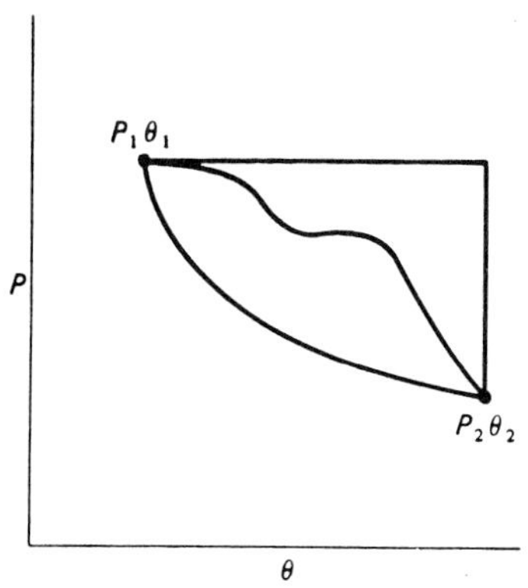

FIGURE 4-4. Various pathways from the state $P_1\theta_1$ to the state $P_2\theta_2$.

If we consider a cycle run in the reverse direction, the device will take up heat from the cold reservoir, work will be done by an external source, and heat will be given to the hot reservoir. Since external work is done on the working substance no violation of the second law is involved in this operation. If Q_2^* is the amount of heat taken from the cold reservoir, and W^* is the work done on the system, the amount of heat delivered to the hot reservoir must be $Q_1^* = W^* + Q_2^*$, if the refrigerator is to act in a cycle. Refrigerators are often rated by a coefficient of performance, which is the ratio of the heat taken from the cold reservoir to the work done, or Q_2^*/W^* in the terms just defined.

Next we introduce the concept of an engine which can be operated reversibly at every stage of the work cycle. In the sense that we have discussed above we must be able to retrace each path in the operation of the working substance. Such a device will be an engine when operated in one direction and a refrigerator when operated in the reverse direction. The identities $Q_1 = Q_1^*$, $Q_2 = Q_2^*$, and $W = W^*$ will obtain because of the requirements of reversibility. It is these identities which are fundamental to two very general theorems which we shall proceed to prove.

D. CARNOT THEOREMS

Theorem I. No irreversible engine operating between two heat reservoirs can be more efficient than a reversible engine operating between the same two reservoirs.

Proof: We shall assume the existence of an engine more efficient than a reversible engine and show that it leads to a contradiction. See fig. 4-5 and table 4-1.

TABLE 4-1

	Reversible engine	Other engine
Heat taken up per cycle	Q_1	Q_1'
Work done	W	W'
Heat given up to the cold reservoir	Q_2	Q_2'
Efficiency	$\eta = \frac{W}{Q_1}$	$\eta' = \frac{W'}{Q_1'}$
Assume $\eta' > \eta$		

Working between the same two reservoirs at θ_1 and θ_2, we will allow the irreversible engine to operate the reversible engine backward as a refrigerator,

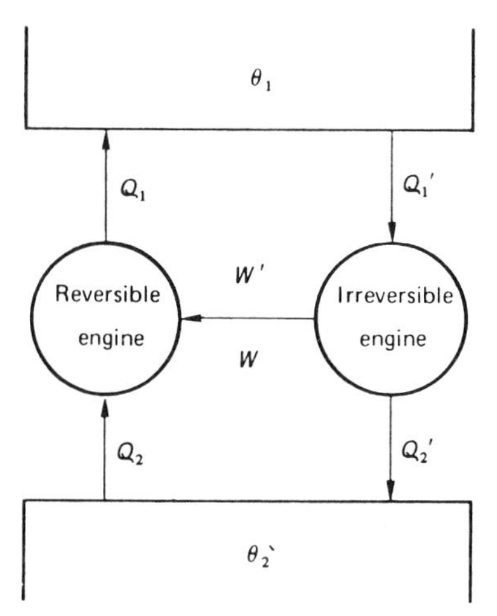

FIGURE 4-5. An irreversible and reversible engine working between the same two heat reservoirs. The irreversible engine drives the reversible engine backward as a refrigerator.

and in that case, $W = W'$, all the work output of the engine is used to drive the refrigerator. The two devices working together therefore do no net work but transfer an amount of heat $Q_1 - Q_1' = Q_2 - Q_2'$ between the reservoirs.

From the assumed efficiencies,

$$\eta' = \frac{W'}{Q_1'} > \frac{W}{Q_1} = \eta \tag{4-5}$$

Since $W = W'$,

$$\frac{1}{Q_1'} > \frac{1}{Q_1}, \quad Q_1 > Q_1' \tag{4-6}$$

The quantity $Q_1 - Q_1'$ is therefore positive. But this is the heat transferred to the hot reservoir. The two engines acting together in a cycle would thus transfer heat from the cold reservoir to the hot reservoir. This, however, is contrary to the refrigerator statement of the second law; hence, η' is not greater than η, which may be formally stated as

$$\eta \geqslant \eta' \tag{4-7}$$

Theorem II. All reversible engines operating between the same two temperatures have the same efficiency.

The proof of this theorem is a simple extension of the proof of the last theorem. We designate the two engines as a and b. We first assume $\eta_a > \eta_b$ and show it leads to a contradiction. We then assume $\eta_b > \eta_a$ and show that this leads to a contradiction. The only remaining possibility is that $\eta_a = \eta_b$. We proceed by allowing engine b to drive engine a backward as a refrigerator (table 4-2).

TABLE 4-2

	Engine a	Engine b
Heat taken in	Q_{a1}	Q_{b1}
Work done	W_a	W_b
Heat discharged	Q_{a2}	Q_{b2}

Assume

$$\eta_b > \eta_a \tag{4-8}$$

Then

$$\frac{W_b}{Q_{b1}} > \frac{W_a}{Q_{a1}}, \qquad W_a = W_b, \quad Q_{a1} > Q_{b1} \tag{4-9}$$

The net effect of the two engines is to transfer an amount of heat $Q_{a1} - Q_{b1}$ from the cold reservoir to the hot reservoir, which is a violation of the second law of thermodynamics; hence,

$$\eta_b \leqslant \eta_a \tag{4-10}$$

By reversing the procedure and allowing engine a to drive engine b as a refrigerator, we can show

$$\eta_b \geqslant \eta_a \tag{4-11}$$

The only way of simultaneously satisfying both equations is to assume

$$\eta_b = \eta_a \tag{4-12}$$

The efficiency of a reversible engine is thus independent of the working substance and independent of the exact mode of operation of the engine; it depends only on the temperatures of the two reservoirs (conclusion from the second law). This dependence of the efficiency on the temperature and the independence on the details of the engine suggests that efficiency might provide an absolute measure of temperature, independent of the detailed nature of the thermometer.

E. THE THERMODYNAMIC TEMPERATURE SCALE

To develop this concept, consider that the temperatures of the two reservoirs are measured on an empirical scale and found to be θ_1 and θ_2. From our previous discussion the efficiency of a reversible engine operating between these two reservoirs is thus a function of θ_1 and θ_2:

$$\eta_{1,2} = \Phi(\theta_1, \theta_2) = 1 - \frac{Q_2}{Q_1} \tag{4-13}$$

The ratio of Q_2 to Q_1 is therefore also a function of θ_1 and θ_2, which we can designate as $\Theta(\theta_1, \theta_2)$.

We now consider three engines, one operating between θ_1 and θ_2, as we have just discussed, and the other two operating between θ_3 as the hot reservoir and θ_1 and θ_2 as the cold reservoirs. We can then write

$$\frac{Q_2}{Q_1} = \Theta(\theta_1, \theta_2), \qquad \frac{Q_1}{Q_3} = \Theta(\theta_3, \theta_1), \qquad \frac{Q_2}{Q_3} = \Theta(\theta_3, \theta_2) \tag{4-14}$$

The three equations can be combined in a simple way to yield

$$\frac{Q_2}{Q_1} = \frac{Q_2/Q_3}{Q_1/Q_3} = \frac{\Theta(\theta_3, \theta_2)}{\Theta(\theta_3, \theta_1)} = \Theta(\theta_1, \theta_2) \tag{4-15}$$

The right-hand equality of eq. (4-15) places certain mathematical restrictions on the function Θ. In general, the requirement is that $\Theta(\theta_i, \theta_j)$ is equal to the product of two functions, one of θ_i and the other of θ_j. Thus $\Theta(\theta_i, \theta_j)$ must be of the form $f(\theta_j)g(\theta_i)$. If we apply this condition to eq. (4-15), we get

$$\frac{Q_2}{Q_1} = \frac{f(\theta_2)g(\theta_3)}{f(\theta_1)g(\theta_3)} = \frac{f(\theta_2)}{f(\theta_1)} \tag{4-16}$$

The Kelvin absolute thermodynamic temperature T will then be defined by the following relation:

$$\frac{T_1}{T_2} = \frac{Q_1}{Q_2} = \frac{f(\theta_1)}{f(\theta_2)} \tag{4-17}$$

The absolute values of T are established by assigning as the temperature of the triple point of water $T = 273.16$ K (in SI units temperature is expressed in kelvins which correspond to degrees kelvin in the older system). The temperature scale so defined has the following properties:

(1) It is independent of the working substance of the thermometer (the engine) because of Carnot's second theorem.

(2) It has an absolute zero since

$$\eta = 1 - \frac{T_2}{T_1} \leqslant 1 \tag{4-18}$$

The inequality is a result of the first law of thermodynamics, since an engine cannot put out more work than its input of heat. Since

$$T_2 \geqslant 0 \tag{4-19}$$

there cannot be sinks with a negative value of T_2, and there is a lowest possible value of T_2, which is zero.

(3) It is a very inconvenient thermometer to use because measuring the efficiency of a perfect reversible engine is a very difficult feat to perform experimentally.

In order to overcome the difficulty in actually using the Kelvin absolute thermodynamic scale, we investigate the relation between it and a specific empirical temperature scale, that of the ideal gas thermometer. When this is done we are able to show the equivalence of the two temperatures. We are thus provided with an experimental method for the measurement of absolute temperature and we are able to tie together all of our various concepts.

The general proof of the equivalence of the two temperature scales, while simple and straightforward, is rather laborious. We will proceed by outlining the method of approach in the following few lines and then presenting the detailed proof which proceeds by constructing an engine from an ideal gas, calculating the efficiency of the engine based on the equations of state of the gas, and determining the quantity

$$\frac{Q_1}{Q_2} = \frac{f(\theta_1)}{f(\theta_2)} \tag{4-20}$$

In actually carrying out the detailed calculation, we arrive at the result

$$\frac{f(\theta_1)}{f(\theta_2)} = \frac{\theta_1}{\theta_2} \tag{4-21}$$

We are therefore able to show, by combining the results of eqs. (4-17) and (4-21), that

$$\frac{\theta_1}{\theta_2} = \frac{T_1}{T_2} \tag{4-22}$$

From this equation we are able to conclude that ideal gas temperature and absolute thermodynamic temperature are proportional. Choosing the reference temperature for both scales as the triple point of water at 273.16 K completes the identity of the two scales.

We now have a temperature scale with the following properties:

(1) It is independent of the working substance.

(2) It has a natural zero.

(3) It is measurable with an empirical thermometer that can be used as a primary standard.

Three separate temperature notions have been introduced: (a) empirical, (b) kinetic theory, (c) absolute thermodynamic. These have been shown in the case of the ideal gas to be identical. Much of the power of thermal physics comes from being able to work back and forth within a conceptual framework which stresses varied aspects of a given problem.

F. EQUIVALENCE OF THE IDEAL GAS AND THERMODYNAMIC TEMPERATURE SCALE

The detailed proof requires us to develop two concepts: (a) the equations of state of an ideal gas, and (b) the nature of a reversible engine using an ideal gas as the sole working substance.

An ideal gas temperature has previously been described as following the equation $PV = nR\theta$. One further feature of ideal gases is necessary for our subsequent treatment. This property was first demonstrated in a free-expansion experiment that utilizes two chambers, one containing an ideal gas and the other under vacuum. The two chambers are enclosed in an adiabatic wall and a hole is opened between them. Initially the gas is at some P_1, V_1, θ_1 and it finally equilibrates at P_2, V_2, θ_2. The energy per mole of a one-component system may be expressed as a function of any two state-variables. So we have

$$U_1 = U(P_1, \theta_1), \qquad U_2 = U(P_2, \theta_2) \tag{4-23}$$

Because of the adiabatic isolation and the absence of any external work,

$$U_1 = U_2 \tag{4-24}$$

The result of the free-expansion experiment is that $\theta_1 = \theta_2$ (empirical generalization); hence,

$$U(P_1, \theta_1) = U(P_2, \theta_1) \tag{4-25}$$

If we represent U as a function of V and θ, we can similarly find

$$U(V_1, \theta_1) = U(V_2, \theta_1) \tag{4-26}$$

Since P_2 and V_2 are arbitrary, as the result is independent of the size of the two chambers, the preceding equations can only be satisfied if the internal energy of a perfect gas is a function of the temperature only. This condition can be expressed as

$$\left(\frac{\partial U}{\partial V}\right)_\theta = 0, \qquad \left(\frac{\partial U}{\partial P}\right)_\theta = 0 \tag{4-27}$$

We could have arrived at the same result from the point of view of kinetic theory. An ideal gas consists of molecules which have no potential energy of interaction except on collision. The total energy is thus the kinetic energy. From Chapter 2, eq. (2-15), we see that, for each molecule,

$$\text{kinetic energy} = \tfrac{3}{2}k\theta \tag{4-28}$$

For a gas of B molecules the total energy is thus

$$U = \tfrac{3}{2}Bk\theta = U(\theta) \tag{4-29}$$

We next introduce the concept of specific heat C_x, which may be defined as

$$C_x = \left(\frac{\partial Q}{\partial \theta}\right)_x \tag{4-30}$$

It is the amount of heat which must be supplied to raise the temperature of the sample by 1 K. In general we determine $\partial Q/\partial\theta$ under two conditions, that of constant pressure or of constant volume. These specific heats are defined as

$$C_V = \left(\frac{\partial Q}{\partial \theta}\right)_V, \qquad C_P = \left(\frac{\partial Q}{\partial \theta}\right)_P \tag{4-31}$$

The first law may be written

$$đQ = dU + P\,dV \tag{4-32}$$

The notation $đQ$ is used to point out that the heat is not an exact differential, but depends upon the path. Where quantities are known to be of this type we will write the differential as $đ$. If eq. (4-31) is substituted into eq. (4-32) we get the following two expressions:

$$C_V = \left(\frac{\partial Q}{\partial \theta}\right)_V = \left(\frac{\partial U}{\partial \theta}\right)_V = \frac{dU}{d\theta} \tag{4-33}$$

$$C_P = \frac{dU}{d\theta} + P\frac{dV}{d\theta} \tag{4-34}$$

since U is a function of θ only. Utilizing the ideal-gas law at constant pressure, we get

$$P\frac{dV}{d\theta} = nR \tag{4-35}$$

Substituting this result into eq. (4-34) we get the result that for ideal gases

$$C_P = C_V + nR \tag{4-36}$$

The ratio of C_P to C_V is usually designated γ.

Consider next the equations for an adiabatic process in an ideal gas. Two equations must obtain, the first indicating the absence of heat flow ($đQ = 0$),

$$đQ = dU + P\,dV = 0 \tag{4-37}$$

and the second indicating an ideal gas,

$$PV = nRT \tag{4-38}$$

We may utilize an accessory condition that for ideal gases U is a function of θ only,

$$dU = \left(\frac{\partial U}{\partial \theta}\right)_V d\theta = C_V\,d\theta \tag{4-39}$$

Expressing the ideal-gas law in differential form we get

$$P\,dV + V\,dP = nR\,d\theta \tag{4-40}$$

We can now combine eqs. (4-37), (4-39), and (4-40) to get

$$C_V \frac{(P\,dV + V\,dP)}{nR} + P\,dV = 0 \tag{4-41}$$

Placing terms over a common denominator and grouping coefficients, we can write the equations as

$$(C_V + nR)P\,dV + C_V V\,dP = 0 \tag{4-42}$$

Since $C_V + nR = C_P$, we can write

$$\frac{C_P}{C_V}\frac{dV}{V} + \frac{dP}{P} = \gamma\frac{dV}{V} + \frac{dP}{P} = 0 \tag{4-43}$$

Integrating this equation leads to

$$\gamma \ln V + \ln P = \text{constant} \tag{4-44}$$

$$\ln V^\gamma + \ln P = \ln PV^\gamma = \text{constant} \tag{4-45}$$

$$PV^\gamma = \text{constant} \tag{4-46}$$

Equation (4-46) thus describes a reversible adiabatic process in an ideal gas.

Consider next the simplest possible reversible engine containing an ideal gas. It consists of a cylinder and a gas communicating with the outside world through a piston, the cylinder having a diathermic wall which may be closed off by an adiabatic wall (fig. 4-3). *A diathermic wall permits the flow of heat but not the flow of matter* (definition). At the beginning of the cycle the gas is at state V_A, θ_A. Work is done by putting the cylinder in contact with an isothermal reservoir at temperature θ_A and allowing a reversible expansion to take place to state $P_B, V_B, \theta_B = \theta_A$. The work having been done, it is necessary to return the system to state P_A, V_A, θ_A. In order to do this, the system is first adiabatically isolated and allowed to reversibly expand until the temperature drops to θ_C, the temperature of the cold reservoir. These steps are shown on a P–V diagram in fig. 4-6. The system is then placed in contact with a reservoir at temperature θ_C and isothermally compressed to $P_D, V_D, \theta_D = \theta_C$. The system is then returned to its original state by adiabatically isolating the cylinder and compressing the gas back to P_A, V_A, θ_A. The total work done in going around the cycle is

$$W = \int_A^B P\,dV + \int_B^C P\,dV + \int_C^D P\,dV + \int_D^A P\,dV \tag{4-47}$$

An inspection of fig. 4-6 shows that this is just the area inside the curves $ABCD$. The device we have just been discussing is called a Carnot engine. It is reversible and operates in a cycle. Heat is taken from a constant high-temperature reservoir and given up to a constant low-temperature reservoir.

The total heat taken from the hot reservoir is the integral of $P\,dV$ from A to B. This is so since θ is constant and U remains unchanged, so that the heat and work must balance in this expansion. This may be formally written as

$$Q_{AB} = \int_A^B P\,dV \tag{4-48}$$

The work done along the isothermals for one mole of gas is

$$W = \int_{V_1}^{V_2} P\,dV = R\theta \int_{V_1}^{V_2} \frac{dV}{V} = R\theta \ln \frac{V_2}{V_1} \tag{4-49}$$

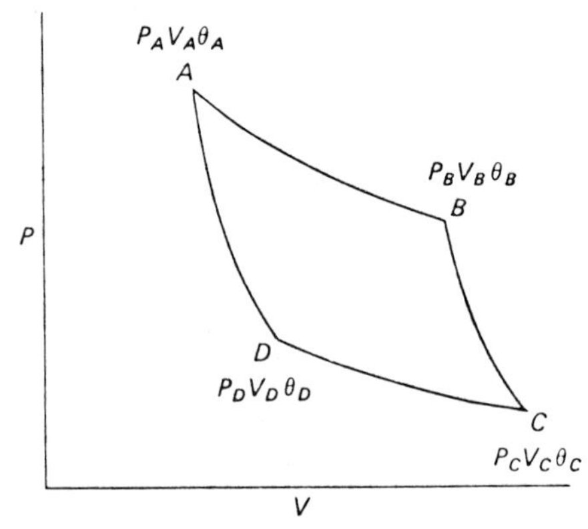

FIGURE 4-6. The working cycle of a Carnot heat engine.

The work done along the adiabatics is

$$W = \int_{V_1}^{V_2} P\,dV = \text{const} \int_{V_1}^{V_2} \frac{dV}{V^\gamma} = \frac{\text{const}}{\gamma - 1}\left[\frac{1}{V_1^{\gamma-1}} - \frac{1}{V_2^{\gamma-1}}\right]$$

$$= \frac{1}{\gamma - 1}[P_1V_1 - P_2V_2] \tag{4-50}$$

The total work in going around the cycle is then

$$W = R\theta_A \ln\frac{V_B}{V_A} + R\theta_C \ln\frac{V_D}{V_C}$$

$$+ \frac{1}{\gamma - 1}[P_BV_B - P_CV_C + P_DV_D - P_AV_A] \tag{4-51}$$

The third term on the right is zero since

$$\begin{aligned}[P_BV_B - P_CV_C + P_DV_D - P_AV_A] &= [R\theta_B - R\theta_C + R\theta_D - R\theta_A]\\ &= R[(\theta_B - \theta_A) + (\theta_D - \theta_C)]\\ &= 0\end{aligned} \tag{4-52}$$

with the last term in brackets vanishing since AB and CD are isothermals so that $\theta_B = \theta_A$ and $\theta_D = \theta_C$. Therefore

$$W = R\theta_A \ln\frac{V_B}{V_A} + R\theta_C \ln\frac{V_D}{V_C} \tag{4-53}$$

The heat taken in is

$$Q_{AB} = R\theta_A \ln\frac{V_B}{V_A} \tag{4-54}$$

The equations of state in going around the cycle are

$$\begin{aligned} P_AV_A &= P_BV_B, & P_CV_C &= P_DV_D\\ P_BV_B^\gamma &= P_CV_C^\gamma, & P_DV_D^\gamma &= P_AV_A^\gamma \end{aligned} \tag{4-55}$$

Eliminating the P's from the preceding equations, we get

$$\frac{V_B}{V_A} = \frac{V_C}{V_D} \tag{4-56}$$

Substituting in eq. (4-53) we get

$$W = R\theta_A \ln\frac{V_B}{V_A} + R\theta_C \frac{V_D}{V_C} = R\theta_A \ln\frac{V_B}{V_A} + R\theta_C \ln\frac{V_A}{V_B} \tag{4-57}$$

This may be rewritten as

$$W = R\theta_A \ln \frac{V_B}{V_A} - R\theta_C \ln \frac{V_B}{V_A}$$

$$= \left(R \ln \frac{V_B}{V_A}\right)(\theta_A - \theta_C) \tag{4-58}$$

The efficiency of the ideal-gas engine is

$$\eta = \frac{W}{Q_{AB}} = \frac{[R \ln(V_B/V_A)](\theta_A - \theta_C)}{R\theta_A \ln(V_B/V_A)} = 1 - \frac{\theta_C}{\theta_A} \tag{4-59}$$

The absolute thermodynamic temperature is defined from the relation

$$\eta = 1 - \frac{T_C}{T_A} \tag{4-60}$$

Therefore

$$\frac{T_C}{T_A} = \frac{\theta_C}{\theta_A} \tag{4-61}$$

This completes the formal proof of the equivalence of absolute thermodynamic temperature and ideal gas temperature.

G. LIFE AT ABSOLUTE ZERO

The concept of an absolute zero of temperature has some immediate biological consequences. Lurking behind many of the advances of modern molecular biology is the very real question of whether, within the foreseeable future, we shall be able to synthesize a living cell. Will it be possible to start out with water, graphite, sulfur, phosphorus, and other vital elements and emerge from the laboratory with an autonomously self-replicating biological organism? One of the approaches to experimentally defining this question has come from studies on the survival of living forms at temperatures near to absolute zero. The significance of low temperatures for this question stems from the existence of pure structure without process in this domain.

An extensive literature exists (see Bibliography) reporting on experiments in which living systems were taken to liquid hydrogen or helium temperatures, kept in that condition at a few degrees from absolute zero for varying lengths of time, brought back to room temperature, and placed under conditions of normal function. A number of forms including bacteria, seeds, and several kinds of microscopic animals survive this treatment with little or no loss of activity. In extreme cases, samples have been kept at a few degrees kelvin for several days, following which the material was found to retain its viability. Rotifers, nematodes, tardigrades, and brine shrimp (Artemia) eggs are among the best characterized examples.

To relate these low-temperature experiments to the synthesis of living systems, we must take the biological concepts of structure and function and examine them at the level of molecular physics. In describing an atomic particle for purposes of statistical mechanical analysis, we specify its state by three position coordinates (x, y, z) and three velocity or, more generally, momentum coordinates (p_x, p_y, p_z). Structure can generally be described in terms of the position coordinates, and process or function corresponds to the momentum. Structure is the result of where the molecules are located, and function is a description of where they are going.

Temperature measures the amount of kinetic energy or the average value of the square of the atomic velocities as the atoms move about in a ceaseless thermal motion. When we reduce the temperature of a sample to its lowest possible value, all molecular motion ceases except for a small quantum mechanical zero point virbation, which need not concern us at this time. In reaching absolute zero, motion effectively comes to a complete stop and the system loses all memory of previous processes except those that have left a structural trace. In this state, the system can be completely specified by noting the positions of the atoms. In other words, all that exists at this temperature limit is structure; the concept of continuity of function loses meaning. Any organism that could survive near-absolute zero would thus be immortal if maintained at that temperature, because one of the processes that would cease would be death or thermal degradation. This is the factor that some people have unsuccessfully attempted to exploit in the low-temperature storage of human corpses.

The fact that, for technical reasons, the experiments described above take the samples within a few degrees of the lower limit rather than absolute zero itself is not of significance in the present context, since the amount of molecular motion is negligible at the temperature studied. For example, at 2 K there is a vanishingly small probability of finding even one vibrationally excited molecule per cell.

When we heat the samples back to room temperature, we do not add back functional information, since heating is a disordering process that simply sets the atoms back in random thermal motion within the constraints imposed by the structure. Yet, the organisms, warmed back to their growth temperatures, commence to function and carry out all their normal processes. The activity that arises is thus entirely the result of configural information, contained in the system near absolute zero, interacting with its new environment.

The conclusion to be drawn from analyzing these low-temperature studies is that the synthesis of a living cell requires only that we construct an appropriate atomic configuration. If the structure is correct we do not have to worry about setting it running; it will set itself running. The problem of cellular synthesis is thus, in kind, no different from the problems of synthesis in ordinary organic chemistry. The difference lies in the fact that the structure is many orders of magnitude more complex than those usually made in the laboratory. The great simplification that emerges from the cryobiology experiments is the rather certain knowledge that the problem of synthesizing a living cell is within the domain of our existing conceptual framework. The task in principle merely requires a team of supersophisticated chemists adept at complex syntheses.

Consider then the hierarchy of chemical structures that go into making up a living cell. First there are the atoms, largely carbon, hydrogen, nitrogen, oxygen, phosphorus, and sulfur, which are assembled into small molecules, many of which are the monomers used in larger structures. The monomers are then combined into high-molecular-weight polymers, which associate by other than valence forces into organelles. The cell is an organized collection of these organelles with an intervening aqueous matrix.

Progress in synthesis has been made at all of these organizational levels. The construction of ordinary small-molecular-weight molecules from simple precursors is the task of organic chemistry, and it is generally believed that any stable, relatively small organic molecule can be synthesized if enough effort is expended. The joining together of monomers to form polymers has taken giant strides in recent years as exemplified by the synthesis of insulin, ribonuclease, and specific high-molecular-weight polynucleotide chains. The construction of organelles from macromolecules is a newer area of research. Yet, already it has been possible to separate small ribosome subunits into ribonucleic acid and 21 types of protein that can be purified and subsequently reassembled into active, functioning organelles. At the organelle and cell levels, microsurgery experiments on amoebas have been most dramatic. Cell fractions from four different animals can be injected into the eviscerated ghost of a fifth amoeba, and a living, functioning organism results. The amoeba does not seem to require the detailed placement of each organelle; their presence appears to be sufficient. Thus, at every level of the hierarchy there is evidence that the structures can be assembled. The smallest living cells from the genus *Mycoplasma* are about 5000 times as large as a single ribosome, but present results suggest that this is a difference of size rather than a difference of kind with respect to the difficulty of synthesis.

Of course, it would be premature to assume that we fully understand all of the structures, even in the smallest organism. Many organelles are primarily built up out of membranes, and the nature of the protein–lipid interaction in these structures is far from being completely understood. The molecular machinery involved in transport processes and various energy transductions has not been isolated, and our understanding of the detailed physical chemistry of these processes is the subject of an ongoing deabte, even among the specialists. A great deal must yet be done to reduce our understanding of biological energy transformations to physics and chemistry. Nevertheless, there seems little reason to doubt that these aspects of cell function lie within the domain of molecular physics.

There are those whose philosophic orientation toward the origin of life would have to be modified following the successful synthesis of a living cell. From the point of view of the present paradigm, the low-temperature experiments should already have forced a shift in viewpoint. Within existing molecular biology, the synthesis of a living cell would not occasion any reorientation of theory, whereas the failure to synthesize a living cell when techniques appeared to have been adequately developed, would force a reexamination of the physical foundations of biology.

TRANSITION

Since thermodynamics arose at the time of the industrial revolution, many of the basic laws were formulated in terms of heat engines and refrigerators. Of special importance is the reversible engine which can be run backward as a refrigerator. Various forms of an empirical generalization known as the second law of thermodynamics have been formulated and these can be shown to be equivalent. A series of theorems (Carnot theorems) describes certain limitations on the efficiency of ideal reversible engines. Using this formalism it is possible to define an absolute thermodynamic temperature scale which is independent of the working substance. It is further shown that this scale is identical with the ideal-gas temperature measure. One feature which emerges is the existence of an absolute zero of temperature which has certain biological implications. The theory as formulated thus far has certain mathematically awkward inexact differentials of work ($đW$) and heat ($đQ$). The next step is to develop mathematical techniques to eliminate these quantities from the equations. This involves the introduction of a new function, the entropy, and this will be our next task.

BIBLIOGRAPHY

Luyet, B. J., and Gehenio, P. M., "Life and Death at Low Temperatures." Biodynamica, Normandy, Missouri, 1940.

A detailed review of the early literature on cryobiology.

Rose, A. H. (ed.), "Thermobiology." Academic Press, New York, 1967.

A series of papers on the effects of heat and cold on a wide variety of organisms.

Skoultchi, A. I., and Morowitz, H. J., Information storage and survival of biological systems at temperatures near absolute zero, *Yale J. Biol. Med.* 37, 158 (1964).

Zemansky, M. W., "Heat and Thermodynamics." McGraw-Hill, New York, 1968.

A detailed discussion of the fundamentals of thermodynamics. This work contains an especially good treatment of experimental background and engineering applications. Includes a wide variety of special topics.

5 Entropy

A. STATE FUNCTIONS

The entropy function has been an enigma to students of thermodynamics since its introduction many years ago. Originally postulated to provide a function of state related to the heat flow $đQ$, it was soon discovered to provide the most general and succinct statement of the condition of equilibrium in any system. With the rise of kinetic theory, entropy, through the Boltzmann H-theorem, provided the most universal notion of mechanical irreversibility. The development of statistical mechanics used entropy to bridge the gap between the statistical formalism and phenomenological thermodynamics. Finally, the modern theory of information relates entropy to information and ultimately to the observer. All of these considerations make entropy central to biology and provide strong incentive to penetrate the conceptual difficulties.

Before discussing entropy *per se* we will briefly review the mathematics of a function of two variables $f(x, y)$. Such an expression is a function of state in that its value depends only on the values of the independent variables of the system x and y and is independent of the path taken to arrive at the state x and y. The differential of $f(x, y)$ can be written

$$df = \frac{\partial f}{\partial x} dx + \frac{\partial f}{\partial y} dy \tag{5-1}$$

$$df = \mathscr{K}(x, y)\, dx + \mathscr{L}(x, y)\, dy \tag{5-2}$$

A path between x_1, y_1 and x_2, y_2 can be represented by a relation

$$\Omega(x, y) = 0 \tag{5-3}$$

where Ω is some function of x and y plus constants. For instance, if the path is a straight line, $\Omega(x, y)$ is of the form $y - ax - b$. Since f is a state function, eq. (5-1) can be integrated directly,

$$\int_{x_1 y_1}^{x_2 y_2} df = f(x_2, y_2) - f(x_1, y_1) \tag{5-4}$$

The integral must be independent of the path since it depends only on the values of the function at the termini of the paths. A direct consequence of (5-1) and (5-2) and the theory of partial differentials is that

$$\frac{\partial \mathscr{K}}{\partial y} = \frac{\partial^2 f}{\partial x \, \partial y} = \frac{\partial \mathscr{L}}{\partial x} \tag{5-5}$$

Many of the most important relations of thermodynamics are derived by the use of this relation.

If we are given the differential of a function in the form

$$\text{differential} = \mathscr{K}^*(x, y)\, dx + \mathscr{L}^*(x, y)\, dy \tag{5-6}$$

and relation (5-5) does not hold, the differential is said to be inexact; i.e., there is no f^* such that $df^* = \mathscr{K}^*\, dx + \mathscr{L}^*\, dy$, and we can only integrate under very special conditions. The general form of the integral must then be written:

$$\text{integral} = \int_{x_1 y_1}^{x_2 y_2} \mathscr{K}^*(x, y)\, dx + \int_{x_1 y_1}^{x_2 y_2} \mathscr{L}^*(x, y)\, dy \tag{5-7}$$

In order to perform the integration we require an additional relation of the form (5-3). Such an expression permits us to solve for y in terms of x and is, as we have noted, a path in x–y space. The integral will depend on this specific path between the limits.

It will be useful to consider a simple thermodynamic example of a state function and a function which does not meet the requirements. Start with an ideal gas for which we may write the following:

$$V = f(P, T) = \frac{RT}{P} \tag{5-8}$$

In P–T space we may consider two paths from P_1, T_1, to P_2, T_2, which are shown in fig. 5-1.

Since V is a function of state, we can write

$$dV = \left(\frac{\partial V}{\partial T}\right)_P dT + \left(\frac{\partial V}{\partial P}\right)_T dP = \frac{R}{P}\, dT - \frac{RT}{P^2}\, dP \tag{5-9}$$

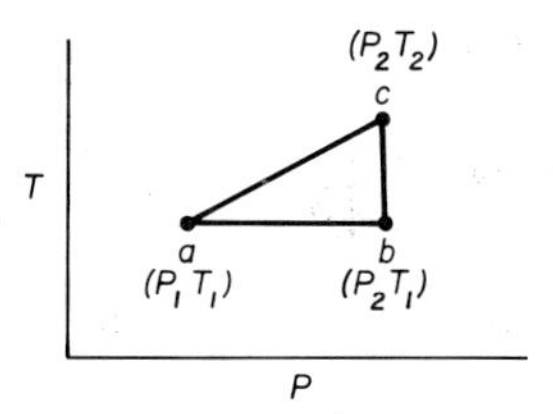

FIGURE 5-1. A graph in P–T space showing alternate pathways between two points.

First integrate dV along the path ac and then along abc. The first path may be represented as

$$T - T_1 = \frac{\Delta T}{\Delta P}(P - P_1) \tag{5-10}$$

Therefore, along the route of integration

$$dT = \frac{\Delta T}{\Delta P}\, dP \tag{5-11}$$

Substituting (5-10) and (5-11) into (5-9) and integrating, we get

$$\Delta V = \frac{RT_2}{P_2} - \frac{RT_1}{P_1} \tag{5-12}$$

Next we integrate (5-9) along the path ab, where T retains a fixed value T_1:

$$\Delta V' = \int_a^b - RT_1 \frac{dP}{P_2} = \frac{RT_1}{P_2} - \frac{RT_1}{P_1} \tag{5-13}$$

In integrating from b to c, P remains constant, P_2, and

$$\Delta V'' = \int_b^c \frac{R\, dT}{P_2} = \frac{RT_2}{P_2} - \frac{RT_1}{P_2} \tag{5-14}$$

The total ΔV is obtained by summing (5-13) and (5-14) and the result is identical with (5-12). The function V which was established as a state function in eq. (5-8) has values which are indeed independent of the path of integration.

Next consider the work function $P\, dV$ and note that, for one mole of an ideal gas,

$$đW = P\, dV = d(PV) - V\, dP = R\, dT - \frac{RT}{P}\, dP \tag{5-15}$$

Along the path ac,

$$đW = R\left[dT - \left(T_1 - \frac{\Delta T}{\Delta P} P_1\right)\frac{dP}{P} - \frac{\Delta T}{\Delta P}\, dP\right] \tag{5-16}$$

which, on integrating, leads to

$$\Delta W = R\left(\frac{\Delta T}{\Delta P} P_1 - T_1\right)\ln\frac{P_2}{P_1} \tag{5-17}$$

We now perform the integral along the path *abc* in two parts *ab* and *bc*:

$$\Delta W_{ab} \quad \int_A^B -\frac{RT}{P}\,dP = -RT_1 \ln\frac{P_2}{P_1} \tag{5-18}$$

$$\Delta W_{bc} = R\,\Delta T \tag{5-19}$$

The total integral is

$$\Delta W_{ac} = R\left[\Delta T - T_1 \ln\frac{P_2}{P_1}\right] \tag{5-20}$$

Equations (5-17) and (5-20) lead to different values for ΔW, so that we have demonstrated that $đW$ is not an exact differential and W is not a function of state. We could have arrived at that result more quickly by noting that

$$đW = R\,dT - \frac{RT}{P}\,dP = \mathscr{K}^*\,dT + \mathscr{L}^*\,dP \tag{5-21}$$

$$\frac{\partial \mathscr{K}^*}{\partial P} = 0, \qquad \frac{\partial \mathscr{L}^*}{\partial T} = -\frac{R}{P} \tag{5-22a}$$

and therefore

$$\frac{\partial \mathscr{K}^*}{\partial P} \neq \frac{\partial \mathscr{L}^*}{\partial T} \tag{5-22b}$$

Returning to eqs. (5-2) and (5-4) we introduce the closed line integral $\oint$ around a path which is a closed loop beginning and ending at the point x_1, y_1. We may then write for any state function

$$\oint df = \int_{x_1 y_1}^{x_1 y_1} df = f(x_1, y_1) - f(x_1, y_1) = 0 \tag{5-23}$$

The property of having the closed line integral equal to zero is a characteristic feature of state functions and may be used to identify them. The vanishing of the closed line integral is equivalent to the function being a function of state only. Refer to fig. 5-2 and consider some arbitrary point x_2, y_2 on the closed path. The vanishing of the closed line integral implies the following equation, since the total line integral can be divided into a sum of successive integrals:

$$\oint df = \int_{x_1 y_1}^{x_2 y_2} df + \int_{x_2 y_2}^{x_1 y_1} df = 0 \tag{5-24}$$

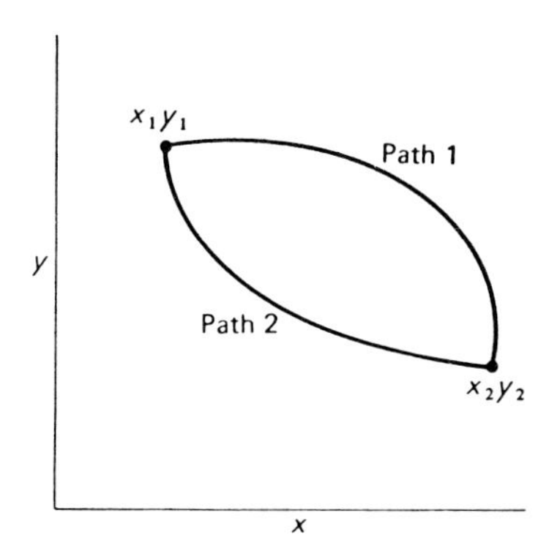

FIGURE 5-2. Two possible paths of integration from the point x_1y_1 to the point x_2y_2.

By a straightforward transformation of the limits, eq. (5-7) can be rewritten as

$$\int_{x_1y_1}^{x_2y_2} df - \int_{x_1y_1}^{x_2y_2} df = 0 \tag{5-25}$$

The first term is the integral from x_1, y_1 to x_2, y_2 by path 1, while the second term is the integral from x_1, y_1 to x_2, y_2 by path 2. We may rewrite this equation in the following form:

$$\underset{\text{path 1}}{\int_{x_1y_1}^{x_2y_2} df} = \underset{\text{path 2}}{\int_{x_1y_1}^{x_2y_2} df} \tag{5-26}$$

The vanishing of the line integral is thus equivalent to the integral being independent of the path. This rather abstract mathematical point has been labored because we shall shortly use this result to show that entropy is a state function.

B. INTEGRATING FACTORS

One further mathematical note will complete our introduction to integrals of dQ. If we have an inexact differential of the form

$$đf(x, y, z) = \mathscr{K}(x, y, z)\,dx + \mathscr{L}(x, y, z)\,dy + \mathscr{M}(x, y, z)\,dz \tag{5-27}$$

it may be possible to multiply $đf$ by an integrating factor $\mathscr{I}(x, y, z)^{-1}$ so that

$$\frac{đf(x, y, z)}{\mathscr{I}(x, y, z)} = dg(x, y, z) \tag{5-28}$$

where dg is an exact differential. This may be illustrated with an example:

$$đf = dx + \frac{x}{y}\,dy + \frac{x}{z}\,dz \tag{5-29}$$

Assume $\mathscr{J}$ equals $1/yz$, then

$$dg = yz\,dx + xz\,dy + xy\,dz \tag{5-30}$$

and

$$g = xyz + \text{constant} \tag{5-31}$$

There is, in general, no uniqueness to the integrating factor (although it is not arbitrary). We could have chosen $\mathscr{J}$ equal to x, in which case

$$dg = \frac{dx}{x} + \frac{dy}{y} + \frac{dz}{z} \tag{5-32}$$

and

$$g = \ln xyz + \text{constant} \tag{5-33}$$

It has been shown by mathematical analysis that an integrating factor exists only when the equation $đf = 0$ has a solution.

C. INTEGRALS OF $đQ$

In describing integrals of $đQ$ we shall start with a rather intuitive physical approach and then move toward a more abstract mathematical formulation.

We have already shown that U is a function of state while Q and W are not. Recalling the first law for a simple one-component system, we see that for reversible transformations

$$dU = đQ - P\,dV \tag{5-34}$$

While $đW$ is not an exact differential, it is the product of a state variable P and an exact differential dV(change of volume). The question arose early in the development of the theory, "Could $đQ$ be represented in terms of state variables and exact differentials?" If this were possible, it would imply important mathematical advantages in dealing with eq. (5-34). Our method of approaching the problem is to employ optimism; that is, we assume that we can represent $đQ$ as the product of some function of state variables (an integrating factor) and an exact differential, and proceed to find the properties of the relevant quantities; i.e., we assume that

$$đQ = (\text{function of state}) \times dS \tag{5-35}$$

where dS is an exact differential. If we can find appropriate functions, then our optimism is justified and we can use the new function S. S is called the entropy and we must now prove that it exists and has the requisite properties. While this may seem a curious method of approach, it has two virtues:

first, it works, and second, it can ultimately be justified by the more exacting Principle of Caratheodory, which will be discussed later in this chapter.

If we assume eq. (5-35) we can write dU for a one-component system as

$$dU = \left(\frac{\partial U}{\partial S}\right)_V dS + \left(\frac{\partial U}{\partial V}\right)_S dV \tag{5-36}$$

We have the following requirements to satisfy in order for equations (5-34) and (5-35) to be in agreement:

$$\left(\frac{\partial U}{\partial S}\right)_V dS = đQ \tag{5-37}$$

$$\left(\frac{\partial U}{\partial V}\right) = -P \tag{5-38}$$

We still have considerable arbitrariness left in the choice of the function S and we narrow down that choice by trying to establish the function so that T plays a role analogous to $-P$ in eqs. (5-34) and (5-36). This then would require

$$\left(\frac{\partial U}{\partial S}\right)_V = T \tag{5-39}$$

Equations (5-36)–(5-39) would lead to the following expression for the entropy:

$$dS = \frac{dU + P\,dV}{T} = \frac{đQ}{T} \tag{5-40}$$

We may regard eq. (5-40) as the defining equation for a function S which satisfies conditions (5-37) and (5-39). The choice of T as $(\partial U/\partial S)_V$ defines S, but does not guarantee either that it exists or that it is a state function. Proof that S is a state function then would complete the problem and allow us to use the results for subsequent analysis.

D. *PROPERTIES OF THE ENTROPY FUNCTION*

For perfect gases we know U as a function of T and we have a relation between P, V, and T, so that we can substitute directly into eq. (5-40), obtain the entropy function, and show that it has all the desired properties. This is done in Appendix III. In the more general case we do not have U as a function of the state variables, so we take a less direct approach. The proof we shall offer is not completely rigorous, but is intuitively satisfactory.

S will be shown to be a state function by demonstrating that $\oint dS = 0$ [see eq. (5-26)]. Consider an arbitrary closed curve in P–V space (fig. 5-3). Divide the area up into a large number of Carnot cycles. For each Carnot cycle

$$\frac{\Delta Q_{\text{in}}}{T_{\text{source}}} = \frac{\Delta Q_{\text{out}}}{T_{\text{sink}}} \tag{5-41}$$

This may be seen from eq. (4-17) since in the Carnot cycle the full energy input is ΔQ_{in} and the amount of energy rejected is ΔQ_{out}. The cycle we are showing is represented in detail in fig. 4-6. Allowing the sign to be part of the ΔQ, we get

$$\frac{\Delta Q_{\text{in}}}{T_{\text{source}}} + \frac{\Delta Q_{\text{out}}}{T_{\text{sink}}} = 0 \tag{5-42}$$

Instead of integrating dS along the actual curve, we integrate along the Carnot cycles approaching the curve. Along the adiabatics, $\Delta Q = 0$.

Along the isothermals, the ΔQ's always appear pairwise, so that divided by the corresponding T's their sum is zero. Therefore along the Carnot cycle curves $\oint dS = 0$. Since these curves can arbitrarily approach the actual curve by taking more and more cycles within the area, we conclude that $\oint dS = 0$ for the actual curve. The entropy S is thus a function of state. Therefore entropy is a property of an equilibrium system. It is not easy to

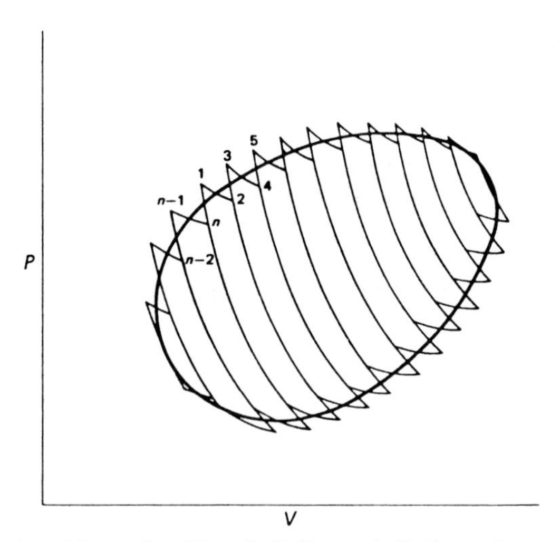

FIGURE 5-3. An arbitrary closed loop in P–V space is divided up into a number of Carnot cycles whose envelope approaches close to the initial curve. The path of integration is from 1 to 2 to 3 to 4 to $\cdots n-1$ to n to 1.

convey a physically intuitive notion of this property in the sense that temperature and volume correspond to abstractions from direct experience. It is perhaps instructive to consider the analogy between the $P\,dV$ and $T\,dS$ terms. In doing work, pressure is the driving force and volume is the extensive property of the system that changes. We could of course write $dV = -đW/P$. Similarly, in heat flow, temperature is the driving force and entropy is the extensive property of the system that changes. Entropy has the units of energy divided by temperature.

A mathematically more precise approach to the problem of the integration of $đQ$ was introduced by C. Caratheodory in 1909. For simplicity we shall present a somewhat restricted version, which may be extended to more general systems. We introduce two proven results of mathematics as starting points:

(1) Any differential expression in two variables has an integrating factor.
(2) For a differential in more than two variables, an integrating factor will always exist if $đf = 0$.

Next consider an equilibrium system of arbitrary complexity and allow a reversible adiabatic process to take place. We can then write

$$đQ = dU + đW = 0 \tag{5-43}$$

For such a system $đQ$ and $đW$ can be expressed in terms of differentials of the state variables and by condition 2 above an integrating factor of $đQ$ exists:

$$dg = \frac{đQ}{\mathcal{I}} \tag{5-44}$$

and dg is an exact differential. An alternative statement of the second law by Caratheodory says that arbitrarily close to any given state of a closed system (closed to the flow of matter) there are an unlimited number of other states which cannot be reached by an adiabatic process. This turns out to provide an equivalent condition to statement 2 and the existence of an integrating factor.

We next consider a global system made up of two subsystems in contact through a diathermic wall. If a reversible adiabatic process takes place, then for each of the subsystems we note that

$$đW = đW_{\text{ext}} + đW_{\text{int}} \tag{5-45}$$

The work term consists of an interaction with the outside world $đW_{\text{ext}}$ and an interaction with the neighboring system $đW_{\text{int}}$. Therefore

$$đW = đW_1 + đW_2 = đW_{1(\text{ext})} + đW_{1(\text{int})} + đW_{2(\text{ext})} + đW_{2(\text{int})} \tag{5-46}$$

But the nature of a reversible process requires that

$$đW_{1(\mathrm{int})} = -đW_{2(\mathrm{int})} \tag{5-47}$$

We can then use the first law to write

$$dU = dU_1 + dU_2 = đQ_1 + đQ_2 + đW_{1(\mathrm{ext})} + đW_{2(\mathrm{ext})} \tag{5-48}$$

Therefore, using (5-43),

$$đQ = đQ_1 + đQ_2 = 0 \tag{5-49}$$

If each subsystem is a pure substance (this is a restriction to simplify the treatment) then

$$h_1(P_1V_1) = f(\theta) = h_2(P_2V_2) \tag{5-50}$$

The term f is a function of the empirical temperature only and eqs. (5-50) are equations of state for the two subsystems. Condition 1 (two-parameter case) now provides that each subsystem will have an integrating factor and condition 2 (vanishing of the differential) provides that the global system has an integrating factor. Combining (5-44) and (5-49) we then get

$$\mathcal{J}\, dg = \mathcal{J}_1\, dg_1 + \mathcal{J}_2\, dg_2 \tag{5-51}$$

A detailed analysis of the properties of eq. (5-51) (which we are not presenting in this treatment) shows that the simultaneous existence of $\mathcal{J}_1$, $\mathcal{J}_2$, and $\mathcal{J}$ requires that

$$\mathcal{J}_1 = T(\theta)\mathcal{G}(g_1), \qquad \mathcal{J}_2 = T(\theta)\mathcal{G}(g_2), \qquad \mathcal{J} = T(\theta)\mathcal{G}(g_1 g_2) \tag{5-52}$$

where $T(\theta)$ is a function of the empirical temperature only, which is the same for the global system and the two subsystems. We may then define the following functions:

$$S_i = \int \mathcal{G}_i(g_i)\, dg_i \tag{5-53a}$$

and hence

$$\mathcal{J}_i\, dg_i = T(\theta)\, dS_i = dU_i + đW_i. \tag{5-53b}$$

Globally this leads to

$$dS = dS_1 + dS_2 = \frac{đQ}{T(\theta)} \tag{5-54}$$

Equation (5-53) shows that S is a state function since it depends only on g. Equation (5-54) establishes that the integrating factor for $đQ$ must be a function of the empirical temperature only. We can designate this function $T(\theta)$ as the Caratheodory absolute temperature since it is the same for all systems. Suppose one of the subsystems above is one mole of an ideal gas,

$$dS_1 = \frac{đQ_1}{T(\theta)} = \frac{dU_1 + đW_1}{T(\theta)} \tag{5-55}$$

Since for such a substance U is equal to $\frac{3}{2}R\theta$ per mole [from eq. (4-29)], $đW = P\,dV$, and $PV = R\theta$, we can rewrite (5-55) as

$$dS_1 = \frac{\frac{3}{2}R\,d\theta + R\theta\,dV/V}{T(\theta)} \tag{5-56}$$

By inspection of the above equation we see that $T(\theta) = \theta$ gives us the simplest possible integrating factor. Thus θ, the ideal gas temperature, serves as an integrating factor for $đQ$ and is identically equal to the Caratheodory absolute temperature. We have already shown θ to be equal to the Kelvin absolute thermodynamic temperature. Since $T(\theta)$ applies for all systems, the restriction to ideal gases is removed and we have an independent approach to absolute temperature. Equation (5-56) allows us to establish the entropy of an ideal gas in terms of the state variables T and V (see Appendix III).

Since entropy is a state function, relations can be formulated between entropy and other state variables as indicated below:

$$S = S(U, V) = S(P, V) = S(P, T) = S(V, T) = S(P, U) = S(U, T) \tag{5-57}$$

Next consider the relationship between entropy and the refrigerator statement of the second law of thermodynamics. If we transfer heat $đQ$ from a colder body at temperature T_2 to a hotter body at temperature T_1, the total entropy change is

$$dS = -\frac{đQ}{T_2} + \frac{đQ}{T_1} = đQ\frac{(T_2 - T_1)}{T_1 T_2} \tag{5-58}$$

since $T_1 > T_2$,

$$dS < 0 \tag{5-59}$$

But, if no work is done on the system, this heat transfer is contrary to the refrigerator statement of the second law, which therefore requires in cases involving only heat flow, that $T_2 > T_1$ and

$$dS \geqslant 0 \tag{5-60}$$

It turns out that the increase of entropy accompanying all real processes in a total system is an alternative and very general statement of the second law of thermodynamics.

Let us next consider the entropy changes accompanying the operation of real engines. First recall that for an ideal engine

$$\frac{Q_1}{T_1} = \frac{Q_2}{T_2} \tag{5-61}$$

The entropy change in one cycle is the sum of the entropy changes of the engine and the reservoir. Since the engine returns to its original state, its entropy change is zero. The net change is that of the reservoirs, and that is

$$\Delta S = -\frac{Q_1}{T_1} + \frac{Q_2}{T_2} = 0 \tag{5-62}$$

The zero sum in eq. (5-62) follows from eq. (5-61). For a real engine the efficiency is less than for an ideal reversible engine. Designating the heat taken in and discharged by a real engine as Q_1' and Q_2', we have as an expression of the efficiency

$$\eta' = 1 - \frac{Q_2'}{Q_1'} \tag{5-63}$$

However, Carnot's theorem requires that the efficiency of a real engine be less than that of an ideal reversible engine, which we have shown is $1 - (T_2/T_1)$. Then,

$$1 - \frac{Q_2'}{Q_1'} < 1 - \frac{T_2}{T_1} \tag{5-64}$$

If we rearrange the terms in eq. (5-64), we get

$$\frac{Q_2'}{T_2} > \frac{Q_1'}{T_1} \tag{5-65}$$

Consequently, the entropy change accompanying a cycle of a real engine is

$$\Delta S = -\frac{Q_1'}{T_1} + \frac{Q_2'}{T_2} > 0 \tag{5-66}$$

The operation of a real engine is accompanied by an entropy increase. Reexamining eq. (5-58) we can see that, if T_2 were greater than T_1, the entropy would increase. But since it is always observed that heat flows from a hot body to a cold body, heat flow is always accompanied by an entropy increase.

A somewhat different view of entropy comes from a consideration of the entropy of mixing of two gases. Consider two chambers, one of volume V_A containing n_A moles of perfect gas A and the second of volume V_B containing

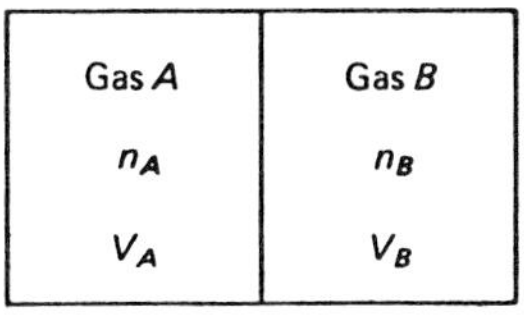

Gas A	Gas B
n_A	n_B
V_A	V_B

FIGURE 5-4. Two compartments containing different gases. The pressures are equal in the two compartments.

n_B moles of ideal gas B, with the volumes chosen so that $P_A = P_B$ as in fig. 5-4. If we adiabatically isolate the system and open a hole between the two chambers, the gases will mix. There will be no external work and ΔQ will be zero; U will remain unchanged and, accordingly, since U is a function of T only, T will remain unchanged. The mixing of the two gases under these conditions is clearly an irreversible process.

We next consider a reversible path from the same initial to the same final state. To do this, we introduce another idealization, the concept of *a perfect selective membrane which will be impermeable to one species of molecule and will freely pass the other species* (definition). Such membranes can be approached experimentally, and indeed the phenomenon of osmosis, which we treat in Chapter 15, depends on the existence of such structures. We will utilize two idealized rigid, selective membranes (fig. 5-5), the first freely permeable to A molecules but impermeable to B molecules, the second freely permeable to B molecules but impermeable to A molecules. Replace the center barrier by the two membranes, the A-permeable one to the left of the B-permeable one. Since both membranes are rigid, connect them to external pistons. Connect the whole system to an isothermal reservoir at temperature T and allow the right-hand piston to move to the right reversibly to the edge of the chamber. Gas A exerts a pressure P_A on the membrane and gas B exerts no pressure, since the membrane is freely

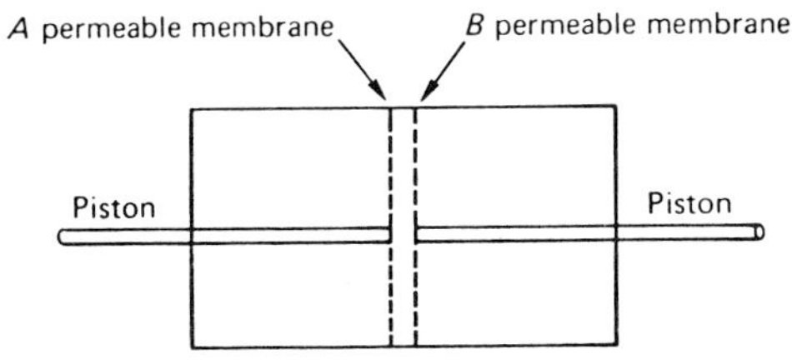

FIGURE 5-5. The same chamber shown in fig. 5-4. The barrier is replaced by two membranes attached to pistons. One membrane is permeable only to A molecules and the second permeable only to B molecules.

permeable to B. The amount of work done is

$$W_A = \int_{V_A}^{V} P\,dV = \int_{V_A}^{V} \frac{n_A RT}{V}\,dV = RTn_A \ln \frac{V}{V_A} \tag{5-67}$$

where $V = V_A + V_B$. Similarly, the amount of work necessary to bring the left-hand piston to the edge of the chamber is

$$W_B = RTn_B \ln \frac{V}{V_B} \tag{5-68}$$

In both cases, heat from the reservoir is being converted into external work. Since the final state is the same as in the previous case, we get

$$\Delta U = \Delta Q - \left(RTn_A \ln \frac{V}{V_A} + RTn_B \ln \frac{V}{V_B} \right) = 0 \tag{5-69}$$

We rearrange terms in eq. (5-69), solve for ΔQ, and compute ΔS, which is simply $\Delta Q/T$ (since the process is isothermal),

$$\Delta S = R\left(n_A \ln \frac{V}{V_A} + n_B \ln \frac{V}{V_B} \right) \tag{5-70}$$

In the irreversible mixing, $\Delta Q = 0$; hence,

$$\int_{\text{reversible}} \frac{đQ}{T} = \Delta S_{\text{reversible}} = \Delta S_{\text{irreversible}} > \int_{\text{irreversible}} \frac{đQ}{T} \tag{5-71}$$

Formula (5-71) turns out to be of considerably more general validity than the single case just given and will get further attention later. Formula (5-70) is usually written in the form

$$\Delta S = -R\left(n_A \ln \frac{n_A}{n_A + n_B} + n_B \ln \frac{n_B}{n_A + n_B} \right) \tag{5-72}$$

Note that, if molecules A and B were identical, it would have been impossible to have a membrane that could select between them. Consequently, the distinguishability of the molecules makes it possible to get the work $W_A + W_B$ out of the system. Mixing the molecules lowers this potential to do work and causes an entropy rise in the system.

In a very general way, entropy measures the orderliness of a system. Spontaneous processes tend toward equilibrium, which is the state of maximum disorder consistent with the constraints of the system. This state of maximum disorder is also the state of maximum entropy, a concept which will later become clearer in the discussions of entropy from a statistical-mechanical and informational point of view.

The global formulation of the second law of thermodynamics may now be stated as follows:

(*a*) There exists a function of state, the entropy, which is defined for reversible transformations by $dS = đQ/T$.

(*b*) In all reversible transformations the entropy universe (system plus reservoirs) remains constant.

(*c*) All irreversible processes are accompanied by an increase of entropy of the system and the surroundings.

(*d*) The last two statements indicate that total entropy remains constant or increases but never decreases.

(*e*) The equilibrium state is the state of maximum entropy of the system plus the surroundings. This is the most general statement of equilibrium.

Part (*a*) has already been demonstrated. Part (*b*) can be shown in the following way. In reversible transformations, the entropy term always involves transfer of heat between a system and reservoir, so that

$$dS = dS_{\text{system}} + dS_{\text{reservoir}}$$

Since heat is transferred, $đQ_{\text{system}} = -đQ_{\text{reservoir}}$ and since the process is reversible, system and reservoir must be at approximately the same temperature, $T_{\text{system}} - T_{\text{reservoir}} = dT$; thus

$$dS = -\frac{đQ}{T + dT} + \frac{đQ}{T} = \frac{đQ\, dT}{T^2} \cong 0 \qquad (5\text{-}73)$$

Thus dS is the product of two differentials and is consequently vanishingly small, particularly since the more nearly the process approaches reversibility, the more nearly does dT approach zero.

Part (*c*) can be demonstrated by considering a system which undergoes an irreversible adiabatic process from equilibrium state I to equilibrium state II. We shall assume an entropy decrease and show that it leads to a violation of the second law of thermodynamics. Assume that $S_{\text{I}} > S_{\text{II}}$. We now return the system from state (*b*) to state (*a*) along some reversible path. Along this path

$$S_{\text{I}} - S_{\text{II}} = \int_{\text{II reversible}}^{\text{I}} \frac{đQ}{T} > 0 \qquad (5\text{-}74)$$

This condition is necessary since, by assumption, $S_{\text{I}} - S_{\text{II}}$ is positive. There must therefore be a flow of heat from reservoirs into the system. Since we return to the same U_{I} by a path involving heat flow into the system, there must be net work done in going around the path. Since ΔU for the cycle is zero, we can sum up the ΔQ and ΔW terms as follows:

$$\Delta Q_{\text{return}} - \int_{\text{II irreversible}}^{\text{I}} P\,dV - \int_{\text{I irreversible}}^{\text{II}} P\,dV = 0 \tag{5-75}$$

Equation (5-75) can be rewritten in the following form:

$$\Delta Q_{\text{return}} = \Delta W_{\text{net work around cycle}} \tag{5-76}$$

A device such as we have just described can act cyclically and convert heat into work with no other effect, which is a violation of the engine statement of the second law; hence, $\Delta S = S_{\text{II}} - S_{\text{I}} \geqslant 0$ for irreversible processes and $S_{\text{II}} > S_{\text{I}}$. The spontaneous process always leads to a state of higher entropy.

E. THE MEASURABILITY OF ENTROPY

The entropy of a pure substance can always be determined relative to some arbitrary standard state. If the standard state is defined by some values of the pressure and temperature, P_0 and T_0, any arbitrary state can be represented by some values of P and T. Since entropy is an extensive function, and depends on the amount of material in the system, the entropy of pure substances is generally determined as entropy per gram or entropy per mole. We will adopt the convention of using entropy per mole. The starting point is one mole of pure substance (one chemical species only and one phase only) at P_0 and T_0. A reversible path must be followed to P, T. First reversibly heat the system at constant pressure from T_0 to T and then place it in contact with a reservoir at temperature T and expand or contract the system until the pressure goes from P_0 to P. The change in entropy is therefore

$$S - S_0 = \int_{T_0}^{T} \frac{đQ}{T} + \int_{P_0}^{P} \frac{đQ}{T} \tag{5-77}$$

Equation (5-77) can be rewritten in the following form:

$$S - S_0 = \int_{T_0}^{T} \frac{(\partial Q/\partial T)_P\,dT}{T} + \frac{1}{T}\int_{V_0}^{V} \left(\frac{\partial Q}{\partial V}\right)_T dV \tag{5-78}$$

V_0 and V are the volumes corresponding to P_0 and P at the temperature T. *The quantity $(\partial Q/\partial T)_P$ is the specific heat at constant pressure, C_P, while $(\partial Q/\partial V)_T$ is the heat of compression at constant temperature, h_c,* (definitions). Both of these quantities can be measured by calorimetric methods and values for pure substances are available in the literature. These empirically determined parameters can be inserted into eq. (5-79):

$$S - S_0 = \int_{T_0}^{T} \frac{C_p\,dT}{T} + \frac{1}{T}\int_{V_0}^{V} h_c\,dV \tag{5-79}$$

The quantities C_p and h_c, as we have already noted, can be determined as functions of temperature and pressure. Equation (5-79) can be evaluated by numerical integration, so that the entropy of a pure substance relative to a standard state can always be determined if the appropriate calorimetry has been done.

In many applications both P and P_0 are taken as atmospheric pressure, so that eq. (5-79) reduces to

$$S = S_0 + \int_{T_0}^{T} \frac{C_p\,dT}{T} \tag{5-80}$$

F. THE THIRD LAW OF THERMODYNAMICS

The third law of thermodynamics states that the entropy of all pure substances at absolute zero is zero. This law may be viewed as a defining statement for the previously arbitrary standard state for entropy. It is probably best to regard the third law as a postulate of thermodynamics, as it later takes on special meaning in light of statistical mechanics. Utilizing the third law, eq. (5-80) becomes

$$S = \int_{0}^{T} \frac{C_p\,dT}{T} \tag{5-81}$$

With the use of this equation plus measured values of the specific heat it is possible to calculate the absolute entropy of any pure substance. The values of entropy given in thermodynamic tables are calculated this way. In Chapter 17 we shall briefly outline experimental methods of measuring specific heats.

G. ENTROPY AND CHEMICAL REACTIONS

Thus far our thermodynamic analysis has been in terms of pure substances or, in one restricted case, in terms of two ideal gases. The next stage is to consider chemical reactions and the associated energy and entropy changes. The method of approach is to go back to the first law of thermodynamics in the form $dU = dQ - dW$. In this form the law states that the change of total energy of a system is the amount of heat crossing the boundaries of the system plus the amount of work crossing the boundaries of the system. Suppose now that a small amount dn_i moles of a pure substance crosses the boundary into the system. The change of energy of the system due to this process is

$$dU = \left(\frac{\partial U}{\partial n_i}\right)_{S,V} dn_i \tag{5-82}$$

This energy will be additive to the energy flows from the other processes, so that we may write the first law for a reversible transformation as

$$dU = T\,dS - P\,dV + \left(\frac{\partial U}{\partial n_i}\right)_{S,V} dn_i \tag{5-83}$$

We will define the coefficient of dn_i as the chemical potential of the ith substance μ_i,

$$\mu_i = \left(\frac{\partial U}{\partial n_i}\right)_{S,V} \tag{5-84}$$

If we now generalize eq. (5-83) for the possibility of many substances, we get

$$dU = T\,dS - P\,dV + \sum \mu_i\,dn_i \tag{5-85}$$

The first law is still a conservation law indicating that the change of energy in a system is the sum of all the energy flows across the boundaries. The existence of more than one chemical species introduces a new factor to be considered, the possibility of chemical reactions. Under such considerations the n_i are not conserved for the system and the surroundings; introduction of dn_i moles of a substance into a system does not necessarily increase the amount of that substance in the system by an amount dn_i, as chemical reactions may act as internal sources or sinks for the ith substance. Similarly, chemical reactions may act as sources or sinks for thermal energy and potential energy. All this does not affect the validity of eq. (5-85), but merely stresses that $\sum \mu_i\,dn_i$ is not an exact differential, just as $đQ$ and $đW$ are not exact differentials. Another way of viewing this is that energy is conserved, mass is conserved, indeed for ordinary chemical reactions (excluding nuclear reactions) the number of atoms of each type is conserved, but the form of the energy is not conserved and the chemical form of the molecules is not conserved. Changes in chemical form are accompanied by changes in energy form, so that the conservation laws as well as the relations governing change of form are related. With this in mind the notion of U being a state function can be generalized and we can write

$$U = U(S, V, n_1, n_2, \ldots, n_i, \ldots, n_N) \tag{5-86}$$

In any chemical reaction taking place within the system the dn_i are not independent but are related by the stoichiometry of the reaction. Consider the following chemical transformation taking place in a system in which there is no flow of matter from the outside:

$$\nu_A A + \nu_B B \leftrightharpoons \nu_C C + \nu_D D \tag{5-87}$$

Equation (5-87) is a representation of any arbitrary chemical reaction in which two reagents react to give two products. The symbols ν_A, ν_B, ν_C, and ν_D are the stoichiometric coefficients which are usually small integers. The capital letters A, B, C, and D are simply the symbols designating the chemical species and in actual cases are represented by chemical formulae. An actual example of an equation of this type is

$$1BaCl_2 + 2NaOH \rightleftharpoons 1Ba(OH)_2 + 2NaCl$$

with

$$\nu_{BaCl_2} = 1, \qquad \nu_{NaOH} = 2, \qquad \text{etc.}$$

In order for ν_A moles of A to be used up, they must react with ν_B moles of B to produce ν_C moles of C and ν_D moles of D. Therefore,

$$-\frac{dn_A}{\nu_A} = -\frac{dn_B}{\nu_B} = \frac{dn_C}{\nu_C} = \frac{dn_D}{\nu_D} \tag{5-88}$$

We can set the expressions in eqs. (5-88) equal to a new function $d\xi$, where ξ is defined as the degree of advancement of a chemical reaction. We can then write

$$dn_A = -\nu_A\,d\xi, \qquad dn_B = -\nu_B\,d\xi, \qquad dn_C = \nu_C\,d\xi, \qquad dn_D = \nu_D\,d\xi \tag{5-89}$$

The number of independent variables has been reduced from four to one because of stoichiometric constraints on the one chemical reaction which is taking place.

The entropy change accompanying a chemical reaction obeys the general conditions regarding entropy that we have just outlined. To look at an example, start with a vessel of constant volume containing molecules of A and B only and in contact with a reservoir at temperature T. Next allow the reaction shown in eq. (5-87) to proceed to equilibrium. The entropy change may be split into two parts—that in the vessel and that in the reservoir. In the vessel the entropy change will be

$$\Delta S = S_{\text{final}} - S_{\text{initial}} \tag{5-90}$$

If the initial and final states are equilibrium states, an equation of the above form can always be written since S is a state function. If ΔQ is the heat transferred to the reservoir during the chemical reaction, the entropy change of the reservoir is

$$S_r = \frac{\Delta Q}{T} \tag{5-91}$$

The total entropy change of the universe is therefore

$$S_t = S_{\text{final}} - S_{\text{initial}} + \frac{\Delta Q}{T} \tag{5-92}$$

If the reaction is carried out along a reversible path, S_t is zero; otherwise, it will be positive, as indicated in our previous considerations of entropy. The formalism of eq. (5-92) provides a useful introduction to free-energy functions and we shall later use it for proving some important theorems about these functions.

H. LIMITATIONS OF THE FORMULATION

The second law of thermodynamics has become such an integral part of physics, physical chemistry, and engineering that it is easy to lose sight of the domain of its validity and try to apply it to problems to which it is inappropriate. Both elementary formulations deal with sources and sinks of a very special character; they are in equilibrium with respect to all modes of energy that are subsequently used in doing work. The sources and sinks are in effect thermal. This viewpoint was indeed the proper one for the industrial revolution where all the engines employed operated by burning fuel and creating high-temperature reservoirs such as boilers, hot gases in cylinders, and similar devices. The stored chemical energy was converted to thermal energy before it was used to perform mechanical work. Under these conditions the Carnot scheme allows one to analyze maximum efficiencies and other parameters. Biological engines are of an entirely different character. They operate approximately isothermally and convert chemical energy directly into mechanical or other energy without going through a heat stage. As such they are not subject to Carnot's efficiency theorem. They do not, of course, violate the second law of thermodynamics; they cannot, however, be treated by heat engine analysis. Keeping this in mind will prove helpful when we come to investigate biological machines and move into a domain where current theory is inadequate.

There also exists a class of conventional equilibrium devices, such as fuel cells, where the efficiency limitations of eq. (4-18) do not hold. The differences can be seen in fig. 5-6.

The Carnot engine which we have just been studying has a maximum efficiency of $(1 - T_{\text{sink}}/T_{\text{source}})$. Actual heat engines used in almost all industrial applications mimic the Carnot idealization by having a heat reservoir which is kept at more or less constant temperature by the combustion of fuel. The maximum efficiency is given as above, where T_{source} is the temperature of the reservoir, and T_{sink} is the exhaust temperature. Since exhaust

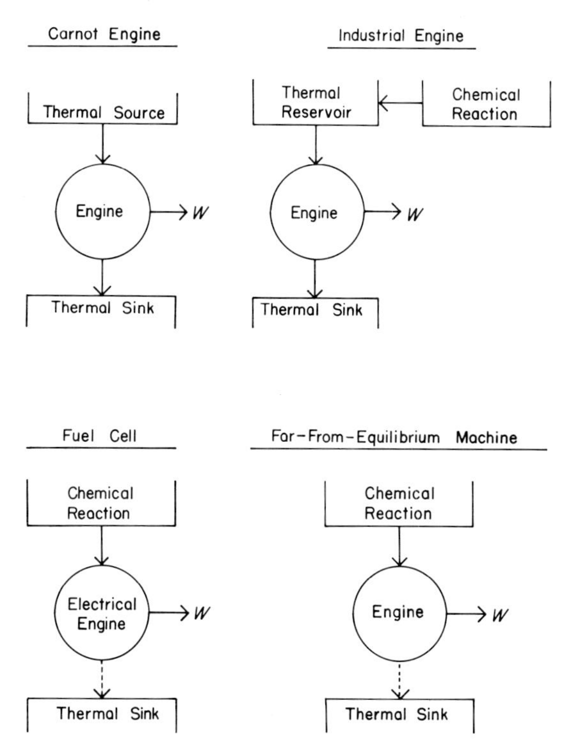

FIGURE 5-6. Types of engines. Carnot and industrial engines are limited by the Carnot heat theorem. Fuel cell driven engines may be approximately isothermal and are limited only by the entropy term in the chemical reactions. For far-from-equilibrium machines the heat loss is written as a dotted line since present theory does not require such a term except from friction or viscous effects.

temperatures are in general controlled by ambient values, there are fundamental limitations on the efficiencies.

The last two engines, shown in fig. 5-6, operate isothermally and involve direct conversion of chemical energy to another form without going through a thermal stage. In an electrolytic cell the oxidation–reduction energy of the chemical reaction is converted directly to electrical energy which operates an engine. The maximum efficiency is, as we shall later demonstrate,

$$\eta = 1 - \frac{T\Delta S}{\Delta U} \tag{5-93}$$

where ΔU and ΔS are the changes of internal energy and entropy of the fuel cell and T is the temperature of the surroundings. All of the first three types of systems must operate slowly and reversibly to achieve maximum efficiency.

A fourth type of chemical-to-work converter is possible which operates in a very far-from-equilibrium domain. Such devices transfer energy before it is converted to heat by molecular collisions and therefore escape the normal thermodynamic restrictions. A gas phase chemical laser in series with a photocell and an engine would be just such a system. In the chemical laser, the high energy state of molecules is brought about by a chemical reaction, and the energy is rapidly radiated out as a laser beam before it is converted to heat. The beam striking the photocell leads to a flow of current that drives the engine. The entire scheme operates very far from equilibrium and does not, in principle, require any heat loss. Here is a theoretically possible scheme that could in principle lead from fuel to work with very high efficiency. There are undoubtedly a whole class of devices which may act in this far-from-equilibrium fashion, and we have simply not explored the problem in any systematic way. We are limited not only by our resources but by our imaginations.

A fifth possibility is not shown on the drawing, yet needs some consideration. Muscle fibers and bacterial flagella are machines which operate by direct conversion of chemical to mechanical energy. They are isothermal (constant-temperature) machines, so they cannot be among the first two types. They must therefore function either as equilibrium devices of type three or as very far-from-equilibrium devices as in type four. It is also possible that they operate on a molecular quantum mechanical domain that is not describable by any of the macroscopic engines. They are not Carnot theory limited, and might possibly achieve high conversion efficiencies. If we could transform the biological principles of energy conversion to industrial use, we might open a whole new technology.

We have now completed our introduction to temperature and entropy from the point of view of thermodynamics and kinetic theory. To give deeper meaning to these concepts and ultimately to relate them to biology, we examine the approaches of statistical mechanics and the insights of information theory.

TRANSITION

Entropy is introduced by seeking an integrating factor for the inexact differential $đQ$. This problem may be solved by a number of different mathematical methods and the simplest factor is found to be the reciprocal of the absolute thermodynamic temperature. The exact differential that results, $đQ/T$, then defines the entropy, a function which has been of great importance in the structure of thermodynamics as well as its biological applications. The second law can be reformulated much more concisely in terms of the entropy that is shown to be measurable in terms of experimental quantities such as specific heat and temperature. The zero point of entropy is defined by the third law of thermodynamics. A discussion of some limitations of the formulism completes our introduction to macroscopic thermodynamics and sets the stage for the more microscopic view of statistical mechanisms. Before going to postulates of the statistical approach, we will develop some aspects of the information theory approach which will lend insight to subsequent developments.

BIBLIOGRAPHY

Edsall, J. T., and Wyman, J., "Biophysical Chemistry." Academic Press, New York, 1958.

Chapter 4 is a 104-page treatment of thermodynamics with special reference to biochemical problems. This chapter also includes a detailed discussion of the method of Caratheodory.

Margenau, H., and Murphy, G. M., "The Mathematics of Physics and Chemistry." Van Nostrand-Reinhold, Princeton, New Jersey, 1965.

Chapter 1 of this book is a thorough discussion of the mathematical methods underlying thermodynamics.

McClare, C. W. F., Chemical machines, Maxwell's demon and living organisms, *J. Theor. Biol.* 30, 1–34 (1971).

A discussion of the meaning of work and heat at the cellular and microscopic level.

6 | *Information*

A. PROBABILITY

This chapter represents a transition in our thinking from the gross macroscopic behavior of matter to a consideration of the submicroscopic details underlying the phenomenology. A number of conceptual tools must be introduced from probability theory, combinatorial analysis, and information theory. When these have been reviewed it will be possible in the next chapter to go on to an introduction of elementary statistical mechanics.

We begin with the notion of examining the properties of aggregates. In thermal physics, in biology, and in many other areas we are frequently called on to deal with collections of quantities, whether they be molecules, experimental data, sequences of computer readouts, or any other sequence of numerical values. Often properties of the aggregates rather than the individual elements may be of interest. Also, it may happen that, although we are unable to obtain information about the elements, we may still deal effectively with the collection. Statistical mechanics represents such a case, where we are unable to say much about the individual molecules, but are able to make predictions about properties of a group of molecules such as pressure, temperature, and density.

In cases where we may assign numerical values to each of the elements, we may utilize directly certain properties of the aggregate. If we designate the numerical value of the ith element in a sequence as x_i, the mean, or average value of x, of a collection of N elements is defined by the following equation:

$$\bar{x} = \frac{\sum_{i=1}^{N} x_i}{N} \tag{6-1}$$

and the mean square deviation Δ^2 is defined by

$$\Delta^2 = \frac{\sum_{i=1}^{N} (x_i - \bar{x})^2}{N} \tag{6-2}$$

The collection x_i may form either a continuous or a discrete set of numbers. The set of x_i may consist of certain discrete values $x_1, x_2, x_3, \ldots, x_N$ or continuous values within the domain $x_0 \leqslant x \leqslant x_1$. If the set is discrete, we may designate the fraction of the elements with the value x_j as $f(x_j)$. The subscript i numbers the elements in a sequence, while the subscript j refers to all elements in the sequence with a value x_j regardless of where they occur. We may then write

$$f(x_j) = N_j/N \tag{6-3}$$

$$\sum_j f(x_j) = 1 \tag{6-4}$$

where N_j is the number of elements having the value x_j. If the x_i may have continuous values, then $f(x)\,dx$ is taken to represent the fraction of elements having values between x and $x + dx$. It then follows that

$$\int_{x_0}^{x_1} f(x)\,dx = 1 \tag{6-5}$$

The functions $f(x_j)$ and $f(x)$ are of a somewhat different character since $f(x_j)$ is dimensionless and $f(x)$ has dimensions of x^{-1}.

In actual practice, if the x_i represent experimental data, the collection is always discrete, as every instrument will have a limiting sensitivity, and can be read only to a certain number of significant figures. This is more clearly seen in reading digital display devices than in squinting at a pointer on a meter and trying to squeeze out one more significant figure.

The function $f(x_j)$ or $f(x)$ is usually designated as a distribution function. Figure 6-1 represents such a discrete distribution function. If the function $f(x_j)$ has only one maximum, the value of x_j at which it occurs is known as the mode. A distribution with a single maximum either can be symmetrically distributed or can be skewed in its distribution about that maximum. *A distribution with two maxima is termed bimodal, three maxima trimodal, etc.* (definition).

A number of mathematical methods have been devised to deal with aggregates, among them probability analysis, statistics, theory of errors,

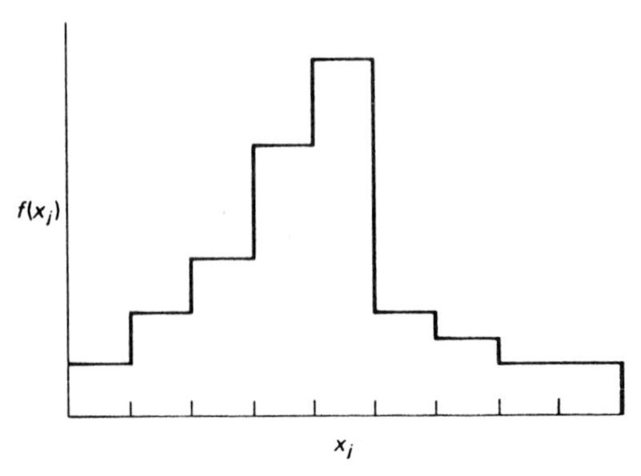

FIGURE 6-1. Distribution function. The probability that a quantity will have a value x_j is plotted as a function of the value.

and information theory. Each of them provides certain background information which is useful in understanding the basic methods of statistical mechanics. We shall therefore pick out and discuss a few topics which will be useful in subsequent chapters.

Probability theory seems to be historically rooted in games of chance, so the reader will realize that we are following a time-honored tradition when we introduce the subject by talking about throwing dice. We will begin by asking the question, "What is the chance of throwing a three with a die?" We may answer this question in one of two ways. First, we may reason that there are six possible numbers that may be thrown and, given an "honest" die and an "honest" thrower, each number will have an equally likely chance of turning up; hence, in a long series of throws we will expect a three to appear one-sixth of the time. This corresponds to the *a priori* definition of probability, a definition going back to Laplace. Alternatively, we might throw the die a large number of times, record the results, and measure the ratio of threes to the total number of throws, and take the limit of this ratio for a long sequence of observations as the chance of throwing a three. This procedure corresponds to the *a posteriori* method of measuring probabilities.

In practical application the first definition is troublesome because of the problem of knowing that we have an "honest" die and an "honest" thrower (the physics and psychology of the situation are in fact too difficult). The second definition is incomplete in being unable to say how many observations are necessary to establish the limit. In addition, it will not serve us where we are unable to measure the elements of a sequence, but may deal only with averages. This will prove to be the case in statistical mechanics, where the situation is handled by assuming certain states to be equally probable and justifying the assumption by the correspondence of theory and experiment.

The question of when we may use probability theory in dealing with an aggregate is a very important one. Below are presented two sequences of numbers:

Sequence A: 1, 2, 3, 4, 5, 6, 1, 2, 3, 4, 5, 6, 1, 2, 3, 4, 5, 6, 1, 2, 3, 4, 5, 6, . . . , etc.

Sequence B: 1, 4, 3, 1, 4, 1, 6, 1, 3, 6, 2, 5, 1, 5, 2, 4, 1, 6, 2, 5, 1, 1, 3, 5, 3, 6, 2, 5, 2, 4, 3, 4, 3, 1, 5, 4, 6, 3, 2, 5, 5, 6, 4, 1, 2, 5, 5, 2, 3, 3, 6, 2, 5, 1, 2, 2, 2, 3, 4, 5, 6, 6, 3, 2, 5, 4, 3, 2, 5, 5, 6, 3, 4, 5, 6, 1, 1, 4, 6, 5, 4, 1, 3, 4, 3, 3, 2, 6, 4, 5, 3, 5, 6, 1, 2, 4, 2, 5, 2, 3, 6, 5, 2, 5.

The first sequence was obtained in an obvious manner, the second by the somewhat less obvious manner of throwing dice. Because the first sequence was formulated by a rule, we can unambiguously state the next element, whereas in the second we can only state the probability of the next element having some particular value. The latter is known as a random sequence, and probability is usually limited to such cases. The notion of randomness is a very important one in physics, yet difficult to describe. Often a process is so complicated, or we are so ignorant of the boundary conditions, or of the laws governing the process, that we are unable to predict the result of the process in any but a statistical fashion. Randomness is, in a certain sense, a consequence of the ignorance of the observer, yet randomness itself displays certain properties which have been turned into powerful tools in the study of the behavior of systems of atoms. In quantum mechanics randomness becomes even more fundamental and probability distributions are the irreducible minimum descriptions that are allowed.

To simplify our discussion of probability, we shall consider at the moment only independent events; that is, given a sequence of elements, we assume that the probability of any one element having a particular value is independent of the values of the other elements in the sequence. Consider first two elements 1 and 2, with the probability of the first element having a specified value x being $\mathscr{P}_1(x)$ and that for the second element being $\mathscr{P}_2(x)$. Now, in considering the pair there are four possibilities:

I. Both elements will have the value x.
II. Neither element will have the value x.
III. Element 1 will have the value x, but element 2 will not.
IV. Element 2 will have the value x, but element 1 will not.

The axioms of probability theory give the following probabilities to each of the four cases:

I. Both: $\mathscr{P}_{\text{I}} = \mathscr{P}_1(x)\mathscr{P}_2(x)$.
II. Neither: $\mathscr{P}_{\text{II}} = [1 - \mathscr{P}_1(x)][1 - \mathscr{P}_2(x)]$.

III. 1 but not 2: $\mathscr{P}_{\text{III}} = \mathscr{P}_1(x)[1 - \mathscr{P}_2(x)]$.
IV. 2 but not 1: $\mathscr{P}_{\text{IV}} = [1 - \mathscr{P}_1(x)]\mathscr{P}_2(x)$.

Since the two elements must fit one of the four possibilities, the sum of $\mathscr{P}_{\text{I}}, \mathscr{P}_{\text{II}}, \mathscr{P}_{\text{III}}, \mathscr{P}_{\text{IV}}$ must be unity, as can be verified by adding them. In our use of probability theory we shall proceed from the rules listed above.

After first reviewing the elementary laws of probability, we shall investigate certain of the properties of random systems which are important in statistical mechanics.

B. INFORMATION THEORY

In addition to probability and statistics, a more recent method of handling aggregates is information theory. While, historically speaking, statistical mechanics is many decades older than information theory, there is certain justification in treating information theory first. The later theory is more general, applies to any kind of aggregate, and within its context we can develop, in a familiar fashion, certain abstract concepts which are of importance in statistical mechanics. In particular, the entropy function of statistical mechanics can be more easily appreciated after having first comprehended the analogous information function.

Information theory arose in the communications industry, where it was necessary to quantitatively measure the amount of information transmitted in a communication channel. With such a measure it would then be possible to optimize the information communicated per unit energy expenditure or per unit time or per unit cost or any other parameter of practical importance in the design of communication networks. The resulting measure of information has had unforeseen consequences on the thinking of a number of disciplines.

Information theory starts with a message, a linear array of symbols which, in normal language, may be a set of words, syllables, letters, or phonemes. The theory is quite general and finds application in any linear array of symbols, which may be musical notes, experimental observations, amino acids in a linear peptide, or any other defined sequence.

The notion of information bears very heavily on an overlapping subject, the relation between physical science and psychology. This is a very ill-understood area and confusion can easily arise. An attempt is made to avoid this problem by designing an information measure that is independent of the truth or meaning of the message but is just related to the formal aspects of transmitting and receiving a set of symbols.

The logic of our approach may be difficult to follow since information is not a physical quantity in the sense that mass, charge, or pressure are physical quantities. Information deals with the use of a set of symbols by an observer. Since it does not measure anything physical we are free to choose an information measure that fulfills our requirements. The definition is therefore at first arbitrary and the choice is based on developing a concept of information that is measurable and makes contact with intuitive notions. The original definition arising from the needs of the communications industry was, to use P. W. Bridgman's words, "of such unblushing economic tinge." What in the end turns out to be surprising is that the definition which was introduced is found to relate to the entropy concept in interesting and very fundamental ways. What appears to lie behind all this is the fact that our view of the physical universe is not a cold, lifeless abstraction but is structured by the human mind and the nature of life itself. While it has long been fashionable to inquire how biology depends on the underlying physics, we are being forced to inquire how physics depends on the underlying biology of the mind.

The approach used is to define a measure of the information of an element in a sequence of elements or message. The measure is then demonstrated to accord with some of our everyday ideas about information and to have certain analytical properties which makes it useful. The information measure is ultimately shown to have certain formal relations to the entropy measure of statistical mechanics. Lastly, it is noted that when the information measure relates to which of the possible quantum states a system is in, then the relationship between information and entropy becomes more than just a formal analogy.

The information content of a symbol is then defined as

$$I = \ln_2 \mathscr{P}_r / \mathscr{P}_s \tag{6-6}$$

I is the information content of the symbol, $\mathscr{P}_s$ is the probability that the symbol was sent, and $\mathscr{P}_r$ is the probability that the symbol that was sent was correctly received. The logarithm to the base two was originally chosen because relays and electronic devices used in communication networks had two states, on and off. With such components a binary mathematics (Boolean algebra) was the method of choice. With the equipment now in use there is less reason to use arithmetic to the base two, except that, considering the body of existing theory, there seems little reason to change. Note that for any logarithmic base

$$\ln_\alpha x = \zeta \ln_2 x \tag{6-7}$$

with appropriate algebraic manipulation we can show that

$$\zeta = \frac{1}{\ln_2 \alpha} = \ln_\alpha 2 \tag{6-8}$$

It is frequently useful to convert from the $\ln_2$ used in information theory to $\ln_e$, which is the normal exponential base used in calculus, in which case

$$\zeta = \frac{1}{\ln_2 e} = \ln_e 2 = 0.693 \tag{6-9}$$

Next rewrite eq. (6-6) as

$$I = \ln_2 \mathscr{P}_s + \ln_2 \mathscr{P}_r \tag{6-10}$$

The quantity $\mathscr{P}_r$, the probability that the message has been received once it has been sent, relates to the fidelity of the communication channel between sender and receiver. For an ideal channel $\mathscr{P}_r = 1$, $\ln(\mathscr{P}_r) = 0$, and this term makes no contribution to the information measure.

Information, as we have pointed out, is not a physical quantity, but, in order to transmit symbols over a communication channel, we must convert them to some physical form such as electrical pulses, flashes of light, sound waves, or other forms. In the encoding, transmission, and decoding of these signals, errors may occur due to mechanical failure, poorly designed equipment, extraneous signals being introduced from other sources, or from thermal noise. For all real channels $\mathscr{P}_r < 1$, $\ln(\mathscr{P}_r) < 0$; the failure of the channel to transmit the message with 100% fidelity reduces the information content of the message. The lower the value of $\mathscr{P}_r$, the larger the negative value of $\ln(\mathscr{P}_r)$ and the more information is lost. This clearly agrees with our intuitive notions of information. A poor transmission line corresponds to low $\mathscr{P}_r$ and reduces the amount of information transmitted. In many applications, particularly those which will be of importance to us in relating information theory to thermal physics, $\mathscr{P}_r \cong 1$ and we can drop the second term, reducing eq. (6-10) to

$$I = -\ln_2 \mathscr{P}_s \tag{6-11}$$

This expression is the real heart of information theory and the one that we must relate to our intuitive notions of information. The quantity $\mathscr{P}_s$ applies to the entire range of symbols that can comprise the message; that is, $\mathscr{P}_s$ is a normalized probability distribution of the message source. Designate the probability of the jth type of symbol as $\mathscr{P}_j$, then

$$\sum_{j=1}^{N} \mathscr{P}_j = 1 \tag{6-12}$$

Equation (6-12) means that N possible symbols can be sent; that is, the effective alphabet or dictionary contains N symbols. Since an element in the sequence must contain one of the N possibilities, the sum of the probabilities must be one, as in eq. (6-12). Since all probabilities are positive and each is less than one, the information content of a symbol in eq. (6-11) is always positive. If $\mathscr{P}_s$ approaches one, I approaches zero; that is, if we were certain what a message was going to be ($\mathscr{P}_s = 1$), we would get no additional information by receiving the message.

Suppose $\mathscr{W}$ possible messages can be sent and they are all equally probable. Then $\mathscr{P}_s = 1/\mathscr{W}$ and eq. (6-11) becomes

$$I = -\ln_2 \frac{1}{\mathscr{W}} = \ln_2 \mathscr{W} \tag{6-13}$$

Equation (6-13) says that the more possible messages that can be sent, the greater the information that we have when we receive one of them. The greater the number of messages that can be sent, the more possibilities can be distinguished and the greater the information in knowing which of those possibilities is in fact being indicated by the message.

Returning to eq. (6-11), let us examine a concrete case, where the elements in the sequence are the letters of the alphabet; that is, let us assume that the message is written in ordinary English. The 26 letters have a far from equal probability distribution, as is shown in table 6-1.

The conclusion from our definition is that an infrequent letter carries far more information per letter than a common letter. This is a conclusion that has been known for a long time by devotees of crossword puzzles, who know that a rare letter narrows the choice of words far more than a common letter and therefore conveys more information about the missing word. By the

TABLE 6-1

Probability Distribution and Information Measure of the Letters of the English Alphabet

Letter	$\mathscr{P}_i$	$-\ln_2 \mathscr{P}_i$	Letter	$\mathscr{P}_i$	$-\ln_2 \mathscr{P}_i$
a	0.0805	3.6716	n	0.0719	3.8364
b	0.0162	6.0080	o	0.0794	3.6917
c	0.0320	5.0161	p	0.0229	5.5037
d	0.0365	4.8279	q	0.0020	9.0565
e	0.1231	3.0525	r	0.0603	4.0927
f	0.0228	5.5101	s	0.0659	3.9635
g	0.0161	6.0171	t	0.0959	3.4166
h	0.0514	4.3256	u	0.0310	5.0625
i	0.0718	3.8384	v	0.0093	6.8170
j	0.0010	10.0665	w	0.0203	5.6792
k	0.0052	7.6365	x	0.0020	9.0565
l	0.0403	4.6799	y	0.0188	5.8248
m	0.0225	5.5295	z	0.0009	10.2202

same reasoning the value of a letter in "Scrabble" is related to its information content.

Equation (6-13) suggests another approach to the problem of information. Suppose the elements in the sequence are words instead of letters. Then, assuming equal probabilities of using words, $\mathscr{W}$ represents the total vocabulary of the sender. The information measure says that someone with a large vocabulary can send more information per word than someone with a small vocabulary.

If we allow the elements in the sequence to be letters, then the information content is independent of the vocabulary of the sender; if we allow it to be words, then it is dependent on the vocabulary of the sender. Thus the same message will have a different information content according to how we evaluate it. This seeming paradox relates to the fact that information by our definition is not dependent on meaning or truth but is a property of the probability distribution of eq. (6-11). The rules that go into grouping symbols can alter the probability distributions and hence alter the information measure. Information, as we have pointed out, is not a physical quantity that has an absolute measure; it is rather a property of a sequence of elements and depends on certain arbitrary choices for classifying and grouping elements. Information measure thus depends on the classifying and grouping procedure. This involves the observer or experimenter in a very intimate way. We are already familiar with the observer being intimately involved in his observation. The uncertainty principle in quantum mechanics and the transformation of special relativity are examples of this. Information theory begins with the observer; after all, information is ultimately a property of the human mind. The theory then starts with a certain arbitrariness in the choice of what kind of information the observer wants and how he chooses to coarse-grain or fine-grain his observations. Thus in evaluating written English the primary elements may be letters, phonemes, words, or sentences. The information measure will be dependent on the choice of element. When the information is about physical systems rather than arbitrary sequences of elements, then the relation of information to measurable quantities must be examined. This will be done in Chapter 7.

The next step in the development of information theory is the proof of the additivity of the information of two elements in a sequence, and the derivation of the expression of the average information of an element in a long sequence.

Consider a message that contains two symbols A and B which are completely independent. If the probability of sending A is $\mathscr{P}_A$ and the probability of sending B is $\mathscr{P}_B$, the probability of the joint message AB is given by our probability axioms as $\mathscr{P}_A\mathscr{P}_B$. The information content of the message AB

is thus

$$I_{AB} = -\ln_2 \mathscr{P}_A \mathscr{P}_B = -\ln_2 \mathscr{P}_A - \ln_2 \mathscr{P}_B = I_A + I_B \tag{6-14}$$

The information content of the combined message is thus the sum of the information contents of the two individual messages.

The average information content of a message of independent, uncorrelated elements is the sum of the information divided by the number of elements. For a message in which the jth element occurs N_j times,

$$I = \sum_j \frac{N_j I_j}{N} = \sum_j \frac{N_j}{N}(-\ln_2 \mathscr{P}_j) \tag{6-15}$$

As the message gets sufficiently long, the *a posteriori* measure of probability indicates that

$$\frac{N_j}{N} \rightarrow \mathscr{P}_j \tag{6-16}$$

Equation (6-15) then becomes

$$\bar{I} = -\sum \mathscr{P}_j \ln_2 \mathscr{P}_j \tag{6-17}$$

Equation (6-17) is the expression for information that we shall use most frequently.

Suppose in a given situation we have incomplete knowledge about the probability distribution, but we do know some constraining relations. The least committal statement we can make about the probabilities is to choose that distribution which maximizes I. That is, we assume that a series of measurements would yield the maximum amount of information, and this is equivalent to assuming that we have the least amount of information before the measurements are made. Such an approach, which is philosophical in nature, provides one possible way of exploring incompletely determined or incompletely described systems.

TRANSITION

Information theory, which arose in the communication industry, provides a measure related to a random sequence of symbols. Such an abstraction applies to physical systems because the detailed microstate keeps changing in a nonpredictable way so that the system is describable by a sequence of symbols representing its time course at the atomic level. The formalism developed in information theory will be related to the distributions which will be derived using statistical mechanics in the next two chapters.

BIBLIOGRAPHY

Lindsay, R. B., and Margenau, H., "Foundations of Physics." Dover, New York, 1963.

Chapter IV entiled Probability and Some of Its Applications is a concise review of the relation of probability theory and its physical applications.

Quastler, H., "Information Theory in Biology." Univ. of Illinois Press, Urbana, 1953.

Contains a survey of the original applications of information theory to biology.

Quastler, H., A primer of information theory, Office of Ordinance Research, U.S. Army, Technical Memorandum 5G1, 1956.

A very lucid introduction to the elementary constructs of information theory.

Shannon, C., and Weaver, W., "The Mathematical Theory of Communication." Univ. of Illinois Press, Urbana, 1967.

This short book contains Shannon's original paper on information theory.

Watanabe, M. S., "Knowing and Guessing." Wiley, New York, 1969.

A very deep formal treatment, which attempts to unify many aspects of physics, philosophy, communication theory, cybernetics, and statistics.

7 Statistical Mechanics

A. THE STATISTICAL POINT OF VIEW

In thermodynamics we deal with macroscopic measurements, using measuring instruments which are large compared with individual molecules and which have response times long in comparison to the times of individual molecular processes. Small measuring probes of the order of 1 mm contain about 10^{20} atoms, while a measurement which takes 1 msec is long in comparison with molecular vibration times, which are of the order of 10^{-8} sec or less. Thus macroscopic measurements represent averages over a very large number of molecular events. This type of averaging suggests one of the most powerful methods of thermal physics, predicting the macroscopic properties of systems by averaging over all possible solutions of the mechanical system viewed at the atomic level. This general method is designated as statistical mechanics, which by its applications to macromolecules has already made major contributions to biological thought.

To discuss the general ideas, visualize a box of dimensions L_x, L_y, and L_z which contains N atoms of an ideal monatomic gas and is in contact with a thermal reservoir at temperature T. The system can be completely described by specifying the position (x, y, z) and momentum (p_x, p_y, p_z) of each of the N particles. Thus $6N$ coordinates are required to completely specify the system at the atomic level. If we construct an abstract $6N$-dimensional space, the instantaneous state of the system can be determined as a single point in this space since a point involves assigning a value to each one of the $6N$ variables. The behavior of the system with time can be completely described by the trajectory of the point in the $6N$-dimensional space.

Any macroscopic measurement is a time average over a long segment of this trajectory. We cannot, of course, determine this trajectory from first principles, since that would involve an N-body problem where N is very large and thus provides insuperable mathematical difficulties.

A given macroscopic system can be visualized in an alternate way. We begin by forming an ensemble; that is, by taking a very large number of identical macroscopic systems and placing them in contact with an isothermal reservoir. The meaning of identical is that all of the systems have the same macroscopic properties (composition, volume, temperature) although they may certainly vary with respect to the actual molecular details. Each system in the ensemble may be represented by a point in our $6N$-dimensional space and the entire ensemble will be represented by a cluster of points. At equilibrium that cluster can be represented by a time-independent density distribution function. We can now calculate any macroscopic average over the ensemble in the following way. With each point in the $6N$ space we may associate a value of the given macroscopic property $\mathfrak{M}$ $(x_1y_1z_1, x_2y_2z_2, \ldots, p_{x1}, p_{y1}, p_{z1}, \ldots, p_{xN}, p_{yN}, p_{zN})$. This follows directly from mechanics. In a small volume dV around any point there are $\rho\, dV$ members of the ensemble, where ρ represents the density distribution function. The average of $\mathfrak{M}$ over the whole ensemble is then

$$\overline{\mathfrak{M}} = \frac{\int_V \mathfrak{M}\rho\, dV}{\int \rho\, dV} \tag{7-1}$$

The idea of an ensemble is of course a conceptual one since such a large group of systems are being envisioned that we could not actually form it experimentally. The ensemble is thus an abstraction which turns out in some cases to be a remarkably efficient computing device. Since we have defined the member systems by properties such as composition and temperature which have a well-defined meaning only at equilibrium, the representation that we are using is confined to that condition. This differs from the trajectory formalism, which has equal validity for equilibrium and nonequilibrium systems.

Contact is made between the trajectory representation in $6N$-dimensional phase space for equilibrium systems and the ensemble representation by a postulate called the "ergodic hypothesis," which states that for large ensembles and long time trajectories the two representations approach each other; that is, given a small volume of phase space dV, the normalized probability of an ensemble point being in that volume is the same as the fraction of time in which the phase point of a long trajectory falls in that volume. Thus, if we average over the ensemble, we will be able to predict the values to be obtained in a macroscopic measurement.

Before proceeding it might be well to consider a simple analogy which will illustrate the difference between the trajectory and ensemble approaches. Suppose we wish to ask the question: What is the average distance of a worker bee from the beehive during food procurement? We could capture a bee, mount a tiny identifying reflector on him, release him, and take several hours of motion pictures. Each frame would give r, the distance as a function of the frame number or time, and we can compute

$$\frac{1}{a}\sum_{i=1}^{a}|r_i|$$

where i is the index number of the photographic frame and a frames are examined. Alternatively, we can take an appropriate aerial photograph and determine the positions of all of the bees visible in the photograph. If the jth bee is a distance r_j from the hive, then the mean distance is

$$\frac{1}{b}\sum_{j=1}^{b}|r_j|$$

where we average over b bees. The first approach gives us a trajectory average, while the second gives us an ensemble average. How close the two would be would depend on the validity of the ergodic theorem for this case.

While analogies of the beehive sort give some feeling for the ensemble approach, they fail to encompass one distinct property unique to molecular systems: macroscopically indistinguishable systems which differ in their detailed mechanical description. It is this feature which makes it impossible for us to go back and forth in any simple way between mechanics and thermodynamics, and forces us ultimately to statistical method. With this in mind we can note that the trajectory approach is straightforward from the point of view of mechanics; it assumes the detailed description, the time course of all the atoms and molecules in the system. For real molecular systems this approach is impossibly complicated, so that we invent the ensemble with the assumption that *at equilibrium* the ensemble average of any measurable property will be the same as the time average of that property for a given system. These ideas should become more concrete when we later actually apply them to physical problems.

The ergodic hypothesis has been the subject of a large amount of profound consideration from the points of view of physics, mathematics, measurement theory, and philosophy of science. We will avoid the subtleties and disputes in this field by taking a very pragmatic point of view toward the ensemble approach to statistical thermodynamics. We will assume two postulates whose ultimate validity rests solely on the agreement between the predictions of the theory and macroscopic measurements.

(1) *The macroscopic properties of equilibrium systems may be predicted by averaging the property in question over an ensemble of all microstates of the system that are consistent with some set of macroscopic constraints* (postulate).

(2) *In performing the averaging, it is usually possible to approximate the average value of a property by its value in the most probable microstate* (postulate).

This is equivalent to replacing mean values by modal values. This second postulate can in fact be "proved" for the actual distributions of statistical mechanics by a rather sophisticated mathematical technique known as the Darwin–Fowler method of steepest descent. (See Tolman in the Bibliography.)

These two statements on ensemble averaging allow us to make predictive statements provided we can formulate an ensemble of microstates. In classical mechanics the ensemble forms a continuum over all possible coordinates and momenta of the atoms subject to the constraints of the system. This continuum approach leads to certain indeterminacies in evaluating thermodynamic functions. Quantum mechanics seems to form a more natural basis for statistical thermodynamics because the theory itself generates an ensemble of discrete states. Indeed, using the wisdom of hindsight, it has been pointed out that the founders of statistical mechanics might have been alerted to the existence of quantum mechanics by the problem of being able to choose an elementary volume in phase space. We will therefore carry out the actual formal analysis of statistical theory in terms of quantum states, which will necessitate the briefest review of some of the elementary notions of quantum mechanics. We will shortly make some of these notions more concrete by considering the quantum mechanical problem of the particle in the box. This particular example will later be extended to a collection of particles in a box, which is, of course, the quantum-mechanical analog of an ideal gas.

B. THE QUANTUM MECHANICAL DESCRIPTION OF SYSTEMS

It should be understood that we cannot here attempt to explain or justify quantum mechanics, but only to present the results so that the notions of quantum state and energy eigenvalue will be more familiar when we utilize them in setting up the fundamental equations of statistical mechanics. The reader interested in more detailed quantum mechanical background should consult a book like that by M. W. Hanna (see Bibliography).

The formulation of quantum mechanics we shall use was carried out by Erwin Schrödinger, a man who later produced a probing examination of the relationship between biology and the statistical aspects of thermal physics. The starting point of this view of quantum mechanics is the equation for the total energy of a system which we have represented as eq. (1-8) of Chapter 1. The equation states that the sum of the potential and kinetic energies of a system equals the total energy. The left side of the equation, *the sum of the two energies, is called the Hamiltonian* (definition) and is represented by the symbol $\mathscr{H}$. If we designate the total energy of the system by the symbol E, we can state the energy conservation principle in the very simple form

$$\mathscr{H} = E \tag{7-2}$$

This equation is the starting point for quantum mechanics, which states that for atomic systems the simplified form of the energy equation must be modified. We postulate that the equation $\mathscr{H} = E$ must be changed to an operator equation of the form

$$\mathscr{H}\psi = E\psi \tag{7-3}$$

where $\mathscr{H}$ and E are mathematical operators of a form we will describe and ψ is a function whose values describe the behavior of the system. It should be understood that there is no intuitive way of justifying the use of an operator equation or the actual operators used. The ultimate justification is that the formalism predicts with high precision the observable behavior of atomic and molecular systems.

A mathematical operator, such as multiplicative constants, $\dfrac{d}{dx}, \dfrac{d^2}{dx^2}$ operates on a function f to yield a new function,

$$\text{constant} \times f, \quad \frac{df}{dx}, \quad \frac{d^2f}{dx^2}$$

It is a very convenient formalism in a number of areas of theoretical physics. Returning to eq. (7-3), $\mathscr{H}$ is the quantum-mechanical operator that is derived from the classical Hamiltonian, which, as we have noted, is a function equal to the total energy of a system, i.e., the sum of the kinetic and potential energies. Classically, we may write the Hamiltonian for a single particle, $\mathscr{H}$, as follows:

$$\mathscr{H} = \tfrac{1}{2}mv^2 + \mathscr{V} \tag{7-4}$$

Frequently problems are formulated in terms of momentum, the product of mass times velocity, instead of in terms of velocity. This turns out to be an especially convenient form for quantum-mechanical problems. Under these circumstances, using p for the momentum, eq. (7-4) becomes

$$\mathscr{H} = p^2/2m + \mathscr{V} \tag{7-5}$$

To put the theory in quantum-mechanical notation, the variables of eq. (7-5) must be transformed into quantum-mechanical operator form. *This is done by a rule which states that each momentum p_x is replaced by an operator*

$$-\frac{ih}{2\pi}\frac{\partial}{\partial x}$$

(postulate). The operator involves h, which is Planck's constant; i, the square root of minus one (quantum mechanics involves the mathematical methods of complex variables); and $\partial/\partial x$, the partial derivative with respect to x. Equations (7-3) and (7-5) become in expanded quantum-mechanical form

$$\frac{-h^2}{8\pi^2 m}\left(\frac{\partial^2\psi}{\partial x^2} + \frac{\partial^2\psi}{\partial y^2} + \frac{\partial^2\psi}{\partial z^2}\right) + \mathscr{V}(x, y, z)\psi = E\psi \tag{7-6}$$

The successive application of the operator p in the previous equation may be viewed as

$$\begin{aligned} \frac{p_x{}^2}{2m}\psi &= \frac{1}{2m}\left(-\frac{ih}{2\pi}\frac{\partial}{\partial x}\right)\left(-\frac{ih}{2\pi}\frac{\partial}{\partial x}\right)\psi \\ &= \frac{1}{2m}\left(-\frac{ih}{2\pi}\frac{\partial}{\partial x}\right)\left(-\frac{ih}{2\pi}\frac{\partial\psi}{\partial x}\right) \\ &= \frac{1}{2m}\left(-\frac{ih}{2\pi}\right)^2\frac{\partial^2\psi}{\partial x^2} = -\frac{h^2}{8\pi^2 m}\frac{\partial^2\psi}{\partial x^2} \end{aligned} \tag{7-7}$$

Let us again note that we are merely giving an abbreviated statement of a very complicated and sophisticated theoretical approach. The reader is being asked to accept the formalism as a mathematical method which gives the right results. The justification for presenting this material here is the hope that it will convey some notion of a quantum state and an energy eigenvalue, concepts which are the foundations of modern statistical thermodynamics.

The function $\psi(x, y, z, t)$ which is a solution to eqs. (7-3) or (7-6) is related related to the probability of finding the particles at x, y, z at time t. In general ψ is a complex number $a + ib$. *For each such number there exists a complex conjugate $\psi^* = a - ib$ and the product $\psi\psi^* = a^2 + b^2$ is the probability of finding the particle* (postulate). *The function ψ must be continuous, single-valued, and the integral of $\psi\psi^*\,dV$ over all available space must be unity* (postulate); that is, the probability of the particle being somewhere is unity.

As can be seen from eq. (7-6), ψ is related to E and the restrictions we have just placed on ψ lead to E having discrete allowable values. The existence of the discrete values is the quantization of quantum mechanics; the energy can have only a certain discontinuous set of values and must therefore change by certain quantized amounts. The state ψ_i corresponding to a given allowable energy E_i is designated an eigenstate. The important point is that the system can only exist in a series of discrete states each of which is characterized by an energy eigenvalue. Statistical mechanics consists of a certain kind of enumeration of these states over an appropriate ensemble.

The example we shall use to illustrate the preceding is a particle of mass m in an ideal one-dimensional box. The particle can only move in the x direction between $x = 0$ and $x = L$, where there are infinitely high walls. The potential energy $\mathscr{V}(x)$ is shown in fig. 7-1. Inside the box $\mathscr{V}(x)$ is everywhere zero and at both walls $\mathscr{V}(x)$ rises to infinity. Inside the box the fundamental equation of quantum mechanics (7-6) becomes

$$\frac{-h^2}{8\pi^2 m}\frac{\partial^2\psi}{\partial x^2} = E\psi \tag{7-8}$$

The particle cannot actually be located at either wall, since its energy would go to infinity at those two points. We therefore require that $\psi = 0$ at $x = 0$ and $x = L$. The solution of eq. (7-8) that has the properties we desire is

$$\psi = A \sin \alpha x \tag{7-9}$$

where

$$\alpha^2 = \frac{8\pi^2 mE}{h^2} \tag{7-10}$$

The reader can of course verify that eq. (7-9) is a solution of eq. (7-8) by substituting expressions (7-9) and (7-10) into (7-8). This solution always

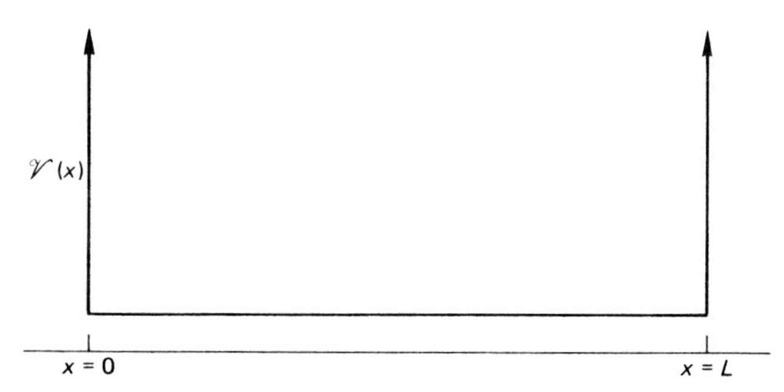

FIGURE 7-1. The potential energy curve for a particle in a one-dimensional box. The potential is constant within the box and rises vertically at the walls.

vanishes at $x = 0$. In order for it to vanish at $x = L$, we require that

$$\alpha L = n\pi \tag{7-11}$$

The quantity n may be any integer. That is, the sine of $n\pi$ is always zero for any integral value of n. Equations (7-10) and (7-11) can be combined to give the following expression:

$$\frac{n^2\pi^2}{L^2} = \alpha^2 = \frac{8\pi^2 mE}{h^2} \tag{7-12}$$

We now see that, from eq. (7-12), E cannot take on any values but is given by

$$E = \frac{n^2 h^2}{8mL^2} \tag{7-13}$$

where n takes on only integral values. The quantity n is designated the quantum number and the $E(n)$ are the eigenvalues of the energy.

The particle in the box is characteristic of a broad range of quantum-mechanical problems where the solutions form a quantized set. If we return to the ensemble approach, we may now formulate the problem in the following way. We wish to investigate a macroscopic system that is characterized by specifying the composition and volume, and placing the system in contact with an isothermal reservoir. In principle (but not in practice) we could write down a complete equation of the form of (7-3) for any system. The solutions would then indicate all possible states consistent with the constraints. A given system would be described as completely as we are able to if we specify which quantum state it is in; that is, which ψ function describes it.

C. THE ENSEMBLE APPROACH

For an ensemble of N systems, each macroscopically identical, the most complete description we can give is to specify which quantum state each member system is in. Since we are only interested in averages over the entire ensemble, the quantities we desire to know are how many of the N systems are in each quantum state. For the jth state we will designate this number as N_j. The task of statistical mechanics is to determine the N_j or the ratios N_j/N, which we will designate as f_j, the probability that a system will be in the jth state.

To proceed with the analysis, the ensemble is rearranged in the following way: all of the members are removed from contact with the isothermal reservoir, placed in contact with each other, and enclosed in an adiabatic envelope. This rearrangement is a thought experiment, a conceptual device, since we never actually experimentally assemble the ensembles of statistical analysis. From the point of each individual system the boundary conditions remain unchanged since the $N - 1$ remaining systems constitute a large isothermal reservoir at average temperature T, as they were all in contact with a reservoir of temperature T at the time of isolation. It is now possible to write two equations of constraint for the ensemble:

$$\sum_j N_j = N \tag{7-14}$$

$$\sum_j N_j E_j = E_T \tag{7-15}$$

It should be noted that a given system remains in a given quantum state for a very short time and then, by interaction with the wall or internal interaction, goes over into another quantum state. The set of N_j used in this analysis refers to the instantaneous distribution over the ensemble. The first equation describes the conservation of systems, while the second reflects the fact that the adiabatic enclosure results in the total energy within the enclosure being a constant.

We next introduce one of the fundamental postulates of statistical mechanics: *for the adiabatically isolated system all possible quantum states of that system have the same a priori probability*. This postulate refers to the large system which is adiabatically isolated, not to the small subsystems which can exchange energy with each other. A state of the large system is completely specified if we specify the quantum states of each of the small systems which comprise it. The assignment of quantum states to the small systems must be consistent with eqs. (7-14) and (7-15).

D. THE MOST PROBABLE DISTRIBUTION

In general we are interested in averages over the ensemble, so that it is not important to know which individual systems are in any given quantum state, but only how many systems are in that quantum state. What we ultimately wish to know are the average values of N_j for the ensemble.

In general any set of N_j's can occur in many different ways since different sets of assignments of j's to the individual systems can result in the same N_j. That is, suppose 3 systems of the ensemble are in the jth state. This can happen by having the 1st, 2nd, and 3rd systems in that state or the 1st, 19th, and 23rd, or the 4th, 17th, and 95th, or any other such set. The question we then ask is how many ways can we have N_1 systems in the first state, N_2 systems in the second state, N_3 in the third state, etc., independent of which N_1 systems are in the first state, etc. This is an old problem in combinatorial analysis and the answer for $\mathscr{W}$, the number of ways, is

$$\mathscr{W} = \frac{N!}{N_1!N_2!N_3!\cdots N_s!} = \frac{N!}{\prod_{i=1}^{s} N_i!} \tag{7-16}$$

The symbol $N!$ is read "N factorial" and designates the product $N \times (N-1) \times (N-2) \times (N-3) \cdots 3 \times 2 \times 1$. The quantity s represents the number of possible states of the system. For a discussion of the combinatorial aspects of statistical mechanics, see the book by Gurney in the Bibliography.

Since all distributions consistent with eqs. (7-14) and (7-15) have equal *a priori* probabilities, the most probable set of N_j's is the one that can occur in the largest number of ways, i.e., the one with the largest value of $\mathscr{W}$. We have already indicated that the average state of the system approximates the most probable state of the system. Our task then becomes the maximization of $\mathscr{W}$ subject to the constraining eqs. (7-14) and (7-15). For reasons of analytical convenience we will deal with $\ln \mathscr{W}$ rather than $\mathscr{W}$ itself. However, the maximum of the logarithm of a function occurs at the same value of the variables as does the maximum of the function, so the essential problem is unchanged by the use of logarithms. We rewrite eq. (7-16),

$$\ln \mathscr{W} = \ln\left(\frac{N!}{\prod_j N_j!}\right) = \ln N! - \ln \prod_j N_j!$$

$$= \ln N! - \sum \ln N_j! \tag{7-17}$$

The notation $\prod_j N_j!$ is the serial product of $N_1!$ times $N_2!$ times $N_3!$, etc., up to $N_s!$. The second term in eq. (7-17) comes about as the logarithm of a product is the sum of the logarithms. Factorials are very difficult to deal with analytically, so that eq. (7-17) presents a considerably difficult mathematical problem. However, for large values of N and N_j we are able to simplify the expressions by the use of a device known as Stirling's approximation, which states that for large x

$$\ln x! \cong x \ln x - x \tag{7-18}$$

This relationship is extremely important in the mathematical simplification of statistical mechanics and is proved in Appendix IV. When we use this approximation in eq. (7-17) it becomes

$$\ln \mathscr{W} = N \ln N - N - \sum N_j \ln N_j + \sum N_j \tag{7-19}$$

Substituting eq. (7-14) into eq. (7-19) yields

$$\ln \mathscr{W} = N \ln N - \sum N_j \ln N_j \tag{7-20}$$

The mathematical problem now before us is to find a set of values N_j which satisfy eqs. (7-14) and (7-15) and maximize $\ln \mathscr{W}$ in eq. (7-20). There is a technique known as Lagrange's method of undetermined multipliers which has been designed just for problems of this nature and we shall proceed with a brief exposition of the formal mathematical problem. An alternative and more geometric approach to Lagrange's method is given in Appendix V.

Start with a function $f(x_1, x_2, \ldots, x_s)$ which is to be maximized subject to two constraining equations

$$g(x_1, x_2, \ldots, x_s) = 0, \qquad h(x_1, x_2, \ldots, x_s) = 0 \tag{7-21}$$

Finding an extremum (maximum or minimum) of the function f means that df must be set equal to zero:

$$df = \left(\frac{\partial f}{\partial x_1}\right) dx_1 + \left(\frac{\partial f}{\partial x_2}\right) dx_2 \cdots \left(\frac{\partial f}{\partial x_s}\right) dx_s = 0 \tag{7-22}$$

The equations of constraint can also be written in differential form:

$$\begin{aligned} dg &= \left(\frac{\partial g}{\partial x_1}\right) dx_1 + \left(\frac{\partial g}{\partial x_2}\right) dx_2 + \cdots \left(\frac{\partial g}{\partial x_s}\right) dx_s = 0 \\ dh &= \left(\frac{\partial h}{\partial x_1}\right) dx_1 + \left(\frac{\partial h}{\partial x_2}\right) dx_2 + \cdots \left(\frac{\partial h}{\partial x_s}\right) dx_s = 0 \end{aligned} \tag{7-23}$$

We may eliminate two variables by multiplying dg by α and dh by β and adding these equations to eq. (7-22):

$$\begin{aligned} &\left(\frac{\partial f}{\partial x_1} + \alpha \frac{\partial g}{\partial x_1} + \beta \frac{\partial h}{\partial x_1}\right) dx_1 + \left(\frac{\partial f}{\partial x_2} + \alpha \frac{\partial g}{\partial x_2} + \beta \frac{\partial h}{\partial x_2}\right) dx_2 \\ &\quad + \left(\frac{\partial f}{\partial x_3} + \alpha \frac{\partial g}{\partial x_3} + \beta \frac{\partial h}{\partial x_3}\right) dx_3 + \cdots \\ &\quad + \left(\frac{\partial f}{\partial x_s} + \alpha \frac{\partial g}{\partial x_s} + \beta \frac{\partial h}{\partial x_s}\right) dx_s = 0 \end{aligned} \tag{7-24}$$

α and β are at the moment undetermined multipliers, but we shall choose their values so that at the extremum of f they will make the first two coefficients of eq. (7-24) vanish:

$$\frac{\partial f}{\partial x_1} + \alpha \frac{\partial g}{\partial x_1} + \beta \frac{\partial h}{\partial x_1} = 0, \qquad \frac{\partial f}{\partial x_2} + \alpha \frac{\partial g}{\partial x_2} + \beta \frac{\partial h}{\partial x_2} = 0 \tag{7-25}$$

Equation (7-24) then becomes

$$\sum_{j=3}^{s} \left(\frac{\partial f}{\partial x_j} + \alpha \frac{\partial g}{\partial x_j} + \beta \frac{\partial h}{\partial x_j}\right) dx_j = 0 \tag{7-26}$$

The remaining dx_j are all independent, since we have gone from three equations in s variables to one equation in $s-2$ variables. Since each dx_j can be varied independently, the only way that (7-26) can be satisfied, therefore, is that

$$\frac{\partial f}{\partial x_j} + \alpha \frac{\partial g}{\partial x_j} + \beta \frac{\partial h}{\partial x_j} = 0, \qquad j = 3, 4, \ldots, s \tag{7-27}$$

Equations (7-25) and (7-27) now give s equations in s variables $x_1, \ldots, x_s$ and two undetermined multipliers α and β. We can therefore solve for each x_j as a function of α and β. These solutions can be substituted back in the equations of constraint (7-21) and we end up with two equations in two variables α and β which can then be solved. Substituting the values of α and β back into the solutions of (7-27) determines the values of x_j which solve the original problem.

E. THE MAXWELL–BOLTZMANN DISTRIBUTION

The method of Lagrange may now be applied to maximize In $\mathscr{W} = N \ln N - \sum N_j \ln N_j$ subject to $\sum N_j = N$ and $\sum N_j E_j = E_T$.

Before getting lost in a sea of equations it is well at this point to review verbally what we have done. First, we have recognized that in principle the most detailed physical description we can give of a system is to note which quantum state it is in. Second, we note that any macroscopic measurement is a time average over a very large succession of quantum states. Third, we set up an ensemble, which is a calculating device to approximate the very large succession of quantum states. This calculating device consists of a large array of simultaneous systems (subject to the same boundary conditions) all packed in an adiabatic box. We then assert that all possible quantum states of the big box are equally likely and ask what this implies about the individual systems. We find that three things are implied: (1) conservation of systems [eq. (7-14)]; (2) conservation of energy [eq. (7-15)]; and (3) the large system being in its most probable configuration [a maximization of eq. (7-20)]. We then develop a mathematical method, that of undetermined multipliers, which allows us to solve for all three conditions simultaneously. We are now ready to apply that mathematical method to the actual ensemble equations.

We start out by seeking an extremum to ln $\mathscr{W}$ and we write

$$d \ln \mathscr{W} = -\sum \ln N_j \, dN_j - \sum dN_j = 0 \tag{7-28}$$

This is exactly analogous to eq. (7-22). The analogs to eq. (7-23) are the derivatives of the equation of constraint and are presented below:

$$\sum dN_j = 0 \tag{7-29}$$

$$\sum E_j dN_j = 0 \tag{7-30}$$

We multiply by the undetermined multipliers and then add the equations [as in eq. (7-24)]. The sum is

$$\sum_j [(\alpha - 1) + \beta E_j - \ln N_j] \, dN_j = 0 \tag{7-31}$$

This condition requires that

$$\alpha - 1 + \beta E_j - \ln N_j = 0 \tag{7-32}$$

Equation (7-32) can be solved directly for N_j:

$$N_j = e^{\alpha - 1} e^{\beta E_j} \tag{7-33}$$

In principle, the N_j could now be substituted into (7-14) and (7-15) to determine α and β. This is not usually done, since E_T and N are in general not known. We are eliminate α, however, by substituting (7-33) into (7-14):

$$\sum e^{\alpha - 1} e^{\beta E_j} = N \tag{7-34}$$

$$e^{\alpha - 1} = \frac{N}{\sum e^{\beta E_j}} \tag{7-35}$$

We have now expressed α as a function of β and N. Substituting back into eq. (7-33), we get

$$N_j = \frac{N e^{\beta E_j}}{\sum e^{\beta E_j}} \tag{7-36}$$

Since we are interested only in the ratio N_j/N rather than the absolute value of N_j, we may write

$$f_j = \frac{N_j}{N} = \frac{e^{\beta E_j}}{\sum e^{\beta E_j}} \tag{7-37}$$

This leaves β as the sole undetermined parameter. The evaluation of β must be deferred until we come to the comparison of statistical mechanics with thermodynamics.

The next item of business is to derive eq. (7-37) from a different point of view utilizing the notions of information theory. The mathematical manipulations will be very similar to those that we have just gone through, but the slightly different meanings ascribed to the concepts may impart some insight into the methods of statistical physics. Again start with an ensemble of macroscopically identical systems which are characterized by a complete set of quantum states. The information we wish to obtain is what fraction f_j of the members of the ensemble are in the jth quantum state. Since the f_j are a normalized probability distribution, we can write

$$\sum f_j = 1 \tag{7-38}$$

Since the members of the ensemble are each characterized by an internal energy U, we can consider that ensemble members have some average energy,

$$\bar{E} = \sum f_j E_j \tag{7-39}$$

In order to view the problem from the point of view of information theory, assume for a moment that one could examine each member of the ensemble and determine what quantum state it was in. The set of quantum states would constitute a message and the average information content of knowing which quantum state a system is in is given by

$$I = -\sum f_j \ln f_j \tag{7-40}$$

The preceding formula comes directly from eq. (6-17). The least committal statement I can make about the system is that such a determination of which quantum state a system is in would give a maximum amount of information. That is, if I knew nothing more about the system other than eqs. (7-38) and (7-39) then 1 would anticipate that an otherwise unconstrained system would maximize I subject to these constraints. Therefore, the f_j can be determined by finding a maximum of I by the method of undetermined multipliers. The procedure is formally identical to the one we have just carried out and leads to eq. (7-37).

Both the most-probable-state approach and the information-theory approach lead to the same distribution function for quantum states of the system. In the next chapter we shall study the relationship between the E_j and f_j and the measurable parameters of macroscopic thermodynamics.

TRANSITION

The statistical point of view involves the representation of the macroscopic state of the system by an ensemble average of similar macroscopic systems differing in their detailed microstates. The microstates are defined quantum mechanically by a set of quantum numbers and classically by the positions and momenta of the particles. The most probable distribution of an interacting ensemble is defined by the one that can occur in the maximum number of ways by combinatorial analysis. The ensemble is then subject to the constraints of the total energy and total number of members being constants.

The most probable distribution is then found by Lagrange's method of undetermined multipliers. This leads to a distribution function involving a single undetermined quantity. To evaluate this parameter we must make contact between the statistical approach and traditional thermodynamics, which is the first task in the next chapter.

BIBLIOGRAPHY

Gurney, R. W., "Introduction to Statistical Mechanics." Dover, New York, 1966.

Contains a particularly lucid treatment of combinatorial analysis as applied to statistical mechanics.

Hanna, M. W., "Quantum Mechanics in Chemistry." Benjamin, New York, 1969.

A concise treatment of introductory quantum mechanics and its application to chemical systems.

Jaynes, E. T., *Phys. Rev.* **106**, 620 (1957).

This is the original paper in which the information theory approach to statistical mechanics is set forth.

Tolman, R. C., "The Principles of Statistical Mechanics." Oxford Univ. Press, London and New York, 1938.

This is a major work, which elaborates in great detail the foundations of statistical mechanics both from the classical and quantum mechanical points of view.

8 Connecting Thermodynamics and Statistical Physics

A. EQUIVALENT FORMULATIONS

Our presentation of statistical mechanics in the previous chapter provides a method of dealing with ensembles of systems distributed over quantum states; however, until we identify β we have no method of using the theory to make macroscopic predictions. By way of preempting later developments, we might note that nowhere in the theory have we explicitly introduced the temperature of the reservoir T. Since we would anticipate different distributions over possible quantum states as a function of temperature, we might guess that β will be a function of T.

Contact between statistical mechanics and thermodynamics is first made by identifying the internal energy of a system with its average energy over the ensemble,

$$U \leftrightarrow \bar{E} = \frac{E_T}{N} = \sum f_j E_j \tag{8-1}$$

U is the total energy of the system and E_j is the total energy of the system in the jth quantum state. If the macroscopic measurement of U averages over a sufficient number of quantum states, $\bar{E}$ and U become measures of the same quantity. We may then write

$$U = \sum f_j E_j \tag{8-2}$$

For a single homogeneous substance in contact with an isothermal reservoir we may apply the simplest form of the first law of thermodynamics:

$$dU = T\,dS - P\,dV \tag{8-3}$$

Equating eqs. (8-2) and (8-3), we find that

$$T\,dS - P\,dV = \sum E_j\,df_j + \sum f_j\,dE_j \tag{8-4}$$

In simple systems the energy levels E_j are primarily functions of the dimensions of the system, as we have shown in eq. (7-13) of the last chapter. Consider a change in one dimension only. Then we have

$$dE_j = \frac{\partial E_j}{\partial x}\,dx \tag{8-5}$$

In order to make eq. (8-5) somewhat more meaningful, let us consider a specific case of a rectangular box of gas whose sides are parallel to the three axes of a coordinate system (see fig. 8-1). In order to change the dimensions by an amount dx, we must apply a pressure P on the yz face and do an amount of work $P\,dV$. If the box is adiabatically isolated, the change in energy of the system is $P\,dV$, which must be equal to the change of energy in quantum-mechanical terms, which, if the system were entirely in the jth quantum state, is dE_j. Since dV is $yz\,dx$, we can then write

$$-P\,dV = -Pyz\,dx = \frac{\partial E_j}{\partial x}\,dx = dE_j \tag{8-6}$$

However, P in eq. (8-6) is the specific pressure when the system is in the jth quantum state and should consequently be designated as P_j.

The last term in eq. (8-4) then becomes

$$\sum f_j\,dE_j = -\sum f_j P_j\,dV = -dV \sum f_j P_j \tag{8-7}$$

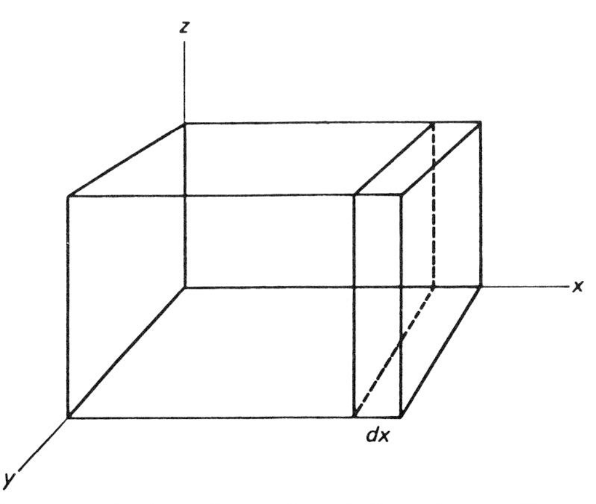

FIGURE 8-1. A rectangular box of gas oriented along the coordinate axes. The problem of interest is the change in energy accompanying a small change in the x dimension of the box.

The quantity $\sum f_j P_j$ is the ensemble average of the pressure, which we can identify with the macroscopic pressure P. Therefore the last term of eq. (8-7) is simply $P\,dV$, and eq. (8-4) reduces to

$$T\,dS = \sum E_j\,df_j \tag{8-8}$$

However, from eq. (7-37) of Chapter 7 we note that

$$E_j = \frac{\ln f_j + \ln \sum e^{\beta E_j}}{\beta} \tag{8-9}$$

Substituting eq. (8-9) into eq. (8-8) we get

$$T\,dS = \frac{\sum \ln f_j\,df_j}{\beta} + \frac{(\ln \sum e^{\beta E_j})}{\beta} \sum df_j \tag{8-10}$$

The second term on the right can be dropped since we have already noted $\sum f_j = 1$, so that $\sum df_j = 0$. Utilizing this result, we can rewrite eq. (8-10) in a somewhat more simplified form as

$$dS = \frac{1}{\beta T} d(\sum f_j \ln f_j) \tag{8-11}$$

The function $\sum f_j \ln f_j$ may be represented as $\mathscr{X}$, in which case eq. (8-11) becomes

$$dS = \frac{1}{\beta T} d\mathscr{X} \tag{8-12}$$

At this point we introduce properties of the entropy function which allow us to solve eq. (8-11) for β and also to solve for the integrated form of S. The first restriction is that the entropy function in thermodynamics must be an exact integral. The mathematical requirement of eq. (8-12) is then

$$\frac{1}{\beta T} = f(\mathscr{X}) \tag{8-13}$$

If eq. (8-13) is satisfied, dS becomes $f(\mathscr{X})\,d\mathscr{X}$, which in principle can always be integrated. One further requirement of the entropy function, that of additivity, allows us to solve eqs. (8-12) and (8-13) and determine β. Additivity means that if we have two systems at the same temperature of entropies S_A and S_B and bring them in contact through a diathermic wall, then the entropy of the combined system S_{AB} is equal to $S_A + S_B$. The specification of β then completes the identification of thermodynamic and statistical theories and provides the generalized background to go on to more specific applications.

We shall not solve the problem in detail, owing to the mathematical complexities necessary for a rigorous solution. For a discussion of this point see T. Hill in the Bibliography. We shall quote the result, which is that the only function $f(\mathscr{X})$ which permits eq. (8-12) to conform to the additivity postulate is that $f(\mathscr{X})$ is a constant, which we shall designate as $-k$. The fact that the additivity condition holds between any two systems requires that k be a universal constant. Application of the formalism of statistical mechanics to ideal gases shows that this universal constant is indeed equal to the Boltzmann constant introduced in Chapter 2. Equations (8-12) and (8-13) then lead to the result

$$\beta = -\frac{1}{kT} \tag{8-14}$$

Equation (7-37) of Chapter 7 now becomes

$$f_j = \frac{e^{-E_j/kT}}{\sum e^{-E_j/kT}} \tag{8-15}$$

and the entropy function S becomes

$$S = -k \sum f_j \ln f_j \tag{8-16}$$

B. THE PARTITION FUNCTION

The denominator of eq. (8-15) is a particularly important quantity in statistical mechanics and is usually designated the partition function Z, which may formally be defined as

$$Z = \sum e^{-E_j/kT} \tag{8-17}$$

The function Z is a hybrid notion combining the E_j, which are eigenvalues from quantum mechanics, with T, the temperature from phenomenological thermodynamics, and k, the quantity which was introduced in the connecting of statistical mechanics with classical thermodynamics. We shall prove that, once Z is known, all the thermodynamic information about a system can be obtained. This is a particularly important notion since it demonstrates the consistency of our various approaches. Start by writing the probability f_j in terms of Z and E_j:

$$f_j = \frac{e^{-E_j/kT}}{Z} \tag{8-18}$$

We then take the logarithm

$$\ln f_j = \frac{-E_j}{kT} - \ln Z \tag{8-19}$$

Substitute eq. (8-19) into the expression for entropy given in eq. (8-16),

$$S = -k \sum f_j \left(-\frac{E_j}{kT} - \ln Z \right) \tag{8-20}$$

C. HELMHOLTZ FREE ENERGY

Rearranging terms in eq. (8-20), we get

$$-kT \ln Z = \sum f_j E_j - TS \tag{8-21}$$

The quantity $\sum f_j E_j$ has already been identified as U, the internal energy. *The right-hand side of the equation* $(U - TS)$ *will be designated* A, *the Helmholtz free energy* (definition). From eq. (8-21) we can then write

$$A = -kT \ln Z = U - TS \tag{8-22}$$

Thus if Z is known, A is known; that is, the thermodynamic function is directly available from the partition function, since T and S are state functions. Taking the differential, we get

$$dA = dU - T\,dS - S\,dT \tag{8-23}$$

Substituting the appropriate form of dU for a multicomponent system $(T\,dS - P\,dV + \sum \mu_i\,dn_i)$, we get

$$dA = -P\,dV + \sum \mu_i\,dn_i - S\,dT \tag{8-24}$$

Since A is a state function, we can also write dA in the form of an exact differential

$$dA = \left(\frac{\partial A}{\partial V}\right)_{n_i,T} dV + \sum \left(\frac{\partial A}{\partial n_i}\right)_{V,T,n_j} dn_i + \left(\frac{\partial A}{\partial T}\right)_{n_i,V} dT \tag{8-25}$$

From (8-24) and (8-25) we get

$$P = -\left(\frac{\partial A}{\partial V}\right)_{n_i,T}, \qquad \mu_i = \left(\frac{\partial A}{\partial n_i}\right)_{V,T,n_j}, \qquad S = -\left(\frac{\partial A}{\partial T}\right)_{n_i,V} \tag{8-26}$$

Thus if we know $Z(V, n_i, T)$, we can derive all other thermodynamic information.

In developing statistical mechanics, we started out with a quantum-mechanical description of the individual members of the ensemble, so that the partition function we arrived at depends upon the eigenvalues. Since the quantum-mechanical description is the best representation of a system which we have, our approach has been fundamentally correct. The difficulty with this method is that for complex systems it is often very difficult to solve for the eigenvalues and the summations do not always lead to convenient analytical expressions. Where the eigenvalues are closely spaced it is sometimes possible to replace the summation of eq. (8-17) with an integral.

D. CLASSICAL STATISTICAL MECHANICS

Alternatively, it is possible to formulate statistical mechanics from a classical point of view and to apply it directly to those areas where classical mechanics gives a reasonably good approximation to quantum mechanics. For classical systems the energy E is a function of the coordinates and momenta of the particles. If we have N particles, we can then formulate a $6N$-dimensional space in which the dimensions represent the coordinates and momenta of all particles in the system. A point in this phase space then represents a complete classical description of an individual system and an ensemble is represented by a cluster of points in this space. If we now divide the phase space up into a group of small subvolumes and assign a number i to each subvolume, then to each subvolume there corresponds an energy E_i,

$$E_i = E(p_i, q_i) \tag{8-27}$$

If we now designate the number of systems of an ensemble whose representative point is in the ith subvolume as N_i, we can then proceed exactly as in Chapter 7, eqs. (7-28)–(7-33), and get the same representation subject to the fact that the exact size of the subvolume is arbitrary. The resultant equations are the same. In computing the partition functions for a system of N identical particles, we can go from the sum to the integral in the following way:

$$Z = \sum \exp - \frac{E_j}{kT} = \frac{1}{N!}\frac{1}{\Delta V}\int \exp - \frac{E(p,q)}{kT}\,dp\,dq \tag{8-28}$$

ΔV is the volume of an element in phase space and cannot be determined from classical mechanics. (From a comparison of classical and quantum theory we know that $\Delta V = h^{3N}$, where h is Planck's constant.) The factor $N!$ enters because in the classical case particles are distinguishable while in quantum mechanics they are indistinguishable. A detailed treatment of this material is found in the book by Hill that we have already noted. To evaluate Z in a specific case we note that for an ideal gas

$$E = \sum_j \frac{p_{xj}^2 + p_{yj}^2 + p_{zj}^2}{2m} \tag{8-29}$$

Equation (8-29) says that all the energy is kinetic, and for each particle is the sum of three terms of the form $\frac{1}{2}mv_x^2$, which may be written in momentum notation as $p_x^2/2m$. We may now substitute eq. (8-29) into eq. (8-28) and integrate. There is one integral of six dimensions for each particle, and the partition function is the product of these for N particles:

$$Z = \frac{1}{N!h^{3N}}\left[\int_{\substack{\text{all possible values}\\ \text{of coordinates}\\ \text{and momenta}}} \exp - \frac{p_x^2 + p_y^2 + p_z^2}{2mkT}\, dp\, dq\right]^N$$

$$= \frac{1}{N!}\left[\left(\frac{2\pi mkT}{h^2}\right)^{3/2} V\right]^N \tag{8-30}$$

The details of this integration are presented in Appendix VI, and eq. (8-30) simply records the results.

We can now proceed to derive some of the thermodynamic properties from Z. The Helmholtz free energy now becomes

$$A = -kT \ln Z = -kTN\left[\ln V + \tfrac{3}{2}\ln\frac{2\pi mkT}{h^2}\right] + kT \ln N! \tag{8-31}$$

Using the formalism developed in eq. (8-26), we get

$$P = -\frac{\partial A}{\partial V} = \frac{kTN}{V} = \frac{nRT}{V} \tag{8-32}$$

Equation (8-32) is the familiar ideal-gas law, which has now been derived from the classical partition function using the formalism of statistical mechanics.

Since there is no interaction potential for molecules of an ideal gas, eq. (8-30) can be regarded as the N-fold product of partition functions for a single molecule, $\mathfrak{z}$, divided by $N!$ because of quantum mechanical indistinguishability:

$$Z = \frac{\mathfrak{z}^N}{N!} \tag{8-33}$$

$$\mathfrak{z} = \left(\frac{2\pi mkT}{h^2}\right)^{3/2} V \tag{8-34}$$

E. *VELOCITY DISTRIBUTION FUNCTION*

Consider next the velocity distribution in an ideal gas. The probability that an individual molecule has a velocity between v_x and $v_x + dv_x$, v_y and $v_y + dv_y$, v_z and $v_z + dv_z$, and lies between x and $x + dx$, y and $y + dy$, z

and $z + dz$ is given by

$$d\mathscr{P}(v_x, v_y, v_z, x, y, z) = \frac{m^3}{h^3}\left[\exp - \frac{m(v_x^2 + v_y^2 + v_z^2)}{2kT}\right]\frac{dv_x\, dv_y\, dv_z\, dx\, dy\, dz}{\mathfrak{z}} \tag{8-35}$$

To set up eq. (8-35) we start with the distribution of eq. (8-15) and apply the transition to a classical formalism in the same manner as introduced in (8-28).

In writing eq. (8-35) we have gone from momentum representation to velocity representation. This equation is the classical analog of eq. (8-18) when applied to an ideal-gas molecule. If we wish to consider just the velocity distribution, we integrate the numerator of eq. (8-35) over $dx\, dy\, dz$. This integral is simply the volume V. Substituting for $\mathfrak{z}$ from eq. (8-34) and rearranging, we get

$$d\mathscr{P}(v_x, v_y, v_z) = \frac{\exp[-m(v_x^2 + v_y^2 + v_z^2)/2kT]\, dv_x\, dv_y\, dv_z}{(2\pi kT/m)^{3/2}} \tag{8-36}$$

Equation (8-36) is a velocity distribution function for molecules of an ideal gas. It is usually expressed in a somewhat different fashion. We can ask, what is the probability that the velocity in the x direction lies between v_x and $v_x + dv_x$ independent of the y and z velocities. To obtain this, we integrate eq. (8-36) over y and z. This leads to

$$d\mathscr{P}(v_x) = \frac{\exp(-mv_x^2/2kT)\, dv_x}{(2\pi kT/m)^{1/2}} \tag{8-37}$$

Alternately, we might ask what is the probability that a molecule has a speed v independent of the x, y, and z components of the velocity. Consider a representative point in a phase space of coordinates v_x, v_y, and v_z. We note from geometrical considerations that $v^2 = v_x^2 + v_y^2 + v_z^2$; that is, the square of the velocity is equal to the sum of the squares of its components along the three axes. A particle whose velocity lies between v and $v + dv$ will be represented by a point lying in a spherical shell between v and $v + dv$. The element of volume $dv_x\, dv_y\, dv_z$ is therefore replaced in v notation by $4\pi v^2\, dv$. Rewriting eq. (8-36) in v notation leads to

$$d\mathscr{P}(v) = \frac{[\exp(-mv^2/2kT)]4\pi v^2\, dv}{(2\pi kT/m)^{3/2}} \tag{8-38}$$

Equations (8-37) and (8-38) are the Maxwell velocity distribution and are of fundamental importance in understanding the behavior of matter at the molecular level.

The Maxwell velocity distribution emerges from a rather lengthy and abstract set of arguments in statistical mechanics and thermodynamics. Yet it gives a very detailed prediction of how the velocity distribution of ideal-gas molecules depends on temperature and molecular mass. The experimental validity of the distribution should then provide a rather probing examination of the correctness of the chain of reasoning. If eq. (8-38) corresponds to the experimental distribution of velocities, it will be strong evidence that our method of approach has been essentially correct. If eq. (8-38) fails to correspond, then the entire argument must be examined step by step to locate the difficulty.

In practice the Maxwell velocity distribution has been checked in many ways and found to agree very well with experiment. We will sketch some general features of one type of measurement so as to make the argument more concrete. Suppose we have a vessel containing a gas maintained at temperature T as shown in fig. 8-2. If we place a hole in the vessel so that the leakage rate is small compared with the amount of gas in the vessel, we will not appreciably disturb the equilibrium. If we now place a slit in front of the hole and some distance from it, we should get through the slit a nearly parallel beam of molecules showing a Maxwell velocity distribution. This beam can now be passed through a device shown in fig. 8-3. Two concentric discs are rotated on a common axis. Each disc has a slit, the second slit being displaced by an angle δ from the first slit. The beam from the device in fig. 8-2 enters the analyzer from the left. Those molecules which pass through the first slit are undeviated, and in general hit the second disc. If the time of flight between the discs, τ, is the time for the second disc to move an angular distance δ, the molecules will pass through the second slit. This condition is

$$\tau = \frac{L}{v} = \frac{\delta}{\omega} \tag{8-39}$$

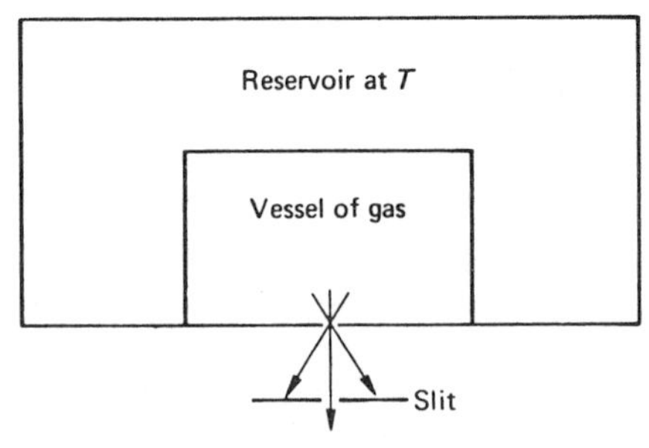

FIGURE 8-2. Oven used to produce a stream of molecules to test the velocity distribution law.

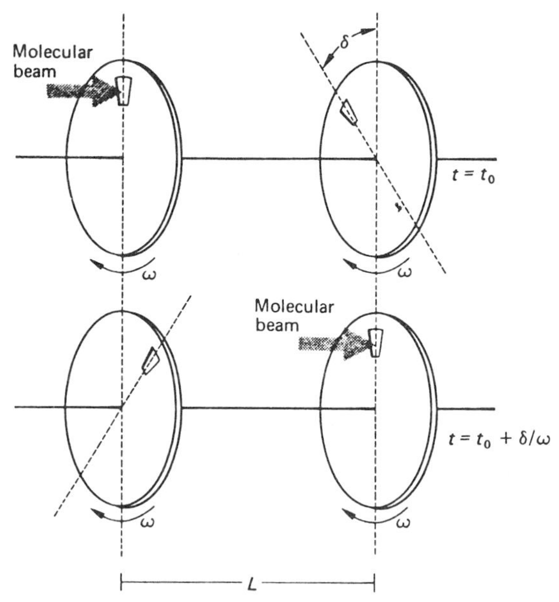

FIGURE 8-3. Velocity spectrometer to analyze the beam produced by the device shown in fig. 8-2. The beam passes through two slits which are separated by a distance L and are separated in time by δ/ω.

The discs rotate at an angular velocity ω, and L is the distance between the discs. Particles will pass through both discs only if they have a velocity

$$v = \frac{\omega L}{\delta} \tag{8-40}$$

The angular velocity ω can be varied and the number of molecules coming through the second slit can be measured as a function of this quantity. The device then acts as a velocity spectrometer and can be used to analyze the beam coming from the apparatus in fig. 8-2. Such equipment has been used to directly measure the velocity distributions of gases, which are found to conform to the Maxwell velocity distribution.

F. INTERNAL DEGREES OF FREEDOM

The ability to relate the statistical approach to thermodynamics is not confined to ideal gases or to translational degrees of freedom. Internal degrees of freedom can be treated as well as a discussion of the interactions between molecules. The extension to internal degrees is relatively simple and we will discuss some examples. Molecular interactions are considerably more difficult to handle and we shall only be able to give a brief introduction into how the problem is treated.

For an individual molecule we can often write the energy as the sum of four terms:

$$\varepsilon_T = \varepsilon_t + \varepsilon_r + \varepsilon_v + \varepsilon_i \tag{8-41}$$

Equation (8-41) states that the total energy is the sum of the translational, rotational, vibrational, and the energy of interaction with other molecules. It is an approximate relation and its validity must be examined for each case in which it is used. All molecules are postulated to be in the electronic ground state. We will for the moment assume the equation and discuss how it can be used to include internal degrees of freedom in the analysis. First form the partition function corresponding to eq. (8-41):

$$\begin{aligned}\mathfrak{z} &= \sum e^{-\varepsilon_T/kT} = \sum_{t,r,v,i} e^{-(\varepsilon_t+\varepsilon_r+\varepsilon_v+\varepsilon_i)/kT} \\ &= \sum_{t,r,v,i} e^{-\varepsilon_t/kT} e^{-\varepsilon_r/kT} e^{-\varepsilon_v/kT} e^{-\varepsilon_i/kT} \\ &= \sum_t e^{-\varepsilon_t/kT} \sum_r e^{-\varepsilon_r/kT} \sum_v e^{-\varepsilon_v/kT} \sum_i e^{-\varepsilon_i/kT} \\ &= \mathfrak{z}_t \mathfrak{z}_r \mathfrak{z}_v \mathfrak{z}_i \end{aligned} \tag{8-42}$$

When the energy is in the form of eq. (8-42) the partition function is the product of four independent partition coefficients, one for each type of additive term. In going to the Helmholtz free energy for the system, we have

$$\begin{aligned} A &= -kT \ln \mathfrak{z} = -kT \ln \mathfrak{z}_t \mathfrak{z}_r \mathfrak{z}_v \mathfrak{z}_i \\ &= -kT \ln \mathfrak{z}_t - kT \ln \mathfrak{z}_r - kT \ln \mathfrak{z}_v - kT \ln \mathfrak{z}_i \end{aligned} \tag{8-43}$$

The partition functions for various internal degrees of freedom can thus be considered separately. For instance, for a diatomic molecule in a dilute gas we can write for its translational energy

$$\varepsilon_t = \frac{\mathfrak{n}^2 h^2}{8mL^2} \tag{8-44}$$

Equation (8-44) comes from considering the molecule as a particle in a cubic box with sides of dimension L. The vibrational energies are also available from quantum mechanics. We can regard the molecule as a harmonic oscillator, two point masses connected by a spring (fig. 8-4). The solution to this problem, which we shall not carry out, is that the energy eigenvalues are

$$\varepsilon_v = \frac{h}{2\pi} \sqrt{\frac{\kappa}{m}} \left(\mathfrak{v} + \frac{1}{2} \right) \tag{8-45}$$

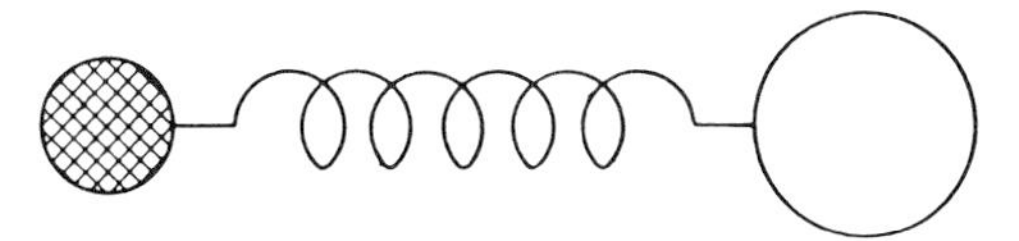

FIGURE 8-4. Representation of a diatomic molecule as two masses connected by a spring.

κ is the spring constant (force constant in the molecular case), and m is the reduced mass of the system and is equal to $m_A m_B/m_A + m_B$. The vibrational quantum number is $\mathfrak{v}$. Similarly, the molecule will have energy associated with its rotation about its center of gravity. The problem of the rigid rotator has also been solved in quantum mechanics, and the eigenvalues are

$$\varepsilon_r = \frac{\mathfrak{r}(\mathfrak{r} + 1)h^2}{8\pi^2 \mathscr{I}} \tag{8-46}$$

The quantity $\mathscr{I}$ is the moment of inertia of the molecule computed about the center of mass and $\mathfrak{r}$ is the rotational quantum number.

It should be noted that eqs. (8-45) and (8-46) can be derived in an exact way from the applications of the Schrödinger formulation of quantum mechanics. We are simply quoting the energy eigenvalues in order to indicate how real molecular situations are treated in statistical mechanics. To gain some more feeling for the method of approach, we shall proceed to show how one of the molecular partition functions can be computed from the energy eigenvalues.

The partition function for vibration can be formulated in the following way:

$$\mathfrak{z}_v = \sum_{\mathfrak{v}=0}^{\infty} \exp\left[-\frac{h}{2\pi} \sqrt{\frac{\kappa}{m}} \left(\mathfrak{v} + \frac{1}{2}\right) \Big/ kT \right] \tag{8-47}$$

Define a quantity ν, the fundamental frequency of the oscillator

$$\nu = \frac{1}{2\pi} \sqrt{\frac{\kappa}{m}} \tag{8-48}$$

This is the frequency of a classical oscillator of reduced mass m and spring constant κ. Define a second quantity x as

$$x = e^{-h\nu/kT} \tag{8-49}$$

Equation (8-47) then becomes

$$\begin{aligned} \mathfrak{z}_v &= x^{1/2}[1 + x + x^2 + x^3 + x^4 + \cdots] \\ &= \frac{x^{1/2}}{1 - x} = \frac{e^{-h\nu/2kT}}{1 - e^{-h\nu/kT}} \end{aligned} \tag{8-50}$$

The partition function for rotation can be computed by somewhat more elaborate processes. The resultant value is

$$\mathfrak{z}_r = \frac{8\pi^2 \mathcal{I} kT}{h^2} \tag{8-51}$$

The total partition function for noninteracting molecules then is

$$Z = \frac{\mathfrak{z}^N}{N!} = \frac{1}{N!}\left(\left[\frac{(2\pi mkT)^{3/2}V}{h^2}\right]\left[\frac{e^{-h\nu/2kT}}{1 - e^{-h\nu/kT}}\right]\left[\frac{8\pi^2 \mathcal{I} kT}{h^2}\right]\right)^N \tag{8-52}$$

where m in eq. (8-52) is the sum of the atomic masses, $m_A + m_B$. Using eq. (8-52) we can now proceed to derive the thermodynamic quantities associated with internal degrees of freedom.

The problem of treating interactions between molecules is considerably more difficult since we can no longer consider partition functions of individual molecules but must compute the function for the whole collection of molecules in a gas. Thus for a gas of N molecules we could write

$$E = NE_T + N\left(\frac{\Lambda N}{V}\right) \tag{8-53}$$

For the first term we sum up the internal terms from the N individual molecules. The second term (applicable in the case of weak interactions only) asserts that the interaction energy for each molecule depends on the concentration of surrounding molecules. Treating actual cases depends on a study of the nature of Λ.

The chief virtue of the methods we have been studying in this chapter is that we can utilize detailed analysis of molecules involving the most precise available quantum-mechanical methods to describe and study the energetics and molecular mechanics. From the very best quantum-mechanical description of individual molecules and their interactions we can "explain" (predict) many thermodynamic properties of macroscopic collections of these molecules.

Very powerful applications have been made of these methods in biology. The helix-coil transition in DNA is perhaps one of the best examples of the important phenomena in biological macromolecules which have been treated with great insight by the methods of statistical thermodynamics. These techniques can be expected to add further understanding of biological macromolecules.

TRANSITION

Identification is made between the average energy of an ensemble member and the internal energy function of thermodynamics. When this is done a number of results follow, the first of which is the evaluation of the undetermined multiplier as $-1/kT$. The relation between the two viewpoints also allows the evaluation of the entropy function in terms of distribution probabilities. We define the partition function which depends only on temperature and eigenenergies. It is shown that all of the thermodynamic information about a system may be derived from this function. Classical statistical mechanics is introduced, and the velocity distribution function of an ideal gas is obtained by this approach. A number of experiments confirm the distribution. The partition function of diatomic models is derived from the solution of the harmonic oscillator and the rigid rotator. With this background we next return to the relation of entropy and information measure.

BIBLIOGRAPHY

Hill, T., "Introduction to Statistical Thermodynamics." Addison-Wesley, Reading, Massachusetts, 1960.

A thorough introduction to statistical mechanics written with the chemist and biochemist in mind. The treatment of difficult sections is very detailed and complete.

Jeans, J., "An Introduction to the Kinetic Theory of Gases." Cambridge Univ. Press, London and New York, 1959.

Contains the theory of dilute gases with particular emphasis on the Maxwell velocity distribution.

9 Information Theory Considerations

A. INFORMATION AND ENTROPY

The analytical form of the entropy function in statistical mechanics, as well as the approach used in our second method of deriving the distribution functions of statistical mechanics, suggests a relationship between entropy and information. Indeed, we may use the ensemble approach to provide a very direct link between the two concepts. Starting with an ensemble, let us serially number each member of the ensemble. The eigenstate of each member can be regarded as a symbol, and the sequence of eigenstates constitutes a message in the sense indicated in our introduction to information theory. The probability of a system being in the jth eigenstate is f_j. We can then pose the question: What is the average information we would have if we knew which eigenstate a system was in? From eq. (6-17) of Chapter 6 we can immediately write down the answer:

$$\bar{I} = -0.693 \sum \mathscr{P}_j \ln_2 \mathscr{P}_j = -\sum \mathscr{P}_j \ln \mathscr{P}_j \tag{9-1}$$

The numerical coefficient is introduced when the logarithm to the base two is converted to the logarithm to the base e.

The entropy of the ensemble just discussed is

$$S = -k \sum f_j \ln f_j \tag{9-2}$$

where $\mathscr{P}_j$ and f_j are the same normalized probability distributions. Combining eqs. (9-1) and (9-2) we get

$$S = 0.693\, k\bar{I} \tag{9-3}$$

The entropy is directly proportional to the information we would have if we knew which microstate the system was in. Entropy thus measures the amount of residual ignorance we have about the microstate of the system, once we know the equilibrium-state parameters. Note once more, entropy is related to the average information *we would have* if we had the maximal possible information about the detailed microscopic state of the system. Thus a system of high entropy is one in which a knowledge of the macroscopic state of the system tells very little as to which of the possible microstates it is in. Such a system is considered as disordered at the molecular level. However, order is used here in a somewhat subjective sense since the incompleteness is always in the observer's knowledge of which eigenstate a system is in. If we think of the phase-space representation, a high-entropy system is one whose representative point traverses a wide volume of phase space, while the phase point of a low-entropy system tends to be confined to a small subvolume of phase space. This point may be illustrated by our previous discussion of the entropy of mixing. Referring to fig. 9-1, suppose we have a molecule of A in the left-hand box (the x coordinate lying between 0 and 1) and a molecule of B in the right-hand box (the x coordinate lying between 1 and 2). Consider the phase point representing this system in coordinate space. Let the x coordinate of the A molecule be represented on the ordinate and the x coordinate of the B molecule be represented on the abscissa. Before mixing, the phase point is confined to the dashed square in fig. 9-2, while after mixing, the phase point can be anywhere within the larger square. The higher-entropy state after mixing gives us less knowledge about where the A and B molecules actually are.

Consider next the change of distribution of eigenstates with temperature. We start by restating the distribution functions of statistical mechanics,

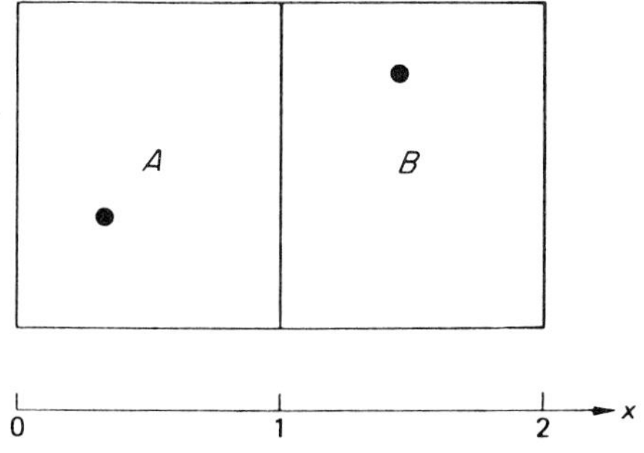

FIGURE 9-1. Two boxes of equal dimensions, one containing a molecule of A and the second containing a molecule of B.

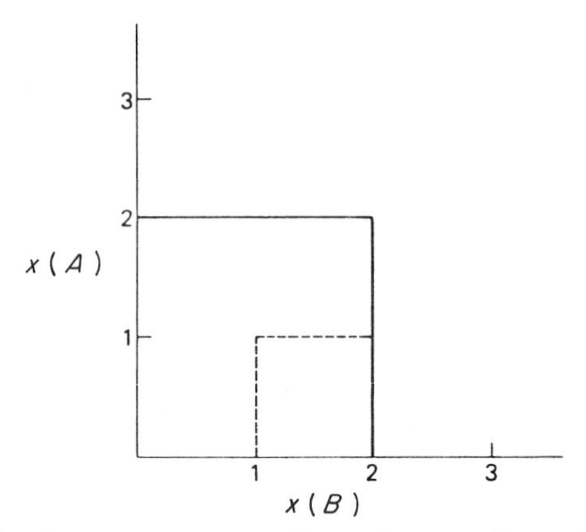

FIGURE 9-2. The phase space corresponding to the x coordinate of molecules A and B. The portion enclosed by the dotted line is the available space before mixing, while the portion enclosed by the solid line is the space after mixing.

from eq. (8-15):

$$f_j = \frac{e^{-E_j/kT}}{\sum_{j=1}^{n} e^{-E_j/kT}} \tag{9-4}$$

The state numbered 1 is the lowest-lying eigenstate. As $T \to 0$, $f_1 \to 1$, $f_j \to 0$ for $j \neq 1$. At absolute zero the probability becomes one that the system will be found in the energetically lowest-lying eigenstate. The entropy in this case becomes

$$S = k[1 \ln 1 + (n - 1)(0 \ln 0)] = 0 \tag{9-5}$$

The quantity in brackets is equal to zero because the logarithm of one is zero and the quantity $x \ln x$ approaches zero as x approaches zero. That is, since we know exactly what microstate the system is in, no further information remains to be obtained about the microstate and the entropy goes to zero.

The vanishing of the entropy at absolute zero has also been introduced on thermodynamic grounds alone and this postulate is known as the third law of thermodynamics. Statistical mechanics provides the basis for understanding this property of the entropy function.

In general, as a system goes from a state of higher entropy to a state of lower entropy, we go from a state of greater uncertainty about the microstate (high I) to a state of lesser uncertainty. We have gained information because in the second state we are less uncertain about the microscopic description of the system than we were orginally. Thus a decrease of entropy corresponds to an increase of information about the microstate.

As a result of the preceding a good deal has been written about the equivalence of information and negentropy (entropy decrease). This has been somewhat confusing because from eq. (9-3) information measure and entropy are linearly related, so that a decrease of one is equivalent to a decrease of the other. The confusion is removed when we remember that entropy is proportional to a function which measures our lack of information about the microstate of a system. A high I means that we would have a great deal of information *if* we knew which microstate a system was in, but this implies that we have very little knowledge as to which state it actually is in. As I decreases (S decreases) we are less unsure as to the state of the system; hence, we have gained real information about the system. Therefore $-\Delta I$ represents the change in our information about the system, and a decrease of entropy does correspond to an increase in our actual knowledge about the microstate of the system. So negentropy is information, but this poetical form of the statement has to be thoroughly understood so as not to be confusing.

B. MAXWELL'S DEMON

The entropy of a system is also related to its capacity to do work, so that in some way information, which is a rather biological or even psychological concept, is related to purely energetic concepts. This problem was seen by Maxwell, who introduced an imaginary construct, subsequently known as a Maxwell demon. The demon is located at a wall between two chambers of equal pressure of ideal gas, the entire system being embedded in an isothermal reservoir. This is seen in fig. 9-3. The demon operates a small trap door between the two chambers. Whenever he sees a molecule approaching the door from the right and no molecule approaching from the left, he opens the door. In time a pressure difference develops and work can be done by moving the wall and having it hooked up to an external mechanical device. The demon then opens the door, the wall is restored to its original position, and the demon begins his selection all over again. The entire device of the demon plus the gas reservoirs operates in a cycle and transfers heat from the reservoir into mechanical work, thus violating the second law of thermodynamics. If one accepts the validity of the second law, one must conclude that the demon cannot exist.

The reason for the apparent contradiction is that we have not examined the physical process by which the demon obtains information about molecules. Two approaches are available: one is to try to examine the details of the measurement to see what limitations are imposed, and the second is to assume the validity of the second law and see what limitations this imposes on the measuring process. We begin with the second approach and consider

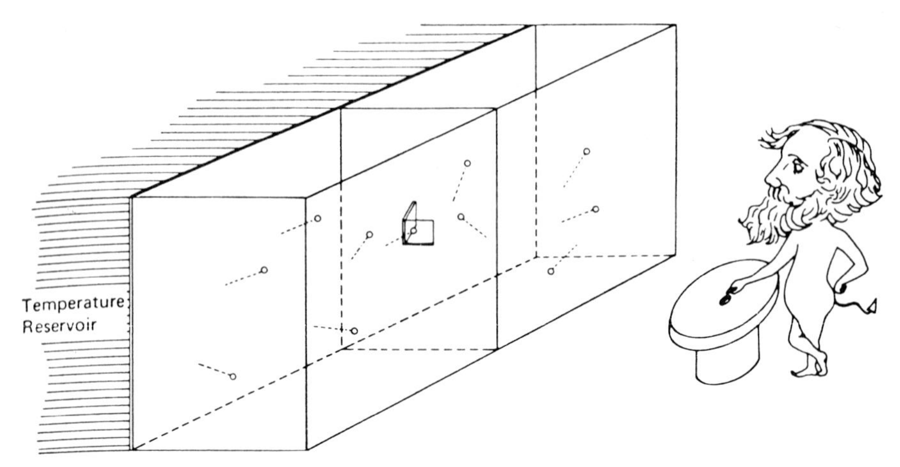

FIGURE 9-3. A Maxwell demon controlling a door between two chambers each initially at temperature T_1 and pressure P_1.

two chambers of equal volume with a wall between them. A molecule is in one of the chambers but we do not know which one (fig. 9-4). The amount of information we have in determining which chamber the molecule is in is

$$I = -\ln_2 \tfrac{1}{2} = \ln_2 2 = 1\,\text{bit} \tag{9-6}$$

The bit is a unit of information which corresponds to the reduction in uncertainty from two possibilities to one. The entropy decrease in knowing which side the molecule is on is the same as the entropy decrease in isothermally compressing a gas of original concentration $1/(2V)$ to a final concentration of $1/V$:

$$S = \frac{1}{T}\int_{2V}^{V} P\,dV = \frac{1}{T}\int_{2V}^{V} \frac{kT}{V}\,dV = -k\ln 2 \tag{9-7}$$

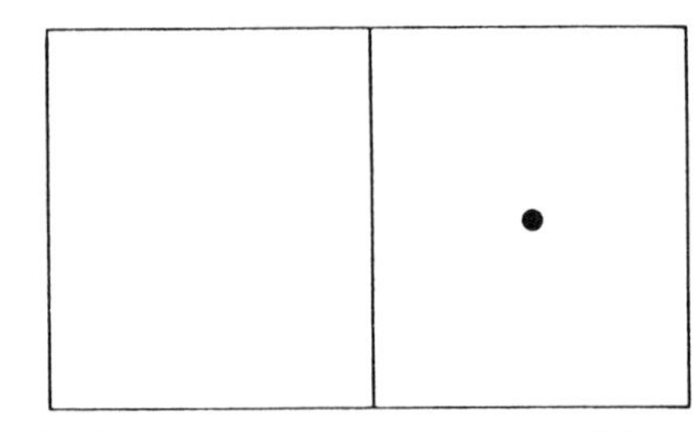

FIGURE 9-4. Two chambers separated by a wall. One of the two chambers is occupied by a molecule.

Thus the determination of which side the molecule is on entails an entropy decrease of $-k \ln 2$. The measurement process must be accompanied by an entropy increase of at least $k \ln 2$ in order to preserve the second law.

Another view of the situation comes from the realization that knowing which side the molecule is on enables us to hook up the system to an external piston and do an amount of work $kT \ln 2$. Such a process could be repeated cyclicly, violating the engine statement of the second law. Information about the microstate enables us to convert thermal energy into work.

Alternatively, Brillouin has shown that work must be expended to obtain information, so that the second law could be derived as a consequence of the theory of measurements. (See L. Brillouin in the Bibliography.)

The situation can be summarized as follows. One of the most fundamental properties of matter on a macroscopic scale is temperature. At a molecular level temperature manifests itself as the kinetic energy of atoms and molecules. Compared to a human observer, molecules are very small, very numerous, and interact very rapidly. These interactions lead to continuous energy exchange between the particles of the system. From a macroscopic point of view we therefore describe the energy as being randomly distributed among the possible degrees of freedom. The Maxwell–Boltzmann distribution function enables us to state the statistical distribution. Indeed, it is possible to prove that there is an average energy of $\frac{1}{2}kT$ associated with each degree of freedom which contributes an energy term proportional to the square of the relevant parameter (such as v_x^2, v_y^2, v_z^2). To convert all of this kinetic energy to work requires a detailed knowledge of each molecule so that it can be hooked up to an external work device. If this knowledge were available, we could continuously convert heat to work. Obtaining detailed molecular knowledge requires measurements, and performing a measurement requires the expenditure of energy (you don't get something for nothing—even information). Hence an entropic price has to be paid to get work out of a heat source. To expand the argument in the preceding paragraph, consider how the demon in fig. 9-3 obtains information about the molecules. If we put a window in the box which is at equilibrium, the observer will see only black-body radiation (see Chapter 11) which will not tell him anything about individual molecules. To make an observation he would have to shine in a light (photons) of higher frequency than the average photon in the box. This requires

$$h\nu > kT \tag{9-8}$$

The entropy associated with transfer of a photon from a source at temperature T_1 to the demon at temperature T is

$$\Delta S = -\frac{h\nu}{T_1} + \frac{h\nu}{T} \tag{9-9}$$

Since the source temperature T_1 must be greater than T if $h\nu > kT$, ΔS is positive and an entropy increase always accompanies a measurement. A more detailed analysis shows that to obtain one bit of information

$$\Delta S > k \ln 2 \tag{9-10}$$

Since $k \ln 2$ is the maximum entropy decrease that the demon could have obtained from the system, the net operation of the demon results in an entropy increase of the universe. Also note that the demon has to use up a photon of energy greater than kT to get back an amount of work $kT \ln 2$ so that nothing is to be gained from the operation. The important fact to note is that the Brillouin analysis comes from a detailed theory of measurement and gives the same answer as the thermodynamic analysis based on the Carnot principle. The theory does not involve the uncertainty principle. The entropic price of information is an independent limitation on the nature of measurements.

The preceding considerations are of minor importance in classical thermodynamics, where one deals with large systems and gross macroscopic properties. They become of greater importance when small systems and nonequilibrium systems are considered. They become of very great importance in dealing with a small, highly organized, far-from-equilibrium system such as a living cell.

The relation between entropy, information, and work thus appears to involve some unusual concepts. Entropy and information about microstates are rather equivalent. Entropy relates the measure to the system and information relates the measure to the observer. That is, however, a rather artificial distinction since thermodynamics always implies the observer as establishing or removing constraints on the system or otherwise manipulating. Microscopic information can also be converted into work and the conversion factor is temperature, the amount of energy stored in microscopic kinetic modes. Alternatively, work can be converted into information since one can do work to place a system in a certain microstate or at least a group of microstates considerably more restricted than the equilibrium ensemble. Continuous work can therefore lead to the self-organization of systems, and this is the principle which underlies the ordering of the biosphere.

Entropy in this formalism measures the average amount of information that the observer is missing because he does not know which quantum state the system is in. Highly disordered states are broadly distributed over many quantum configurations and hence have high entropy. Highly ordered states are narrowly distributed over many fewer quantum states so that an observer has much less uncertainty. These are low entropy arrays.

Realizing that entropy measures the observer's lack of knowledge about the detailed state of the system causes a considerable reorientation in our view of thermodynamics. The observer becomes very much a part of the system, as has happened in relativity and quantum mechanics. This leads to a certain circularity since we wish to reduce much of biology to physics; yet, establishing the foundations of physics we are involved with the observer, who is, after all, an object of biological study. We shall not be able to resolve this circularity, but shall note it as one of the challenges to our deeper understanding of nature.

TRANSITION

Entropy is shown to be proportional to the average information an observer would have if he knew which microstate a system was in. The second law of thermodynamics is shown to be equivalent to the observer lacking sufficient information to convert heat into work. Having established the relation of thermodynamics and statistical mechanics, we now proceed to the formal techniques used to apply this theory to chemical reactions. This is usually done through the use of a number of free-energy functions that we will introduce and discuss in the following chapter.

BIBLIOGRAPHY

Brillouin, L., "Science and Information Theory." Academic Press, New York, 1962.

An introduction to information theory that contains a very detailed analysis of the relation of information to entropy and to physical measurement.

10 Free-Energy Functions

A. GIBBS FREE ENERGY

Our primary concern in this chapter is developing the apparatus which underlies the application of thermodynamics to actual problems in chemistry and biochemistry. For that reason our approach will be somewhat less general as we shift to defining standard states and otherwise developing conventions for use with real systems. Working out these details burdens this chapter with somewhat more than its share of algebraic tedium. It seems a price worth paying for deeper understanding of what is actually done in the use of thermodynamics.

As we have already indicated, the most general description of the equilibrium of a system is that state which maximizes the entropy of the universe, which is interpreted as system plus surroundings. The entropy of the universe is, however, often a difficult quantity to measure and sometimes beset with conceptual difficulties. It is usually possible, under restricted conditions, to find functions, of the state of the system only, which can be related to the global entropy. The most often used of these functions have been given special names and form important tools in solving thermodynamic problems. The formulation of these functions will be carried out by considering a number of equilibrium situations.

First, we discuss a system, closed to the flow of matter, but in mechanical and thermal contact with the rest of the universe under conditions that maintain the system at constant pressure and constant temperature. At equilibrium the entropy is a maximum, so that infinitesimal changes at equilibrium correspond to the mathematical condition

$$dS_u = 0 \tag{10-1}$$

The quantity dS_u is the entropy change of the universe and may be split into two parts, dS_s, the entropy change of the system, and dS_r, the entropy change of the rest of the universe. Equation (10-1) can then be written

$$dS_u = dS_s + dS_r = 0 \tag{10-2}$$

A system at constant pressure and temperature interacts with the outside world by being in contact with an isothermal reservoir through a diathermic wall and being in contact with an isobaric (constant pressure) reservoir through an adiabatic wall (see fig. 10-1). A diathermic wall conducts heat but does not permit the flow of matter. Therefore dS_r can only result from the passage of an amount of heat dQ across the diathermic wall. If dQ_s is the amount of heat change in the system, the heat change of the reservoir is $-dQ_s$ and the entropy change of the reservoir is $-dQ_s/T$. For the system we can write the first law of thermodynamics:

$$dQ_s = dU_s + P\,dV_s \tag{10-3}$$

Equations (10-2) and (10-3) can then be combined in the following way:

$$dS_u = dS_s - \frac{dQ_s}{T} = dS_s - \frac{dU_s}{T} - \frac{P\,dV_s}{T} = 0 \tag{10-4}$$

The condition of equilibrium is now expressed entirely in terms of system parameters S_s, U_s, T, P, V_s. Equation (10-4) may be rewritten

$$T\,dS_s - dU_s - P\,dV_s = 0 \tag{10-5}$$

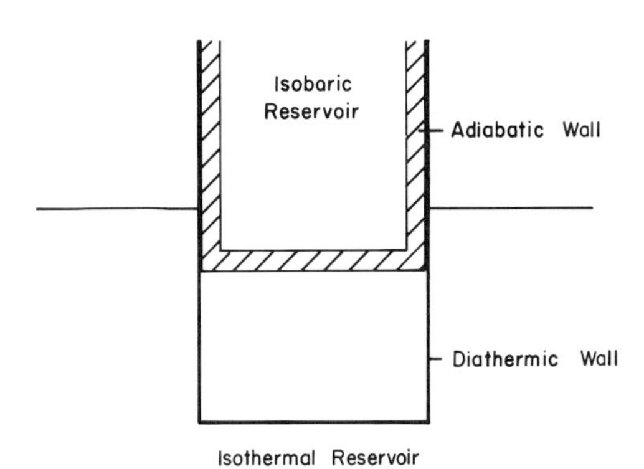

FIGURE 10-1. A chamber in contact with an isothermal reservoir and an isobaric reservoir.

The condition expressed by this equation indicates that there is some function of the system alone which has an extremum at equilibrium, which extremum is equivalent to maximizing the entropy of the universe. We choose a function which we designate G which fulfills these conditions:

$$G = U_s + PV_s - TS_s \tag{10-6}$$

If we now proceed to take the differential of G and apply the isothermal and isobaric conditions $dT = 0, dP = 0$, eq. (10-6) reduces to

$$dG = dU_s + P\,dV_s - T\,dS_s \tag{10-7}$$

Comparison of eqs. (10-7) and (10-5) indicates that the condition $-dG = 0$ is equivalent to the entropy maximization. The condition of equilibrium can then be expressed as the minimum value of G (maximum value of $-G$). G is called the Gibbs free energy and plays a leading role in the biological and biochemical applications of thermodynamics owing to the fact that many reactions take place at constant pressure (atmospheric) and constant temperature (either ambient or, in the case of higher organisms, subject to temperature homeostatis).

B. HELMHOLTZ FREE ENERGY

For systems at constant temperature and constant volume a function of the system other than G is appropriate to describe the equilibrium condition. Equations (10-1) and (10-2) still apply as the most general condition of equilibrium. However, imposing the condition of constant volume modifies eq. (10-3) so that the $P\,dV$ term drops out on the right-hand side, leading to the change in internal energy being equal to the heat term. The equilibrium equations [analogous to (10-4) and (10-5)] now become

$$dS_u = dS_s - \frac{dQ_s}{T} = dS_s - \frac{dU_s}{T} = 0 \tag{10-8}$$

Equation (10-8) may be rewritten as

$$T\,dS_s - dU_s = 0 \tag{10-9}$$

Equation (10-9) corresponds to the minimum of another function A which may be written

$$A = U - TS \tag{10-10}$$

A is called the Helmholtz free energy and has already been introduced in Chapter 8 [eq. (8-22)] in the expression relating the partition function to thermodynamics. Its significance here is that it is a function of the state of the system only whose minimum corresponds to maximum entropy of the universe under conditions of constant temperature and volume.

In addition to the Gibbs and Helmholtz free energies, another energy function of interest is defined and designated as the enthalpy H,

$$H = U + PV \tag{10-11}$$

Referring back to eq. (10-6), we can relate the enthalpy and Gibbs free energy

$$G = H - TS \tag{10-12}$$

To sum up, we note that utilizing the fundamental thermodynamic quantities U, T, S, P, and V we can build up a series of functions which are particularly useful in special applications. The Gibbs free energy has the virtue of being a minimum at equilibrium under the usual conditions for biochemical experiments. The Helmholtz free energy is a minimum under constant-volume isothermal conditions and more importantly is directly available from the partition function of statistical mechanics. The enthalpy function, as we shall shortly see, allows the direct relation between thermochemical data and thermodynamic functions.

The quantities U, G, A, H, S, and V are extensive variables; i.e., they depend on the total size of the system. On the other hand, P and T are intensive variables, whose value is independent of the size of the system. In classical thermodynamics, P and T are global variables which have the same value everywhere within an equilibrium system (in the absence of external fields).

C. GENERALIZED ENERGY FUNCTION

The full meaning of the various free-energy functions only becomes manifest when we know U as a function of the parameters of state or, in the more general case, when we know dU. With the exception of very idealized cases, such as ideal gases, the functional relation between U and the rest of the state parameters is hardly ever known. The changes in total energy dU can, however, be generalized to include a number of other factors. We have previously formulated dU as

$$dU = dQ - P\,dV + \sum_i \mu_i\,dn_i \tag{10-13}$$

The right-hand side refers to ways in which energy can enter or leave the system (thermally, mechanically, or materially). The equation may be extended to include other energy forms as follows:

$$dU = dQ - P\,dV + \sum \mu_i\,dn_i + \mathscr{E}\,de + \mathscr{G}\,dm + \sigma\,d\mathscr{A} + \mathscr{T}\,dl + \Re\,d\mathscr{M} \tag{10-14}$$

The term $\mathscr{E}\,de$ refers to the energy change of the system due to the flow of an amount of charge de driven by an applied potential $\mathscr{E}$. Equations derived from the introduction of this term form the basis of electrochemistry. The term $\mathcal{G}\,dm$ is the change of gravitational energy due to the increase of mass of the system dm in a gravitational potential $\mathcal{G}$. The gravitational potential may be nonuniform within a system, in which case the energy may be changed by altering the position of a mass m in the gravitational field. The term that would then be used would be

$$m\,d\mathcal{G} = m\frac{d\mathcal{G}(x)}{dx}\,dx$$

The use of this type of external mechanical force is of importance in deriving the equations of equilibrium ultracentrifugation and also in meteorological problems involving the earth's atmosphere. The references by Edsall and Wyman and Tanford at the end of this chapter deal with detailed presentations of the problem of the equilibrium distribution of matter in external gravitational and centrifugal fields.

If σ is the surface energy per unit area or surface tension and $d\mathcal{A}$ the change in surface area, then $\sigma\,d\mathcal{A}$ is the change in energy due to extending surfaces. In systems of high surface-to-volume ratio this energy can be an appreciable part of the total energy. The $\sigma\,d\mathcal{A}$ term turns out to be of importance in studying the surface adsorption of molecules.

$\mathcal{T}\,dl$ is a tension elongation term and refers to linear structures in the system, which may be wires, rubber bands, or extensible linear macromolecules. It is a one-dimensional analog of the $P\,dV$ term, tension playing the role of pressure and extension dl playing the role of volume change.

The application of an external magnetic field $\mathfrak{R}$ can also alter the energy of a system. If $d\mathcal{M}$ is the change in the degree of magnetization, this energy is expressed as $\mathfrak{R}\,d\mathcal{M}$.

Equation (10-14) is of course open ended and any number of additional energy terms may be added. This equation forms the starting point for much of the actual analysis of thermodynamics and those terms on the right side are chosen that are pertinent to the problem being studied. For straightforward homogeneous phase chemical thermodynamics, eq. (10-13) provides a suitable starting point. Our first aim is to study the energy changes accompanying chemical reactions taking place at constant temperature and pressure. If we combine eqs. (10-13) and (10-7) and use the fact that for reversible transformations $dQ = T\,dS$, we then get the result that at constant temperature and pressure

$$dG = \sum \mu_i\,dn_i \qquad (10\text{-}15)$$

At equilibrium dG equals zero since G must be an extremum, and we have the following relation:

$$\sum \mu_i\,dn_i = 0 \qquad (10\text{-}16)$$

This equation describes the thermodynamic conditions for chemical equilibrium and indicates how the infinitesimal changes in the equilibrium amount of various chemical species must be related.

Consider a chemical reaction in which a group of reactants are converted into products. The change of Gibbs free energy for the process is the integral of dG from the initial state to the final state. If there is a reversible pathway from reactants to products such that each state is an equilibrium state, then at constant temperature and pressure we have

$$\Delta G = \int_{\text{initial}}^{\text{final}} dG = 0 \tag{10-17}$$

The previous equation holds since $dG = 0$ everywhere along the path. However, since G is a function of state, ΔG will have the same value for all paths from the same initial to the same final state. Chemical reactions being carried out reversibly at constant temperature and pressure therefore take place at constant G.

It is worth considering an idealized physical arrangement for carrying out a reaction at constant G. If the reaction is of the form $A + B \rightleftharpoons C$ we could have a reaction vessel in contact with three isothermal, isochemical potential reservoirs containing A, B, and C, respectively. Contacts are through membranes permeable only to the substances contained in the relevant reservoirs. The chemical potentials μ_A, μ_B, and μ_C are chosen such that ΔG which equals $(\mu_C - \mu_A - \mu_B)$ is zero, and no net reaction takes place. If we now change reservoir A to one of chemical potential $\mu_A + d\mu_A$ the reaction will slowly proceed and C will flow out of the reaction vessel into its reservoir, as A and B flow into the reaction chamber. Alternatively, if we substitute reservoir C by one of chemical potential $\mu_C + d\mu_C$ the reaction will flow in the reverse direction.

The Gibbs and Helmholtz free-energy functions are often referred to as maximum work functions for various reversible isothermal transformations. Consider a system at equilibrium with certain constraints to chemical reactions, such as enzyme inhibitors or walls to prevent mixing of reagents. Remove the constraints and couple the change of the system to the performance of external work. The coupling may be osmotic or electrochemical or any other type that may be operated in a reversible mode. The dW term includes all methods (excluding P–V) of performing external work. We then note that, at constant temperature and pressure,

$$\begin{aligned} dG &= dU + P\,dV - T\,dS = -dW + P\,dV \\ dA &= -dW \end{aligned} \tag{10-18}$$

Over a reversible path $P\,dV$ can be integrated so that we get

$$-\Delta W = \Delta G - \int P\,dV = \Delta A \tag{10-19}$$

The maximum work that can be performed is thus the change in Helmholtz free energy. The change in Gibbs free energy gives the maximum work that can be performed exclusive of P–V work, and would be applicable to a case like an electrical battery where the P–V work is not coupled to the external work device (such as an electric motor) which is operated solely from the $\mathscr{E}e$ energy term.

D. ENTHALPY AND STANDARD STATES

The enthalpy function, H, turns out to be easily accessible to measurement by standard calorimetry. Consider a constant-pressure constant-temperature calorimeter and allow a chemical reaction to take place irreversibly starting with reactants at equilibrium and going to products at equilibrium. The calorimeter can be schematically represented by fig. 10-1. The enthalpy, before and after the reaction, can be represented by

$$H_1 = U_1 + PV_1, \qquad H_2 = U_2 + PV_2 \tag{10-20}$$

The difference in enthalpy is

$$\Delta H = H_2 - H_1 = (U_2 - U_1) + P(V_2 - V_1) \tag{10-21}$$

By the first law we have

$$U_2 - U_1 = \Delta Q - P(V_2 - V_1) \tag{10-22}$$

Equation (10-21) follows from the fact that for the system under consideration energy changes can occur only by heat flow to the thermal reservoir, ΔQ, or work done on a constant-pressure reservoir, $P(V_2 - V_1)$. When we combine eqs. (10-21) and (10-22), we find that ΔQ is equal to ΔH, the heat term measured calorimetrically is equal to the change in enthalpy.

The relation between heat of reaction and enthalpy can be obtained in an alternative fashion by considering what happens when we allow the reaction to take place isothermally, isobarically, and reversibly.

Starting with eqs. (10-11) and (10-13) and assuming the only work term is $P\,dV$, we get

$$dH = dU + P\,dV + V\,dP = dQ - P\,dV + \sum \mu_i\,dn_i + P\,dV + V\,dP \tag{10-23}$$

Introducing the restriction of constant pressure, $dP = 0$, and using eq. (10-15), we then get

$$dH = dQ + dG \tag{10-24}$$

For a chemical reaction going from reactants to products we have

$$\Delta H = \int_{\text{initial}}^{\text{final}} dH = \int_{\text{initial}}^{\text{final}} dQ + \int_{\text{initial}}^{\text{final}} dG \tag{10-25}$$

The first term on the right is the total heat transferred to the reservoir ΔQ, while the second term is ΔG, which we have just shown in eq. (10-17) to be zero under these conditions. Equation (10-25) then reduces to

$$\Delta H = \Delta Q \tag{10-26}$$

Equation (10-26) thus provides us with a state function, the enthalpy whose changes can be measured by direct calorimetry. ΔH for a specific reaction is called the heat of the reaction. In specifying the value of ΔH it is necessary not only to enumerate the reactants and products, but to specify what state they are in, as ΔH, being the difference of two state functions, clearly depends on the detailed nature of the initial and final states. Thus a given chemical reaction has in principle an infinite number of heats of reaction depending on what the initial and final states actually are. To obviate this difficulty and to provide for tabular data, a series of standard states is defined and all standard data (handbook values) are given relative to these standard states. The present conventions are as follows:

(a) The standard temperature is chosen as 298.15 K. For pure substances $dH = dQ = C_p\,dT$. Thus if we wish to calculate enthalpies at any other temperature, we can integrate dH from the standard temperature to the second temperature. This requires that we know the value of C_p as a function of T over the desired temperature range.

(b) The standard state of a solid is its most stable form at 298.15 K and 1 atm pressure (1.013×10^5 N/m^2) unless otherwise specified. For carbon the standard state is chosen to be graphite. The choice could just as well have been diamond (it was in some of the earlier work) and the enthalpy values obtained would vary by the enthalpy difference between graphite and diamond.

(c) The standard state of a liquid is the most stable form at 298.15 K and 1 atm pressure.

(d) Two standard states may be used for gases: (1) One atmosphere pressure at 298.15 K; (2) pressure approaching zero at 298.15 K.

(e) For exothermic reactions ΔH is negative.

The preceding establishes the basis for thermochemistry. The combustion of carbon provides an example:

$$C + O_2 \rightleftharpoons CO_2, \qquad \Delta H_0 = -393.512 \quad \text{kJ/mole} \tag{10-27}$$

The subscript zero on the ΔH indicates the enthalpy difference is measured with reactants and products in standard states. This shorthand notation means that one mole of graphite at 1 atm pressure and 298.15 K combusts with one mole of oxygen gas at 298.15 K and 1 atm pressure to give one mole of CO_2 at 298.15 K and 1 atm pressure. The heat given up to a calorimeter is -393.51 kJ/mole.

The fact that enthalpy is a state function means that enthalpy changes in successive reactions are additive; that is, for the same reactants and products the enthalpy change will be the same regardless of the intermediate pathways. This relationship was established empirically before it was formulated on thermodynamic grounds and was known as Hess's law of constant heat summation. As an example, consider the previous combustion of graphite and allow it to take place in two steps:

$$C + \tfrac{1}{2}O_2 \rightleftharpoons CO, \qquad \Delta H_{1_0} = -110.52 \text{ kJ/mole} \tag{10-28}$$

$$CO + \tfrac{1}{2}O_2 \rightleftharpoons CO_2, \qquad \Delta H_{2_0} = -282.99 \text{ kJ/mole} \tag{10-29}$$

The previous discussion indicates that the overall ΔH in going from graphite and O_2 to CO_2 is the sum of the two intermediate steps, so that the sum of ΔH_{1_0} and ΔH_{2_0} is -393.51 kJ/mole, which is of course equal to the ΔH_0 of the overall reaction as indicated in eq. (10-27). The summation rule is useful in cases where one reaction in a sequence cannot be isolated. Suppose, for instance, that we could carry out the calorimetry on reactions shown in eqs. (10-27) and (10-28) but we were unable to run a controlled reaction (10-29) for calorimetry. We could still determine ΔH_2 as $\Delta H_0 - \Delta H_1$.

E. THE EULER EQUATION

To develop the properties of the free-energy functions further, we need to explore one further property of the internal energy U. We have already noted [see eq. (5-86)] that for simple systems U may be expressed as a function of the entropy, volume, and mole numbers:

$$U = U(S, V, n_i) \tag{10-30}$$

Consider two such identical systems and bring them together. The entropy, volume, and mole numbers will double and so will the energy. This argument can be extended and put in the following mathematical form known as the Euler equation:

$$U(\lambda S, \lambda V, \lambda n_i) = \lambda U(S, V, n_i) \tag{10-31}$$

Stated verbally, this equation says that if a system is uniformally enlarged (or shrunk) by a factor λ, so that the entropy, volume, and all mole numbers change by a factor λ, then the internal energy will also change by a factor λ.

This is a rather obvious additivity property which can be deduced from our previous discussions of the various parameters. Equation (10-3) allows us to formulate the energy functions in very useful ways. If we differentiate the equation with respect to λ, we get

$$\frac{\partial U}{\partial(\lambda S)}\frac{\partial(\lambda S)}{\partial\lambda}+\frac{\partial U}{\partial(\lambda V)}\frac{\partial(\lambda V)}{\partial\lambda}+\sum\frac{\partial U}{\partial(\lambda n_i)}\frac{d(\lambda n_i)}{\partial\lambda}=U(S,V,n_i) \qquad (10\text{-}32)$$

Carrying out the differentiation of the second term in each pair, the equation may be rewritten

$$\frac{\partial U}{\partial(\lambda S)}S+\frac{\partial U}{\partial(\lambda V)}V+\sum\frac{\partial U}{\partial(\lambda n_i)}n_i=U \qquad (10\text{-}33)$$

The parameter λ is arbitrary and can take on any value. In particular, for $\lambda = 1$, eq. (10-33) becomes

$$\frac{\partial U}{\partial S}S+\frac{\partial U}{\partial V}V+\sum\frac{\partial U}{\partial n_i}n_i=U \qquad (10\text{-}34)$$

But we have already shown [eqs. (5-37), (5-39), (5-83), (5-84)] that

$$\frac{\partial U}{\partial S}=T, \qquad \frac{\partial U}{\partial V}=-P, \qquad \frac{\partial U}{\partial n_i}=\mu_i \qquad (10\text{-}35)$$

Hence eq. (10-34) becomes

$$U = TS - PV + \sum \mu_i n_i \qquad (10\text{-}36)$$

The additivity property has allowed us to integrate the internal energy. If we combine eq. (10-36) with eq. (10-6) we see that

$$G = \sum \mu_i n_i \qquad (10\text{-}37)$$

The Gibbs free energy can thus be formulated solely as a function of the chemical potentials and mole numbers. In particular, for one mole of a pure substance in its standard state we may write

$$G_{i0} = \mu_{i0} \qquad (10\text{-}38)$$

The particularly simple forms of eqs. (10-37) and (10-38) are of great value in the applications of thermodynamics to chemical equilibrium, and stress the utility of the choice of Gibbs free energy in handling such problems.

Equation (10-37) may be arrived at in a way which is mathematically less elegant but carries more physical intuition. If we go back to eq. (10-15), which holds at constant temperature and pressure, we see that G can be determined if we have a path of integration. Consider a system whose concentration, temperature, and pressure are kept constant, but whose various components are added in fixed ratios,

$$dn_i = n_{i(\text{final})}\, d\xi \tag{10-39}$$

Equation (10-39) implies that we build up the system by the simultaneous addition of all the components in fixed ratios. While this may seem unusual, it is physically possible and therefore provides a legitimate pathway and allows us to integrate eq. (10-17). Because the μ_i are intensive variables, they are independent of the total size of the system and will depend only on the concentrations, pressure, and temperature. Thus the μ_i are constant during the integration and the only variable along the path is ξ:

$$dG = \sum n_{i(\text{final})}\mu_i\, d\xi \tag{10-40}$$

and

$$\begin{aligned} G &= \int_0^1 \sum n_{i(\text{final})}\mu_i\, d\xi = \sum n_{i(\text{final})}\mu_i \int_0^1 d\xi \\ &= \sum n_{i(\text{final})}\mu_i \end{aligned} \tag{10-41}$$

The Gibbs free-energy function is thus found to be the same by the two methods of determination. Equation (10-41) states that the Gibbs free energy of a system is the sum of the products of the molar chemical potential of each substance times the number of moles of that substance. Starting with eq. (10-41), we can restate the equilibrium conditions,

$$dG = \sum \mu_i\, dn_i + \sum n_i\, d\mu_i = 0 \tag{10-42}$$

But because of eq. (10-15) we can then write $\sum n_i\, d\mu_i = 0$. This is known as the Gibbs–Duhem relation and is a useful device in solving problems of chemical thermodynamics. Equations (10-15), (10-37), and (10-42) all describe chemical equilibrium, and the different forms are useful in different problems.

A somewhat more general approach may be seen by taking the derivative of eq. (10-36) and setting it equal to eq. (5-85):

$$\begin{aligned} T\, ds + S\, dT - P\, dV - V\, dP + \sum \mu_i\, dn_i + \sum n_i\, d\mu_i \\ = T\, dS - P\, dV + \sum \mu_i\, dn_i \end{aligned} \tag{10-43a}$$

then

$$\sum n_i\, d\mu_i + S\, dT - V\, dP = 0 \tag{10-43b}$$

This is a less restrictive form of the Gibbs–Duhem equation.

F. *ACTIVITY*

The molar chemical potential μ is often a difficult quantity to determine, and a related quantity, the activity, is frequently used. The activity of the

ith substance a_i is defined by the following relation:

$$a_i = e^{(\mu_i - \mu_{i0})/RT} \quad \text{(definition)} \tag{10-44}$$

μ_{i0} is the chemical potential of the ith substance in the standard state; R is the gas constant and T the temperature. Equation (10-44) is more usually written as

$$RT \ln a_i = \mu_i - \mu_{i0} \tag{10-45}$$

The activity concept does not adapt itself to any simple intuitive interpretation. The right-hand side of eq. (10-45) represents the energy difference in taking one mole of the ith substance from its standard state to the state under consideration. Dividing $\mu_i - \mu_{i0}$ by RT gives a ratio of the energy change to the thermal energy. In general, activity measures the tendency of the ith substance to function as a reactant in a given chemical reaction. By appropriate choice of standard states the activity can be made numerically equal to the concentration of dilute solutions. Note, however, that the activity is a dimensionless quantity, since the logarithm is equal to the ratio of two energies.

Utilizing the activity concept makes possible a detailed thermodynamic study of equilibrium in chemical reactions. A generalized chemical reaction can be written

$$\nu_A A + \nu_B B \leftrightharpoons \nu_C C + \nu_D D \tag{10-46}$$

The ν's designate the stoichiometric coefficients, while the capitals designate the substances: thus ν_A moles of A react with ν_B moles of B to give ν_C moles of C and ν_D moles of D. The two arrows indicate that the reaction is reversible; that is, we can start with C and D and make A and B. Theoretically, all reactions are reversible in this sense, although there are often great practical difficulties in realizing this reversibility. An even more general representation of chemical reactions is given below:

$$\sum \nu_k X_k \rightleftharpoons \sum \nu_j X_j \tag{10-47}$$

The ν's are stoichiometric coefficients, the X_k's represent reactants, and the X_j's represent product. At equilibrium at constant temperature and pressure we have that, for any infinitesimal changes in the n_i,

$$dG = \sum \mu_i \, dn_i = 0. \tag{10-48}$$

Small fluctuations in n_i at equilibrium are not, however, independent, owing to the stoichiometric constraints [eq. (11-47)], so that each dn_i must be represented by

$$dn_i = \nu_i \, d\xi \tag{10-49}$$

In balanced equations, ν_i must be assigned a minus sign when $i = k$ and a plus sign when $i = j$. The quantity ξ is usually designated the extent of the reaction or the degree of advancement. We next substitute eqs. (10-45) and (10-49) into (10-48). This yields

$$\sum (RT \ln a_i + \mu_{i0})\nu_i \, d\xi = 0 \tag{10-50}$$

Equation (10-50) is a general expression for equilibrium in terms of activities as independent variables. Since $d\xi$ is arbitrary, eq. (10-50) requires the vanishing of its coefficient. This leads to

$$-\sum_k (RT \ln a_i + \mu_{i0})\nu_i + \sum_j (RT \ln a_i + \mu_{i0})\nu_i = 0 \tag{10-51}$$

Equation (10-51) can be rearranged putting all the activity terms on the left-hand side and all the standard-state chemical potential terms on the right-hand side. We use the mathematical fact that the sum of logarithms is equal to the logarithm of the product and we employ the notation of $\prod_i$ representing a sequential product. We may then write

$$RT \ln \frac{\prod_j a_i^{\nu_i}}{\prod_k a_i^{\nu_i}} = \sum_k \mu_{i0}\nu_i - \sum_j \mu_{i0}\nu_i \tag{10-52}$$

The right-hand side now represents the standard-state molar Gibbs free energy of the reactants minus the standard-state free energy of the products, weighted by the stoichiometric coefficients, and is designated $-\Delta G_0$. It is the difference in free energy per unit advancement of the reaction if all components are in their standard states. Equation (10-52) can be rewritten as

$$\frac{\prod_j a_i^{\nu_i}}{\prod_k a_i^{\nu_i}} = e^{-\Delta G_0/RT} \tag{10-53}$$

For the specific reaction shown in eq. (10-46), the equilibrium condition can be expressed as

$$\frac{a_C^{\nu_C} a_D^{\nu_D}}{a_A^{\nu_A} a_B^{\nu_B}} = e^{-\Delta G_0/RT} \tag{10-54}$$

The right-hand side is often designated the equilibrium constant, represented by the letter K,

$$e^{-\Delta G_0/RT} = K \tag{10-55}$$

Equations (10-53) and (10-54) with K substituted for $e^{-\Delta G_0/RT}$ look very similar to the mass-action law except that activities appear on the left side rather than concentrations. Equation (10-54) is the true thermodynamic form of the chemical equilibrium and equations using concentration are in fact merely approximations. Activities are, however, very different from concentrations, and are defined solely by eq. (10-49) or (10-50), along with the specification of standard states. When the proper choice of standard states is made it is often possible to replace activities by concentrations or mole fractions or some other experimentally accessible quantities. In order to gain an understanding of why such approximations can be made [which is not at all obvious from the defining relation of eq. (10-44)], we must actually work through some detailed examples. This is well worth the effort, as activities are often treated in a rather mystical fashion and only by working our way through actual cases can we clear away this aura of mystery.

We will first examine the case of a mixture of two ideal gases in a volume V containing n_1 moles of the first and n_2 moles of the second. We then have

$$U = \tfrac{3}{2}n_1RT + \tfrac{3}{2}n_2RT = \tfrac{3}{2}RT(n_1 + n_2) \tag{10-56}$$

The entropy of the system is the sum of the entropies of each component, which can be obtained by integrating eq. (5-56), plus the entropy of mixing [eq. (5-72)]. We first write the entropy of a single component of ideal gas:

$$S = \tfrac{5}{2}nR \ln T - nR \ln P + n\mathfrak{B}' \tag{10-57}$$

Equation (10-57) is derived in Appendix III starting from the defining relation

$$dS = \frac{dU + dW}{T} \tag{10-58}$$

If we now take n_1 moles of the first gas at pressure P and n_2 moles of the second gas at pressure P and mix them (the final pressure is P), then the entropy consists of the contribution of each component plus the entropy of mixing,

$$\begin{aligned} S = {} & \tfrac{5}{2}n_1R \ln T - n_1R \ln P + n_1\mathfrak{B}' \\ & + \tfrac{5}{2}n_2R \ln T - n_2R \ln P + n_2\mathfrak{B}' \\ & - R\left(n_1 \ln \frac{n_1}{n_1 + n_2} + n_2 \ln \frac{n_2}{n_1 + n_2}\right) \end{aligned} \tag{10-59}$$

The Gibbs free energy of the system G is $U + PV - TS$. We can then use eqs. (10-56) and (10-59) to eliminate U and S and get G as a function of the n_i. In getting the appropriate form of the equation, we substitute $(n_1 + n_2)RT$ for PV and use the definition of the mole fraction of the ith substance $\mathfrak{N}_i$, being equal to n_i divided by the sum of the n_i. This leads to

$$\begin{aligned} G = {} & \tfrac{5}{2}n_1RT - \tfrac{5}{2}n_1RT \ln T + n_1RT \ln P - n_1\mathfrak{B}_1{}'T + n_1RT \ln \mathfrak{N}_1 \\ & + \tfrac{5}{2}n_2RT - \tfrac{5}{2}n_2RT \ln T + n_2RT \ln P - n_2\mathfrak{B}_2{}'T + n_2RT \ln \mathfrak{N}_2 \end{aligned} \tag{10-60}$$

We now define a quantity $f_i(T)$, which is a function of temperature only,

$$f_i(T) = \tfrac{5}{2}RT - \tfrac{5}{2}RT \ln T - \mathfrak{B}_i'T \qquad (10\text{-}61)$$

Using this newly defined term, we can rewrite eq. (10-60) in the following much simpler form:

$$G = \sum_{i=1,2} n_i[RT \ln \mathfrak{N}_i + RT \ln P + f_i(T)] \qquad (10\text{-}62)$$

Equation (10-62) may be generalized to any number of ideal gas components. The form of the equation is suggestive of the integrated form of the free energy in eq. (10-37) ($G = \sum_i \mu_i n_i$) and comparison of the two allows us to write down the detailed expression for the chemical potential

$$\mu_i = RT \ln \mathfrak{N}_i + RT \ln P + f_i(T) \qquad (10\text{-}63)$$

The next steps involve some algebraic manipulation in order to express μ as a function of the concentration C_i, which is by definition n_i/V. The mole fraction $\mathfrak{N}_i$ can therefore be rewritten as $C_i V / \sum n_i$. If we substitute these values in eq. (10-63) and rearrange terms, we get

$$\mu_i = RT \ln C_i + RT \ln \frac{PV}{\sum n_i} + f_i(T) \qquad (10\text{-}64)$$

The quantity $PV/\sum n_i$ is just RT, so we reduce expression (10-64) to

$$\mu_i = RT \ln C_i + RT \ln RT + f_i(T) \qquad (10\text{-}65)$$

Our definition of activity in eq. (10-45) allows us to set μ_i in eq. (10-65) equal to $\mu_{i0} + RT \ln a_i$ and we proceed with this substitution:

$$RT \ln a_i + \mu_{i0} = RT \ln C_i + RT \ln RT + f_i(T) \qquad (10\text{-}66)$$

We are now free to choose a standard state, which we will pick as 298.15 K (we call this T_0) and a concentration of one mole per liter. In the standard state the activity is unity since $\mu_i = \mu_{i0}$ for this condition. Therefore, since ln 1 is equal to zero, we get

$$\mu_{i0} = RT_0 \ln RT_0 + f_i(T_0) \qquad (10\text{-}67)$$

From eqs. (10-66) and (10-67) we can see that at the temperature of the standard state a_i is equal to C_i. It should be noted that the equating of a_i and C_i requires a fairly elaborate argument and depends both on the choice of standard state and the units in which C_i is expressed (moles per liter in the present example).

In general for ideal mixtures it is possible to choose a standard state so that the activity and concentration approach each other over some range of variables. It is instructive to examine one more case, that of ideal solutions. Start with n_1 moles of component 1 and n_2 moles of component 2, each in the pure state. Mix the two components at constant temperature and pressure. From the definition of G and the boundary conditions, we see

$$\Delta G = \Delta H - T\Delta S \tag{10-68}$$

Ideal solutions are characterized by $\Delta H = 0$, no heat of mixing, and the only entropy change being the entropy of mixing. Thus, before mixing, the Gibbs free energy of the system can be expressed as

$$G_0 = \sum_i n_{i0}\mu_{i0} \tag{10-69}$$

This equation assumes all components are in their standard state. After mixing, we can write

$$G = G_0 + \Delta G = \sum n_i\mu_{i0} + \sum n_i RT \ln \frac{n_i}{\sum n_i} \tag{10-70}$$

If we now utilize the fact that G is the sum of $\mu_i n_i$, we can write for each chemical potential

$$\mu_i = \mu_{i0} + RT \ln \mathfrak{N}_i \tag{10-71}$$

The definition of activity allows us to set a_i equal to $\mathfrak{N}_i$, the mole fraction. This approximation is frequently used for systems showing nearly ideal behavior. If we wish to write eq. (10-71) in terms of concentration, we substitute $C_i V$ for n_i as in the previous example. This leads to

$$\mu_i = \mu_{i0} + RT \ln C_i + RT \ln \frac{V}{\sum n_i} \tag{10-72}$$

If we consider a two-component system and allow the solute C_i to go to infinite dilution, the term $V/\sum n_i$ becomes $\bar{V}$, the partial molar volume of the solvent. If we now define a new μ_{i0} equal to $\mu_{i0} + RT \ln \bar{V}$, then we can once again set a_i equal to C_i. Since a large number of substances approach ideal behavior at infinite dilution, this approach is quite general.

Having adopted standard states so as to equate activities with concentrations for ideal behavior, deviations from ideality are then lumped into an activity coefficient. We write

$$a_i = \mathfrak{A}_i C_i \tag{10-73}$$

The quantity $\mathfrak{A}_i$ in general may be a function of concentration. Equation (10-73) preserves the simple form of the relation between activity and concentration but allows more complex situations to be treated.

G. ALTERNATIVE METHODS OF SOLVING A PROBLEM

The chemical potential and the associated concept of the activity provide a method for solving a wide variety of problems, We will use it to solve the classical case of pressure distribution in the atmosphere. We will then treat the same situation by statistical mechanics and finally by hydrostatics in an effort to see the relation between the different approaches. The problem is to find the density or pressure as a function of height for a column of ideal gas of constant temperature T, and pressure P_0 at the bottom of the column. The Gibbs free energy is a function of height if we introduce the gravitational potential as well as the chemical potential.

The use of G as a local variable is a new feature. We have previously dealt with the homogeneous case where all thermodynamic variables characterized the whole system. In these considerations the existence of an external potential which is a function of y, a space variable, has the the effect of partitioning the column into local domains of fixed y.

Returning to eq. (10-14), we note

$$dU = dQ - P\,dV + \sum \mu_i\,dn_i + Mg\,dy \tag{10-74}$$

Divide the column into slabs of thickness dy (fig. 10-2). In order for the Gibbs free energy to be a minimum, the molar free energy must be the same for each slab, for if this were not true, molecules would move from slabs of high free energy to slabs of low free energy. Since we deal with a one-component system, we can write

$$G_i = \mu_i + M_i g y \tag{10-75}$$

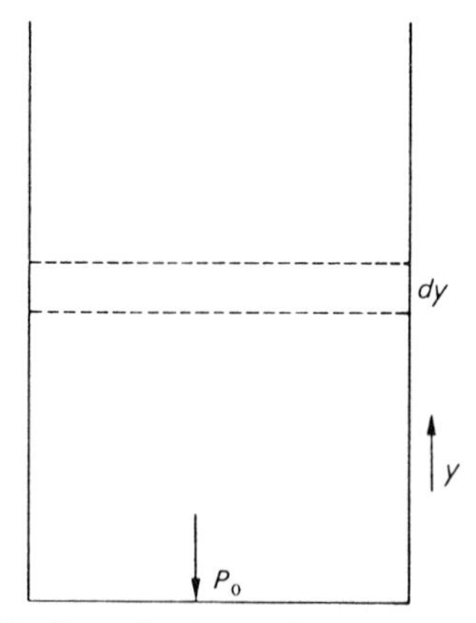

FIGURE 10-2. Standing isothermal column of gas with the pressure P_0 at the bottom of the column. Column may be divided into layers of thickness dy.

The molar free energy at any height is the sum of the chemical potential and the gravitational potential. M_i is the mass per mole or the molecular weight of the ith substance. The gravitational acceleration is given by g and the equilibrium condition for the total free energy to be a minimum is given by

$$\frac{dG_i}{dy} = \frac{d\mu_i}{dy} + M_i g = 0 \tag{10-76}$$

The derivative of G vanishing is the same as G having a constant value for each slab. For an ideal gas we can write

$$\mu_i = \mu_{i0} + RT \ln C_i \tag{10-77}$$

If we substitute eq. (10-77) into eq. (10-76), we can write

$$RT \frac{d(\ln C_i)}{dy} = -M_i gy \tag{10-78}$$

This equation may be directly integrated and solved for the boundary condition that the concentration has a value C_{i0} at the bottom of the column, where $y = 0$. The resulting expression is

$$\frac{C_i}{C_{i0}} = e^{-M_i gy/RT} \tag{10-79}$$

Since we are dealing with ideal gases, we can replace the ratio of concentrations with the ratio of pressures:

$$\frac{P_i}{P_{i0}} = \frac{C_i}{C_{i0}} = e^{-M_i gy/RT} \tag{10-80}$$

The pressure decreases exponentially as a function of height. Equation (10-80) constitutes a complete solution to the problem.

To formulate the problem in classical statistical mechanics, we form an ensemble in which the members are ideal gas molecules in a gravitational field and in contact with a reservoir at temperature T. The potential energy of each ensemble member is mgy, where m is the mass per atom. Applying the Maxwell–Boltzmann distribution law, we see that the number of molecules lying between y and $y + dy$ is

$$dB = \frac{B_0 e^{-mgy/kT}\, dy}{\int_0^\infty e^{-mgy/kT}\, dy} \tag{10-81}$$

The ratio of the number of molecules in a slab at height y to that in a slab at the bottom of the column is

$$\frac{dB(y)}{dB(0)} = \frac{P}{P_0} = \frac{B_0 e^{-mgy/kT}\,dy}{\int_0^\infty e^{-mgy/kT}\,dy} \Bigg/ \frac{B_0 e^{-mg\times 0/kT}\,dy}{\int_0^\infty e^{-mgy/kT}\,dy} \tag{10-82}$$

$$= e^{-mgy/kT}$$

If the numerator and denominator of the exponent are each multiplied by Avogadro's number, we get

$$\frac{P_y}{P_0} = e^{-\mathcal{N}mgy/\mathcal{N}kT} = e^{-Mgy/RT} \tag{10-83}$$

Equation (10-83) follows since $\mathcal{N}m$, Avogadro's number times the mass of a molecule, is the molecular weight, and $\mathcal{N}k$ is R, the gas constant, which is the product of Boltzmann's constant times Avogadro's number. Equation (10-83) is thus identical to eq. (10-80).

The third method of approaching the problem consists of considering the forces on a slab of gas of thickness dy. The three forces are (1) the force due to the pressure on the bottom of the slab, which is balanced by (2) the weight of the slab plus (3) the force due to the pressure exerted on the top (fig. 10-3). For unit area

$$P = dW + P + dP \tag{10-84}$$

Equation (10-84) follows by inspection from fig. 10-2. At equilibrium the forces acting upward on the slab are equal to the downward forces. The change in pressure can be represented by $(dP/dy)\,dy$ while the increment of weight dW is density times acceleration times volume or $\rho g\,dy$. We can

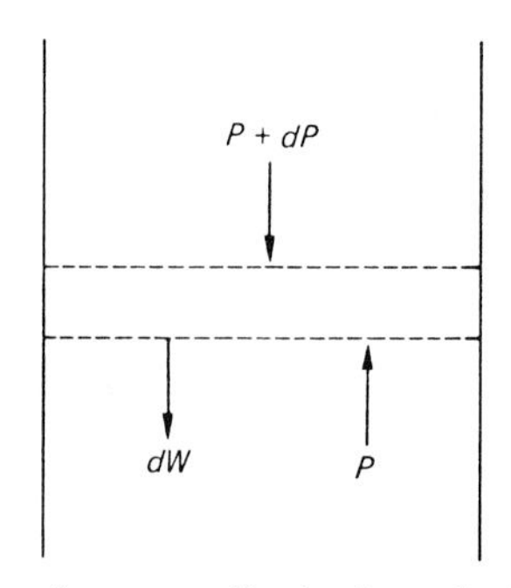

FIGURE 10-3. A layer of gas from a standing isothermal column. At equilibrium the forces down, $P + dP + dW$, are just balanced by the force up, P.

therefore reformulate eq. (10-84),

$$\rho g\, dy = -\frac{dP}{dy}\, dy \tag{10-85}$$

The density is equal to nM/V, the numerator being mass (molecular weight times mole number), while V is the volume. Using the ideal gas law, we note that n/V is equal to P/RT. Applying these factors to eq. (10-85) gives

$$\frac{PMg}{RT} = -\frac{dP}{dy} \tag{10-86}$$

This equation may be integrated subject to the boundary condition that $P = P_0$ at $y = 0$. The integrated equation is

$$P = P_0 e^{-Mgy/RT} \tag{10-87}$$

This is identical to the two solutions already obtained. The ability to derive this formula from varying points of view gives confidence in the conceptual structure we have been developing. It also serves to demonstrate the unity of the subject. Understanding the various methods of solving a problem gives insights into the meaning of the theory and the range of applicability.

H. CHEMICAL REACTIONS

At this point consider a more detailed view of a chemical reaction to get some insight into the meaning of ΔH, ΔG, and ΔS in molecular terms. For simplicity a reaction of the form $A \rightleftharpoons B$ will be looked at (fig. 10-4). The ground vibrational and rotational state of A has an energy ε_{A_0} and upper states have energies of ε_{A_i}. Similarly for B we can define an ε_{B_0} and ε_{B_j}. From

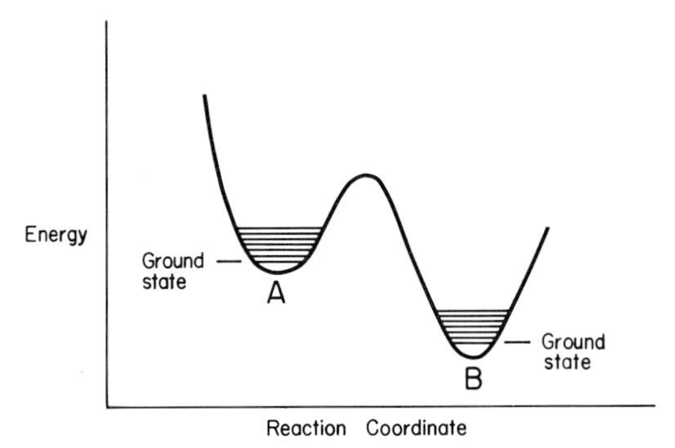

FIGURE 10-4. The energy level diagram of two isomeric species. The concept of a reaction coordinate will be discussed in Chapter 13.

the results of statistical mechanics, the fraction of molecules in isomeric form B is

$$\mathscr{P}(\mathrm{B}) = \frac{\exp(-\varepsilon_{\mathrm{B}_0}/kT) + \sum \exp(-\varepsilon_{\mathrm{B}_j}/kT)}{\mathfrak{z}} \tag{10-88}$$

where $\mathfrak{z}$ is the molecular partition function for the molecule with access to both forms A and B. The ratio of products to reactants is

$$\frac{[\mathrm{B}]}{[\mathrm{A}]} = \frac{\mathscr{P}(\mathrm{B})}{\mathscr{P}(\mathrm{A})} = \frac{\exp(-\varepsilon_{\mathrm{B}_0}/kT) + \sum \exp(-\varepsilon_{\mathrm{B}_j}/kT)}{\exp(-\varepsilon_{\mathrm{A}_0}/kT) + \sum \exp(-\varepsilon_{\mathrm{A}_i}/kT)} \tag{10-89}$$

The numerator is the partition function of species B while the denominator is the partition function of species A, each considered alone. Substituting on the left-hand side from eq. (10-54) (noting that activities and concentrations may be set equal for this case) and on the right-hand side from eqs. (8-1) and (8-22), we see that for isothermal transformations

$$\exp\left(\frac{-\Delta G_0}{RT}\right) = \exp\left(\frac{-\Delta H_0 + T\Delta S_0}{RT}\right)$$

$$= \exp\left(\frac{-\mathscr{N}[(\bar{\varepsilon}_{\mathrm{B}} - \bar{\varepsilon}_{\mathrm{A}}) - T(s_{\mathrm{B}} - s_{\mathrm{A}})]}{\mathscr{N}kT}\right) \tag{10-90}$$

The ΔH term reflects the difference between the average energy of a B molecule and an A molecule. ΔS reflects the difference in entropy due to the distribution among vibrational and rotational states of the two isomeric forms. Both numerator and denominator of the exponent of the middle term have been multiplied by Avogadro's number to convert energies to a molar basis. Remembering that R/k is $\mathscr{N}$, Avogadro's number, we can write

$$\Delta H_0 = \mathscr{N}(\varepsilon_{\mathrm{B}} - \varepsilon_{\mathrm{A}}) = \mathscr{N}\left(\sum_j f_j \varepsilon_{\mathrm{B}_j} - \sum_i f_i \varepsilon_{\mathrm{A}_i}\right) \tag{10-91}$$

The enthalpy change is the actual difference in total energy per mole of products and mole of reactants. Since, in the simple case we are considering, the energies are independent of concentration, the enthalpies are also concentration independent and reflect the intrinsic energy content of each molecule. Note, however, that enthalpy is temperature dependent since the f_i in eq. (10-91) are functions of temperature. In a similar fashion the entropy change, ΔS_0 for the case of concentration-independent energies, reflects the intrinsic entropies of the two species due to their molecular arrangements. Thus, at a first order of approximation, ΔG_0 is concentration independent and measures the contribution of molecular energy states to the Gibbs free energy.

TRANSITION

Most biological and biochemical applications of thermodynamics involve systems at constant temperature and constant pressure. Under such conditions the entropy of the universe is maximized when the Gibbs free-energy function of the system is minimized. Other functions of special value are the Helmholtz free energy and the enthalpy. These functions can be generalized to include all ways by which a system can exchange energy with the environment. The enthalpy can be directly measured from calorimetric experiments. The establishment of standard states is of importance in computing thermodynamic parameters. Relations between extensive variables allows the derivation of the Gibbs–Duhem equation, which is often a useful calculating device. The *activity* is the function most useful in chemical application and, once it is defined in terms of the chemical potential, the equilibrium conditions can be expressed in convenient terms. By appropriate choice of standard states, activity can be expressed as a simple function of the concentration under most conditions. The equations of chemical equilibria can be alternatively derived from statistical mechanical considerations. Thus we can deal with chemical problems but we have not yet developed a formalism for problems in which a system interacts with a radiation field. Since this is a major consideration in global biological applications, we next examine the thermodynamics of radiation.

BIBLIOGRAPHY

Edsall, J. T., and Wyman, J., "Biophysical Chemistry." Academic Press, New York, 1958.

Chapters 4 and 8 deal with a wide variety of thermodynamic problems encountered in biochemistry.

Gibbs, J. W., "The Scientific Papers of J. Willard Gibbs." Longmans, Green, New York, 1906.

The foundations of chemical thermodynamics as expounded by the master.

Lewis, G. N., and Randall, M., "Thermodynamics" (2nd ed., revised by Pitzer, K. S. and Brewer, L.). McGraw-Hill, New York, 1961.

The classical work on the application of thermodynamics to experimental chemistry.

Tanford, C., "Physical Chemistry of Macromolecules." Wiley, New York, 1967.

This work covers thermodynamic background as well as kinetic methods of studying macromolecules.

11 Radiant Energy

A. THE FLOW OF THERMAL ENERGY

Thus far in our discussion of energy we have had little to say about the energy of radiation, or other thermodynamic parameters associated with the emission or adsorption of photons. However, if we wish to deal adequately with photosynthesis, the transfer of energy from the sun to the earth, and problems of thermal regulation, we must develop this area. In our discussion of the measurability of energy in Chapter 3, we noted that the Q term included all energy transfers that could not be accounted for by work terms W. Since this approach defined Q, we did not have to be more explicit about the physical processes involved in the flow of heat. One of the steps in moving toward a more detailed view of thermal physics is to appreciate that there are a number of mechanisms involved in the transfer of heat, or, more properly, in the transfer of thermal energy.

Conventionally, the flow of thermal energy has been accounted for by three processes: conduction, radiation, and convection. Conduction is caused by collisions between molecules. If a temperature gradient exists in a system, the continuing collisions resulting from the random thermal motion will lead to an energy transfer from molecules of higher average velocity (thermal energy) to molecules of lower average velocity. Radiative energy transfer occurs by the emission and adsorption of photons. In a given material, molecules always occur at vibrational, rotational, and electronic levels above the ground state. Some of the decays of these levels are accompanied by the emission of radiation. Similarly, some of the collision processes which occur in the material act to transfer translational energy into internal modes. The net result is that, associated with any object at temperatures other than absolute zero, there is a field of electromagnetic radiation. An adiabatic wall as we have previously defined it must therefore be impervious to the flow of radiant energy. Convection is the flow of thermal energy associated with the flow of matter. The study of convection is thus essential to the interface between thermodynamics and hydrodynamics, and we shall consider this process in more detail in our later discussions of energy flow and irreversible thermodynamics.

While considerations of radiant heat transfer normally deal with radiation in the infrared and microwave region, the theory of electromagnetic radiation is continuous over the entire spectrum, so that we also will consider visible radiation, which eventually provides the background for considerations of photosynthesis. Table 11-1 reviews the classification of radiation into different wavelength regions. We might parenthetically note that two of the regions, visible and heat, are physiologically defined relative to human observers.

TABLE 11-1
Radiation Wavelengths

Type of radiation	Wavelength (nm)
Gamma rays	Less than 0.015
X rays	0.01–15
Ultraviolet	10–380
Visible	380–780
Infrared (heat)	780–40000
Microwave and radio waves	Greater than 4×10^4

The classification might alternatively have been made according to chemical effects. Thus the range below 250 nm is the range of ionization and nonspecific excitation. The 250–1100-nm range is the domain of photochemistry, that is, specific absorption of photons resulting in molecules in highly reactive excited states. Above 1100 nm, the effects are predominantly thermal.

B. BLACK-BODY RADIATION

All material bodies above absolute zero emit radiation which is a function of the temperature and the nature of the material. There are a number of empirical relations regarding the radiation by hot bodies and we will present these before turning attention to the underlying theory. Two essential parameters can be measured about the radiation emanating from an object: the total intensity and the spectral distribution. In 1879, J. Stefan concluded from a series of experiments that *the total radiant energy emanating from the surface of an object was proportional to the fourth power of the absolute temperature* (empirical generalization). This relationship will be derived below. The spectral distribution shifts to shorter wavelengths as the temperature is raised. Thus at 750 K a heated body is visibly red, while at 1275 K it is yellow and at 1475 K it is white hot. These distributions can be seen in fig. 11-1. The response of the eye and the resultant colors cannot be seen from these overall distributions. The wavelength of maximum radiation is approximately related to the temperature by the formula

$$T = \frac{2.9 \times 10^6}{\lambda} \tag{11-1}$$

where T is in absolute temperature and λ_{max} is in nanometers. This result may be regarded as an empirical generalization which was subsequently shown to be derivable from radiation theory. Much of the material in the thermodynamics of radiation appeared first as generalizations and was later independently derived from more reduced postulates.

We next introduce an idealized entity known as a black surface. *A surface is "black" if it absorbs all incident radiation* (definition). Such behavior is

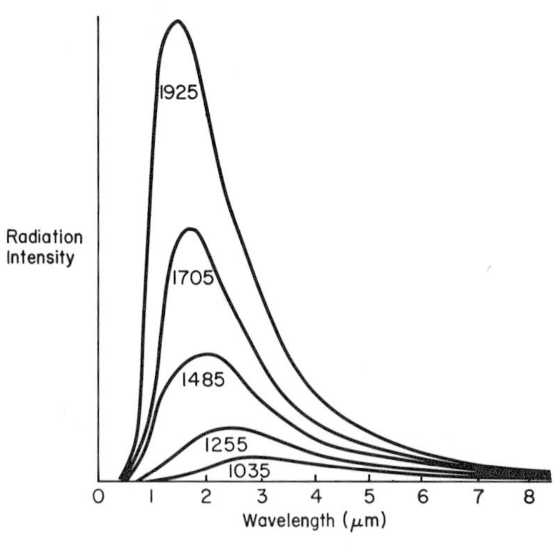

FIGURE 11-1. Black-body radiation intensity as a function of wavelength. The values given are the absolute temperatures of the radiators in kelvins.

approached by materials such as a lampblack. We take a piece of black material and place it in an isothermal reservoir. If we then cut out a cavity from the material and allow it to reach equilibrium, the radiation in the cavity will equilibrate with the black body. The radiation may be sampled by drilling a very fine hole from the surface of the material into the cavity.

The radiation within the hollow we have just described exerts on the walls a pressure whose existence can be inferred from either a classical or a quantum-mechanical point of view. We will adopt the quantum orientation to describe some quantitative aspects. The radiation in the cavity consists of a stream of photons, each with an associated momentum of $h\nu/c$ (from the quantum theory of radiation). If the radiation energy density is u and the average photon frequency is $\bar{\nu}$, there are $u/h\bar{\nu}$ photons per unit volume (each photon has an average energy of $h\bar{\nu}$ by the Planck quantum theory). The number striking unit surface area per unit time is $\frac{1}{6} \times c \times (u/h\bar{\nu})$, where c is the velocity of light. (See Chapter 2 for the discussion of the collision of particles with walls; the same reasoning may be used in discussing photons). As we are dealing with a black body, all the photons striking the surface must be absorbed, but, since the system is at equilibrium, an equal number must be emitted by the wall. This is mechanically equivalent to all the photons striking the wall being elastically reflected with a net change of momentum of $2h\bar{\nu}/c$. The total change of momentum which is the pressure is

$$P = \frac{2h\bar{\nu}}{c} \times \tfrac{1}{6}c \times \frac{u}{h\bar{\nu}} = \frac{u}{3} \tag{11-2}$$

The photon pressure on the walls of the cavity is thus one-third of the energy density. This relationship between P and u had previously been derived by Maxwell from classical electromagnetic theory.

We can now regard the radiation field in the cavity as a thermodynamic system and consider it in the form of a cylinder with a piston. The first law of thermodynamics then may be written

$$dU = dQ - dW = dQ - P\,dV = dQ - \tfrac{1}{3}u\,dV \tag{11-3}$$

where the work term is due to a volume change caused by radiation pressure. The energy density u is a function of temperature alone, since at equilibrium the radiation pressure cannot depend on the volume of the cavity. The total energy U must then be simply uV and the entropy change of the radiation field is given by

$$dS = \frac{dQ}{T} = \frac{d(uV)}{T} + \frac{1}{3}\frac{u\,dV}{T} = \frac{V}{T}\,du + \frac{4}{3}\frac{u}{T}\,dV \tag{11-4}$$

Our previous statement about u allows us to write du as $(\partial u/\partial T)\,dT$ and to reformulate eq. (11-4) as

$$dS = \frac{\partial S}{\partial T}dT + \frac{\partial S}{\partial V}dV = \left(\frac{V}{T}\frac{\partial u}{\partial T}\right)dT + \left(\frac{4u}{3T}\right)dV \tag{11-5}$$

By the equality relation between second partial derivatives we can write

$$\frac{\partial}{\partial V}\left(\frac{V}{T}\frac{\partial u}{\partial T}\right) = \frac{\partial}{\partial T}\left(\frac{4}{3}\frac{u}{T}\right) \tag{11-6}$$

or

$$\frac{1}{T}\frac{\partial u}{\partial T} = \frac{1}{T}\frac{4}{3}\frac{\partial u}{\partial T} - \frac{4}{3}\frac{u}{T^2} \tag{11-7}$$

Rearranging (11-7) leads to

$$\frac{\partial u}{\partial T} = \frac{4u}{T} \tag{11-8}$$

which on integration yields

$$u = \mathfrak{C}T^4 \tag{11-9}$$

Equation (11-9) is known as the Stefan–Boltzmann law. The quantity $\mathfrak{C}$ must be determined experimentally or, as we shall see later, from Planck's radiation law. Since the energy flux into the wall (photon flux $\frac{1}{6}(cu/h\nu)$ times photon energy $h\nu$) is proportional to u, the energy flux out of the wall must also be proportional to u. The flux out is the independent quantity not determined by the equilibrium so that the amount of energy emanating from a black body, even in a nonequilibrium situation, must also be porportional to u. The radiation flux from such an object is given by

$$\text{flux} = \mathfrak{D}T^4 \tag{11-10}$$

The value of $\mathfrak{D}$ has been experimentally determined as $(5.672 \pm 0.003) \times 10^{-3}\ \mathrm{J\,m^{-2}\,deg^{-4}\,s^{-1}}$.

Thus far we have considered only the energy of radiation without attention to spectral distribution. Before taking a more detailed view we introduce a few definitions. *The total surface absorptivity is the fraction of the total incident radiant intensity that is absorbed. The total surface reflectivity is the fraction reflected.* The sum of these two fractions is always unity. Thus surfaces other than ideal black surfaces are characterized by nonzero values of the reflectivity. *The monochromatic surface absorptivity $\mathfrak{H}(\lambda)$ is the fraction of the radiation with wavelengths between λ and $\lambda + d\lambda$ that is absorbed. The monochromatic surface reflectivity is $1 - \mathfrak{H}(\lambda)$. The monochromatic* *emissive power, $\mathfrak{E}(\lambda)$, is the radiant energy flux, per unit area, per unit wavelength range, with wavelengths lying between λ and $\lambda + d\lambda$. The total emissive power $\mathfrak{E}$ is the total energy radiated.*

Therefore,

$$\mathfrak{E} = \int_0^\infty \mathfrak{E}(\lambda)\, d\lambda \tag{11-11}$$

The ratio of the emissive power of a surface to that of a black surface under identical conditions is called the emissivity of the surface. It is a pure number lying between zero and one, and depends on the material composition.

Consider now a cavity cut in a material that is not a black body. If the material is immersed in an isothermal bath we can deduce that the radiation field in the cavity will be identical to black-body radiation at the same temperature. In contact with an isothermal reservoir at temperature T we place two materials, one has black surface properties and the second has a finite reflectivity (fig. 11-2). The two cavities communicate by a narrow channel through which radiation may pass. If the radiation flux at equilibrium were different in the two cavities there would be a continuous net flow of energy from one to the other. Since this is not possible, the radiation fields must be the same with respect to intensity and spectral distribution. This result allows us to derive a relation known as Kirchhoff's law, which states that *at a given temperature all materials have the same ratio of emissive*

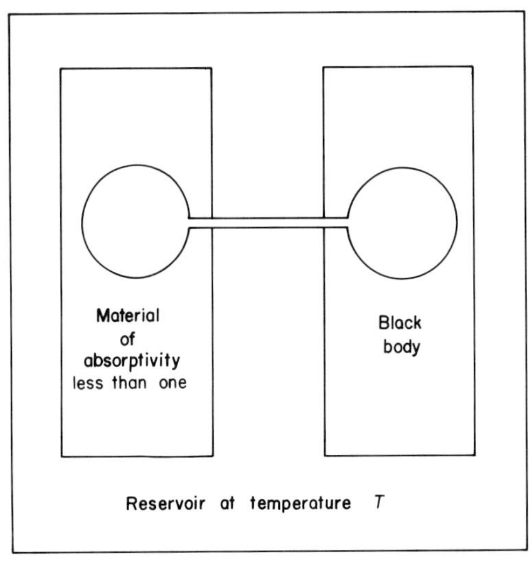

FIGURE 11-2. The radiation equilibrium between materials of different surface absorptivity.

power to absorptivity. This ratio is a universal constant which is a function only of the temperature (conclusion).

Consider the energy flux in both chambers of fig. 11-2. Since we know the value in the right-hand side to be $\mathfrak{D}T^4$, this must also be the value in the other chamber. In the left-hand cavity, the amount of radiant energy absorbed by an area of surface $d\mathscr{A}$ per unit time is then $\mathfrak{H}\mathfrak{D}T^4\,d\mathscr{A}$. Since the system is at equilibrium, the amount of energy radiated $\mathfrak{E}\,d\mathscr{A}$ must be equal to the amount absorbed, so that

$$\mathfrak{H}\mathfrak{D}T^4\,d\mathscr{A} = \mathfrak{E}\,d\mathscr{A} \tag{11-12}$$

$$\frac{\mathfrak{E}}{\mathfrak{H}} = \mathfrak{D}T^4 \tag{11-13}$$

Equation (11-12) is a formal statement of Kirchhoff's law and specifies the universal function of temperature in terms of the Stefan–Boltzmann constant.

C. SPECTRAL DISTRIBUTION

Returning to the radiation field in the cavity, we now wish to focus on its spectral distribution. Consider the quantity u_λ, which is the amount of radiant energy, per unit wavelength per unit volume, with wavelength between λ and $\lambda + d\lambda$. Planck demonstrated that

$$u_\lambda = \frac{8\pi hc}{\lambda^5(e^{h\nu/kT} - 1)} \tag{11-14}$$

The derivation of the equation follows from considering the equilibration of the radiation with a collection of quantized oscillators. In deriving this distribution Planck is credited with originating quantum theory. The derivation of (11-14) will not be carried out here, but may be found in references given at the end of the chapter.

Expression (11-14) is in agreement with the experimental data shown in fig. 11-1. In addition, eq. (11-14) can be integrated over all wavelengths. This yields eq. (11-9) and provides a theoretical value for the constant $\mathfrak{C}$ which is also in agreement with experiment. The function u_λ has a single maximum at a given temperature. This extremum may be determined and leads to the expression given in eq. (11-1).

The radiation emanating from stars has a spectral distribution approximating black radiation. Formula (11-1) or (11-14) can then be used to assign a temperature to the surface of a star based on its spectrum.

Having examined some basic features of the thermodynamics of radiation transfer, we shall now proceed to global bioenergetics which follows from the flow of photons from the 6000-K surface of the sun to the earth.

TRANSITION

The radiation in a system is subject to thermodynamics, and indeed all objects constantly radiate photons, the total flux being proportional to the fourth power of the absolute temperature. In addition, there is a characteristic spectral distribution which is a function of temperature. The distribution function may be derived from a set of assumptions introduced by Planck. The frequency at which the maximum flux occurs is directly proportional to the temperature. An important relation known as Kirchhoff's law states that all materials have the same ratio of emissive power to absorptivity. A background in chemical and radiation thermodynamics enables us to take a detailed look at global bioenergetics. This material begins in the next chapter.

BIBLIOGRAPHY

Cork, J. M., "Heat." Wiley, New York, 1942.

A thorough review of heat and heat transfer. This reviews many of the classical methods of establishing experimental values in thermal physics. Theory is also included.

Planck, M., "The Theory of Heat Radiation." Blakistons, Philadelphia, 1914.

This is the original text in which the modern theory of thermal radiation was first expounded. The work is complete but requires background knowledge in classical electromagnetic theory.

12 The Free Energy of Cells and Tissue

A. SOLAR RADIATION

Almost all of the biological processes taking place on the surface of the earth at the present are powered either by energy coming directly from the sun or by stored energy from past solar radiation. The sun, like all stars, is driven by self-sustaining thermonuclear reactions which are ultimately triggered by the gravitational energy of the contracting sphere. All of this stellar energy is eventually degraded to heat and the solar surface has an effective temperature of 6000 K. Of course, at the core, the sun is considerably hotter, possibly millions of degrees, but the emitted radiation depends on the surface temperature. The major nuclear reaction involves four hydrogen nuclei condensing to form a helium nucleus of lower mass. Because of the very high energy to mass ratio (the velocity of light squared) prodigious amounts of energy are released by this reaction, as has been demonstrated on the earth's surface in the detonation of hydrogen bombs.

From the point of view of terrestrial processes, what interests us about solar radiation is the net flux and the spectral distribution. The flux is measured as that radiation through a plane tangent to the stratosphere and at right angles to a line passing from the center of the sun to the center of the earth. The magnitude of the solar constant is 14×10^2 W/m^2; the total radiation reaching the earth is thus 54.4×10^{23} J/year. The solar distribution is given in fig. 12-1 and table 12-1.

An important feature to note about the solar spectrum is the high fraction of the energy in the visible range, which is also the most interesting range for specific photochemical reactions. Before considering the detailed

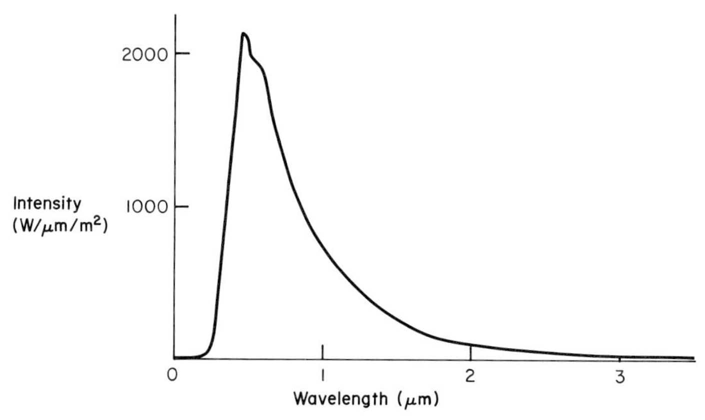

FIGURE 12-1. The spectral distribution of solar energy incident on the earth's atmosphere.

TABLE 12-1
Solar Energy Distribution and Photochemistry

Region	Wavelength range (μm)	Energy range (kJ/Einstein[a])	Fraction of solar spectrum (%)	Molecular changes
Far ultraviolet	0.01–0.2	640–1280	0.02	Ionization
Ultraviolet	0.2–0.38	310–640	7.27	Electronic transitions and ionizations
Visible	0.38–0.78	150–310	51.73	Electronic transitions
Near infrared	0.78–3	40–150	38.90	Electronic and vibrational transitions
Middle infrared	3–30	4–40	2.10	Rotational and vibrational transitions

[a] An Einstein is defined as Avagadro's number of photons.

pathway of the solar energy impinging on the earth, a brief overview will set the scene. About half of the incident radiation reaches the terrestrial surface, the remainder being scattered or reflected by the atmosphere. Most of the flux actually impinging on the surface is converted into heat or reflected. A small fraction is chemically fixed by photosynthetic organisms. This bound energy effectively drives all of the subsequent biological transformations, including the major nutrient cycles. At each stage of the process a fraction of the power is lost as heat, which is then radiated from the planetary surface. Eventually almost all of the incorporated energy flows through and is ultimately lost as heat. In the process of this transit from thermonuclear reactions in the sun, to infrared radiation from the earth to outer space, the flow of energy drives all of the terrestrial biological activity. It is our task to try to eventually understand these processes in terms of thermal physics and quantum chemistry.

With the above in mind we return to the 14×10^2 W/m^2 which impinge on the outer reaches of our planet. As already noted, some of this is returned to outer space without much direct interaction with the planetary surface. The rest hits on the surface and is mostly degraded and emitted as infrared radiation. The heat energy also evaporates water and drives the major meteorological processes. Of the photon flux which reaches the surface, a small fraction, usually less than 4%, and frequently less than 1%, is absorbed by plants, algae, and photosynthetic bateria, and converted into the chemical potential energy of glucose and free oxygen.

B. PHOTOSYNTHESIS AND GLOBAL ECOLOGY

The organisms which carry out photosynthesis are designated primary producers, or the first trophic level. In the terminology of the ecologist, the organisms which live on the primary producers, the herbivores, are classed as the second trophic level and the carnivores that feed on the herbivores are known as the third trophic level. The food chain concept is of course too simple, since actual ecological systems consist of food webs rather than chains. In any case, food flows between trophic levels or, to a more calorimetrically oriented ecologist, energy flows through the trophic network.

Organisms of the first trophic level do not put all of their primary productivity into storage; they must use a fraction of it to carry out their own physiological functions. We thus distinguish between gross primary production, the total amount of energy fixed by photosynthesis, and net primary production, the amount of photosynthesis which goes into plant growth. At all stages of the process, energy conservation must rigorously hold as indicated in our generalization of the first law for nonequilibrium systems.

The process of photosynthesis has three general aspects which we shall outline at this point and later discuss in more detail. They are (1) primary photochemical reactions, (2) the chemical reactions producing energy storage molecules, and (3) physiology and control in first trophic level organisms. The three aspects have been referred to as the physics, chemistry, and biology of photosynthesis. The first steps involve the absorption of photons leading to charge separation (oxidized and reduced centers) in the pigments. These centers participate in the splitting of water into free oxygen, and hydrogen in the form of the energy rich reduced NADP molecule. The second steps convert the energy trapped in the light reaction into the chemical free energy of glucose. The third steps involve the long-term storage and utilization of the energy by the plant or other photosynthetic organism.

The subsequent flow of energy through the ecosystem has led to the model which is summed up in fig. 12-2. This represents the balance for a local system.

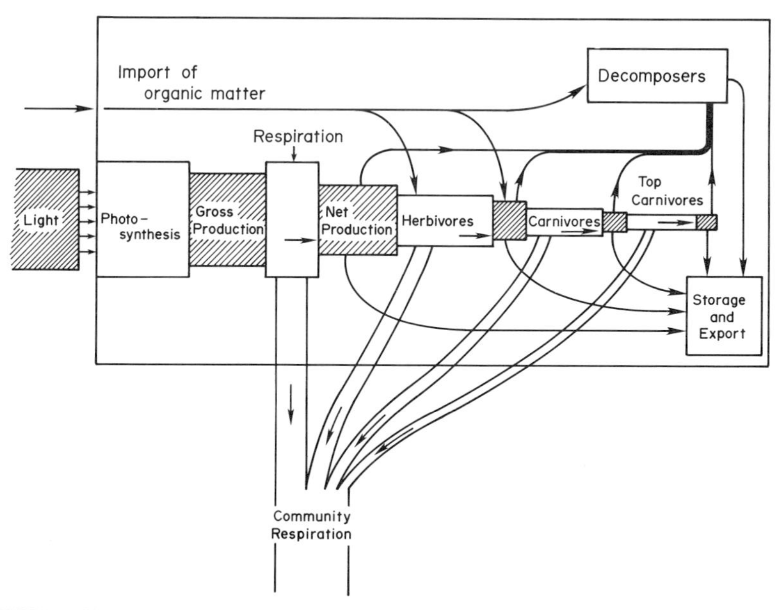

FIGURE 12-2. The flow of energy through an ecological system. Energy enters through sunlight and imported organic matter. It flows through the system as indicated and leaves either as heat (community respiration) or exported organic matter.

A somewhat more condensed global view of the same process is shown in fig. 12-3. This representation shows the carbon cycle as well as the energy flowthrough and stresses the importance of the detritus pool and micro-organisms of decay in the operation of the ecosystem. This feature is indicated by "decomposers" and "storage" in fig. 12-2. Storage is not considered a separate part of diagram 12-3, but is considered as that part of the detritus pool which has not yet been subject to decay. With this brief outline of global processes, we are now in a position to take a more detailed view of the application of thermal physics to energy flow in biological systems.

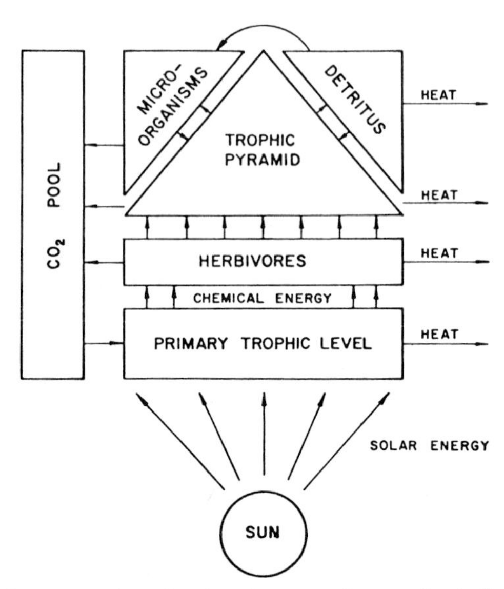

FIGURE 12-3. Representation of ecology in terms of energy flux indicating that ecological processes are driven by the flow of energy from the sun to thermal sinks.

We proceed by restating the concept that biology, as we know it, is dependent on the earth taking in sunlight and radiating infrared energy to outer space. If we were to surround the planet with an adiabatic envelope, all living processes would cease within a short time and the system would begin to decay to some equilibrium state. Solar energy influx is clearly necessary for the maintenance of life. It is just as necessary that a sink be provided for the outflow of thermal energy. If this were not so, the planet would continuously heat up and life would again soon cease to exist. All biological processes depend on the absorption of solar photons and the transfer of heat to celestial sinks.

This principle has been elegantly stated by Albert Szent-Györgi[†] who wrote

> It is common knowledge that the ultimate source of all our energy and negative entropy is the radiation of the sun. When a photon interacts with a material particle on our globe it lifts one electron from an electron pair to a higher level. This excited state as a rule has but a short lifetime and the electron drops back within 10^{-7} to 10^{-8} seconds to the ground state giving off its excess energy in one way or another.

[†] A. Szent-Györgi, *in* "Light and Life" (McElroy and Glass, eds.) p. 7. Johns Hopkins Univ. Press, Baltimore, 1961. Copyright © by Johns Hopkins Univ. Press.

> Life has learned to catch the electron in the excited state, uncouple it from its partner and let it drop back to the ground state through its biological machinery utilizing its excess energy for life processes.

At the risk of repetition, some stress is necessary on the role of the sink in the functioning of the biosphere. The sun would not be a source of "negentropy" if there were not a sink for the flow of thermal energy. The surface of the earth does not gain net energy from the sun but remains at an approximately constant total energy, reradiating as much energy as is taken up. The subtle difference is that it is not energy *per se* that makes life go, but the flow of energy through the system. The concept of flow of energy gives us a powerful basis for physical analysis of a system and has not yet been fully exploited in physics.

Global ecological processes are characterized by major cycles such as the carbon and nitrogen cycles. The existence of flows of this type is as much a feature of geology and meteorology as it is of biology. In geology the energy sources are more complex, consisting of (a) solar radiation, (b) mechanical energy both from gravitational processes on the earth as well as gravitational processes in the solar system such as tides, (c) chemical free energy stored in geological deposits which are above the lowest energy state, (d) nuclear energy, and (e) available thermal energy from the hot core of the earth. These sources serve to drive geological processes as described by the geologist Mason[†]:

> We have been considering the geochemical cycle in terms of the material changes which take place during the various processes. Equally significant, if less studied and less well understood, are the energy changes during the cycle. Geochemical processes operate only because of a flow of energy from a higher to a lower potential or intensity; hence, energy is no less important than matter in the geochemical cycle.

In meteorology, the idealized water cycle is an example of a process driven directly by solar energy. Sunlight falling on bodies of terrestrial water causes vaporization and the rising of water molecules to higher altitudes. The molecules condense into liquid water giving off heat, some of which is radiated to outer space. The condensed water now has mechanical (gravitational) potential energy. The water falls back to earth in the form of rain and the potential energy is converted to heat, some of which is radiated to outer space. The water thus cycles between terrestrial water and clouds. The process depends on the flow of energy from the sun through the earth's surface to outer space. The process is a thermomechanical analog

[†] B. Mason, "Principles of Geochemistry," 3rd ed., p. 291. Wiley, New York, 1966. Copyright © by John Wiley & Sons, Inc.

of the various biological cycles, and indeed closely relates to them since life, as we know it, is intimately connected to the supply of water.

The global ecological system, the biosphere, can then be defined as that part of the terrestrial surface which is ordered by the flow of energy, mediated by photosynthetic processes. The importance of energy flow has long been recognized by ecologists to be the vital step in organizing the ecosystem. Raymond L. Lindeman in his classical paper (see Bibliography) stated, "The basic process in trophic dynamics is the transfer of energy from one part of the ecosystem to another." The entire trophic concept can be formulated in energetic terms as is shown in fig. 12-3. Energy enters the system as photons and is transformed into energetic covalent bonds. All subsequent biochemical changes involve a series of rearrangements which are accompanied by the production of heat. We shall later demonstrate the thermodynamic necessity of this heat loss in terms of entropic considerations. The energy outflow is usually accompanied by the loss of CO_2, water, and nitrogenous compounds. This material then moves through the well-known cycles and eventually back into the biosphere.

C. *THERMODYNAMIC CALCULATIONS*

For a more detailed understanding of the significance of energy flux in the present biosphere, we must examine the actual enthalpy and entropy changes occurring in various biological transformations. These processes can all be encompassed in the scheme involving the synthesis of biomass from the simplest and lowest energy precursors. The model process we shall study is given in the following idealized reaction sequence:

$$CO_2 + N_2 + H_2O + H_2SO_4 + H_3PO_4 \rightarrow \text{Biomass} + O_2 \tag{12-1}$$

Equation (12-1) might at first be regarded as an abstract relation for idealized cycles of carbon and nitrogen, sulfur, and phosphorus. On the other hand, there are living organisms for which this equation is a fairly precise representation of the overall cell growth. Consider, for example, the blue-green algae, phylum Cyanophyta. In the order Nostocinales there are a number of species which are both photosynthetic as well as fixers of free nitrogen. They are mostly included in the genera *Anabaena*, *Cylindrospermum*, *Nostoc*, and *Mastigocladus*. Equation (12-1) is of course incomplete for these organisms in that we have ignored the sodium, potassium, magnesium, manganese, molybdenum, cobalt, and other trace substances. However, from a bulk thermodynamic point of view, the relation approximately describes the overall processes for these extremely versatile cells. The genomes of these prokaryotes are required to encode sufficient information to generate enzymes to catalyze all the reactions shown on the metabolic chart.

A number of other sequences closely related to eq. (12-1) might be studied by using the same methods of analysis. Three of these are

$$CO_2 + HNO_3 + H_2O + H_2SO_4 + H_3PO_4 \rightarrow \text{Biomass} + O_2 \quad (12\text{-}2)$$

$$CO_2 + NH_3 + H_2O + H_2SO_4 + H_3PO_4 \rightarrow \text{Biomass} + O_2 \quad (12\text{-}3)$$

$$C_6H_{12}O_6 + NH_3 + H_2SO_4 + H_3PO_4 \rightarrow \text{Biomass} + CO_2 + H_2O \quad (12\text{-}4)$$

Equations (12-2) and (12-3) are representative of the growth of photosynthetic organisms or symbiotic associations of photosynthesizers with nitrogen reducers. Equation (12-4) is representative of the anaerobic growth of a biochemically versatile bacterium such as *Escherichia coli* on a minimal medium.

We will examine the model processes with respect to various changes in entropy, enthalpy, and free energy. In the course of this analysis, we will find that much of the necessary experimental data is unavailable and, in other case, only approximate values exist. In these cases we will use the best available methods which can be employed to estimate the thermodynamic parameters from a knowledge of molecular structure and additivity of group properties (see Benson, Bibliography). In spite of the limitations we will be able to gain some useful insights into the thermodynamic changes in the major biological cycles, in cell replication, and in the transfer of biomass between trophic levels. This chapter represents an attempt at total thermodynamic bookkeeping in biological systems.

To begin our analysis we must consider the concept of biomass in more specific biochemical terms. If we know the composition of a living organism in terms of the major components (proteins, nucleic acids, fats, and carbohydrates) we can then use this information to deduce the numbers of monomers and small components. However, this figure varies widely: from some adult plants that are almost entirely carbohydrate, to some fish that are primarily protein and fat and contain almost no carbohydrate, to microorganisms that have a considerably higher proportion of their mass in nucleic acids.

In choosing an appropriate distribution to carry out our biomass calculation, we will use compositional data from bateria. This is somewhat arbitrary, but microorganisms represent a large fraction of the total biomass and are ubiquitously distributed. The calculations we are outlining could be carried out for any macromolecular composition and it is clear that, for a detailed view of energetics in general ecology, it will be profitable to examine the free energy of formation for each trophic level.

For present considerations, biomass as used in eqs. (12-1)–(12-4) is high polymer material having the monomeric composition shown in table 12-2. This table was generated by considering overall compositional data on the bacterium *Escherichia coli*. While the detailed values will change with

TABLE 12-2
Chemical Composition of Dry Biomass

Compound	Mole fraction in dry material	Moles/g dry biomass × 10^4
Alanine	0.0552	4.54
Arginine	0.0306	2.52
Aspartic acid	0.0245	2.01
Asparagine	0.0123	1.01
Cysteine	0.0123	1.01
Glutamic acid	0.0429	3.53
Glutamine	0.0245	2.01
Glycine	0.0490	4.03
Histidine	0.0061	0.50
Isoleucine	0.0306	2.52
Leucine	0.0490	4.03
Lysine	0.0490	4.03
Methionine	0.0245	2.01
Phenylalanine	0.0184	1.51
Proline	0.0306	2.52
Serine	0.0368	3.02
Threonine	0.0303	2.52
Tryptophan	0.0061	0.50
Tyrosine	0.0122	1.01
Valine	0.0369	3.02
Hexose	0.1249	10.26
Ribose	0.0544	4.47
Deoxyribose	0.0117	0.96
Thymine	0.0029	0.24
Adenine	0.0170	1.40
Guanine	0.0170	1.40
Cytosine	0.0170	1.40
Uracil	0.0140	1.15
Glycerol	0.0170	1.40
Fatty acid	0.0340	2.80
Phosphate	0.0830	6.80
Other	0.0253	2.08

culture conditions and other environmental factors, the thermodynamic parameters will be fairly insensitive to these small changes. In this table we indicate the number of moles of each constituent necessary to make one gram dry weight of biomass. Utilizing the atomic composition of the components, we can write balanced equations and calculate the amounts of starting material required for the cases represented by eqs. (12-1)–(12-4). The entire analysis must be carried out stepwise to best utilize the currently available thermochemical data on compounds of interest. The first step of the calculation does not represent complete synthesis of biomass, but the synthesis of low molecular weight precursors. Table 12-3 represents the balanced equations for the synthesis of enough precursors to make one gram of biomass.

In Cases I–III (to be presented), oxygen occurs in the products in order to maintain mass balance. The production of oxygen in these cases is also predictable from the general features of the carbon cycle involving the reduction of water in the formation of carbohydrates. Another way of viewing the situation is to note that the precursors are in a highly oxidized state with respect to the final biomass. This accords with a generally held notion that the prebiological synthesis of biological compounds could have proceeded much more easily in a reducing environment. In Case IV, CO_2 and water occur in the products in order to maintain mass balance. Table 12-3 is equivalent to writing eqs. (12-1)–(12-4) in stoichiometric form, which we proceed to do in the case of (12-1). All coefficients are multiplied by 10^4 to avoid negative exponents on the left-hand side of the equations. They therefore have the usual significance of mole numbers:

$$\begin{aligned}&(410.35)CO_2 + (46.64)N_2 + (398.20)H_2O + (3.0)H_2SO_4 + (6.8)H_3PO_4\\&\rightarrow \text{Precursors} \rightarrow 10^4 \text{ g Biomass} + (506.60)O_2 + (82.2)H_2O\end{aligned} \tag{12-5}$$

TABLE 12-3
Mass Balance in the Synthesis of Small Molecule Precursors of Biomass

	Case I, eq. (12-1) (moles × 10^4)	Case II, eq. (12-2) (moles × 10^4)	Case III, eq. (12-3) (moles × 10^4)	Case IV, eq. (12-4) (moles × 10^4)
Reactants				
CO_2	410.35	410.35	410.35	—
$C_6H_{12}O_6$	—	—	—	72.79
N_2	46.64	—	—	—
HNO_3	—	93.28	—	—
NH_3	—	—	93.28	93.28
H_2O	398.20	351.56	258.28	—
H_2SO_4	3.0	3.0	3.0	3.0
H_3PO_4	6.8	6.8	6.8	6.8
Products				
O_2	506.60	618.20	436.64	—
CO_2	—	—	—	26.44
H_2O	—	—	—	178.41
Biomass precursors	Sufficient material to synthesize one gram of biomass (as given in table 12-2)			

The reason that water appears in both reactants and products is that the intermediates are considerably more hydrated than the products due to the molecule of water released in each polymerization step. The final 82.2 moles of water in eq. (12-5) comes from such condensation reactions.

We may now proceed to the calculation of the heats of formation and the entropies of the reactants. In principle, we start with the elements in their standard states and determine the heats of formation and the entropies of the compounds in their standard states. We then consider the thermodynamic changes in going from the pure components to a dilute aqueous solution. This involves a specification of the final volume of the solution, as the entropy of solutes is a function of concentration. In practice much of the thermodynamic data about the reactants is available for compounds whose standard state is given as an aqueous solution of unit molarity and unit activity coefficient. That is, the data are calculated for materials at unit molarity which behave as if they were at infinite dilution. The heats of formation are thus those for infinite dilution and may be used directly. The total entropy must be determined from the tabulated entropies at unit molarity plus the entropy of dilution plus the entropy of mixing. As noted, the entropy depends on the volume of the solution so that volume will be carried as a parameter in the calculations.

The starting states will then be considered as being $T = 298.15$ K, and:

Case I. A dilute aqueous solution of sulfuric acid and phosphoric acid along with containers of nitrogen gas and carbon dioxide at 1 atm.

Case II. A dilute aqueous solution of nitric acid, sulfuric acid, and phosphoric acid along with a container of carbon dioxide at 1 atm.

Case III. A dilute aqueous solution of ammonia, sulfuric acid, and phosphoric acid along with a container of carbon dioxide at 1 atm.

Case IV. A dilute solution of glucose, ammonia, sulfuric acid, and phosphoric acid.

The entropy of dilution is given by

$$\Delta S_d = -R \sum_i n_i \ln(n_i/V) \tag{12-6}$$

where n_i is the number of moles of the ith substance and V is the final volume of the system. Expression (12-6) is obtained from the integration of eq. (5-56) at constant temperature. This leads to ΔS equal to $R\ln(V_2/V_1)$. Since V_1 is one liter (standard state) and V_2 is V/n_i, we get the above expression directly The entropy of mixing from eq. (5-72) is

$$\Delta S_m = -R \sum_i n_i \ln(n_i/\sum n_i) \tag{12-7}$$

for both the liquid and gas mixtures. Their sum is

$$\Delta S_d + \Delta S_m = R\sum_i n_i \ln V - R\sum_i n_i \ln(n_i^2/\sum n_i) \tag{12-8}$$

Note that in calculating these entropies, we do not include the entropy of water since we are dealing with dilute aqueous solutions. The relevant thermodynamic data are given in tables 12-4 and 12-5. We may now utilize tables 12-3 to 12-5 and eq. (12-8) to calculate the standard heats of formation and entropies of the starting materials. The results of these calculations are shown in table 12-6. The original literature sources of all data in the tables are given in the last reference at the end of this chapter.

TABLE 12-4
Thermodynamic Constants of Reactants

Compound	Standard heat of formation in aqueous solution (kJ/mole)	Entropy in aqueous solution (J/mole/K)
$C_6H_{12}O_6$	−1262.01	264.01
H_2O	−285.85	69.95
HNO_3	−206.56	146.44
NH_3	−80.83	110.03
H_3PO_4	−1289.50	176.14
H_2SO_4	−907.50	17.15

TABLE 12-5
Thermodynamic Constants of Gaseous Components

Substance	Standard heat of formation in gas phase (kJ/mole)	Entropy in gas phase (J/mole/K)
O_2	—	205.02
N_2	—	191.50
CO_2	−393.50	213.63

The product of the first stage of the reaction will be a dilute solution of all the small molecule components listed in table 12-2, at a temperature of 298.15 K. In addition, Cases I–III will give rise to free oxygen gas at a pressure of 1 atm and a temperature of 298.15 K. In Case IV the final mixture will be accompanied by free carbon dioxide gas at 1 atm. The group of values required for this reaction stage will then be the heats of formation, entropies, heats of solution, and entropies of solution of all the product molecules.

TABLE 12-6
Heats of Formation and Entropies of Mixtures of Reactants for One Gram of Biomass

Case	Heat of formation (kJ)	Entropy (J/K)
I	−28.68	$12.76 + 0.047 \ln V$
II	−29.27	$13.20 + 0.083 \ln V$
III	−25.43	$12.21 + 0.083 \ln V$
IV	−11.09	$3.89 + 0.146 \ln V$

The primary data for this calculation are given in table 12-7. The heat of formation of each component in the mixture can be obtained by adding the heat of formation of the crystal to the heat of solution at infinite dilution. Thus, the sums of the first and third columns will give the heat of formation of the final mixture. The entropy of the final mixture consists of (a) the entropy of each of the crystalline components plus (b) the entropy of solution at saturation plus (c) the entropy of dilution plus (d) the entropy of mixing. The entropy of the crystalline components can be obtained by direct measurements of specific heats and is given in the second column. The entropy of solution at saturation can be obtained by noting that at equilibrium the free-energy change in going from crystal to solution is zero. Since, however,

$$\Delta G_{sat} = \Delta H_{sat} - T\Delta S_{sat} = 0 \qquad (12\text{-}9)$$

we get

$$\Delta S_{sat} = \frac{\Delta H_{sat}}{T} \qquad (12\text{-}10)$$

where ΔH_{sat} is the heat of solution at saturation. The entropy of dilution is given by

$$\Delta S_d = R \sum n_i \ln \left[\frac{\mathscr{S}_i}{(n_i/V)} \right] \qquad (12\text{-}11)$$

where $\mathscr{S}_i$ is the solubility of the ith component. Then

$$\Delta S_d = R \sum n_i \ln V + R \sum n_i \ln \mathscr{S}_i - R \sum n_i \ln n_i \qquad (12\text{-}12)$$

The entropy of mixing is given by eq. (12-7).

In Cases I–III we need to add the entropy of the oxygen gas formed. In Case IV we need to add the entropy of the CO_2 and the water as well as the heat of formation of these components.

TABLE 12-7
Thermodynamic Constants of the Small Molecular Weight Constituents of Biomass

Compound	Heat of formation (kJ/mole)	Entropy (J/mole/K)	Heat of solution, infinite dilution (kJ/mole)	Heat of solution, saturation (kJ/mole)	Solubility (moles/liter)
Alanine	−560.4	129.2	8.54	9.41	1.665
Arginine	−621.9	250.6	6.28	4.52	
Aspartic acid	−972.6	170.1	25.10	24.27	0.038
Asparagine	−788.6	174.5	24.06	24.06	0.186
Cysteine	−532.2	169.9			
Glutamic acid	−1004.3	188.2	27.32	26.48	0.058
Glutamine	−827.6	195.1			
Glycine	−528.1	103.6	15.69	14.02	2.900
Histidine	−226.7	217.6	13.81	13.43	0.262
Isoleucine	−635.1	207.6		3.51	0.303
Leucine	−635.5	211.7	3.26	3.51	0.181
Lysine	−678.6	205.0	−16.74	−14.64	
Methionine	−754.8	231.4	16.74	16.74	
Phenylalanine	−468.2	213.4		11.80	0.175
Proline	−525.9	134.7	3.14	1.26	6.572
Serine	−726.3	149.3	21.67	21.13	0.464
Threonine	−758.9	166.5			
Tryptophan	−417.6	250.2	5.69	5.69	0.055
Tyrosine	−683.7	213.7			0.0025
Valine	−620.1	178.7	5.98	7.24	0.740
Glucose	−1273.0	212.1	11.00	12.38	3.452
Ribose	−1062.3	177.0			
Deoxyribose	−897.0	177.0			
Thymine	−468.2	137.2			0.032
Adenine	+97.1	151.0			0.0068
Guanine	−182.9	160.2			0.000033
Cytosine	−464.0	120.9			0.064
Uracil	−461.1	122.2			0.032
Glycerin	−665.9	204.4	−5.899	$\Delta S = 22.84$	
Fatty acid	−922.6	350.3			0.0000011
Phosphate	−1266.1	110.5	−8.4	$\Delta S = 65.40$	
Other	−627.0	167.0	8.4	8.37	0.100

A study of table 12-7 shows that we are missing a good deal of the primary data required for an accurate calculation. However, rather than give up entirely we will proceed with a rough calculation based on the following assumptions:

(1) The solubility and heat of solution of the missing amino acids are the same as the average of the amino acids reported.

(2) The average solubility and heat of solution of the missing sugars are the same as for glucose.

(3) The average free energy of solution of the purines and pyrimidines is 26.3 kJ/mole based on the xanthine value. We arbitrarily assume that this is half enthalpic and half entropic.

(4) For stearic acid, a typical fatty acid, we assume a ΔG of solution of 38.5 kJ/mole based on extrapolation of values for smaller carboxylic acids. We arbitrarily assume that this is half ΔH and half $T\Delta S$.

When all these factors are included and the appropriate calculations carried out, results obtained are shown in table 12-8.

The large number of missing data in table 12-7 as well as the uncertain quality of a number of the reported values argues in favor of increased emphasis on experimental thermochemistry of compounds of biological interest.

The next stage in the present calculation involves the condensation of monomers into biomass. This can be represented schematically as

$$\begin{aligned} &4.54 \times 10^{-4} \text{ moles Alanine} \\ &2.52 \times 10^{-4} \text{ moles Arginine} \\ &\quad\vdots \\ &\rightarrow 1\,\text{g Biomass} + 82.2 \times 10^{-4} \text{ moles } H_2O \end{aligned} \tag{12-13}$$

In carrying out the calculation in this way we have grouped two steps into one. The steps are the condensation of monomers into macromolecules (proteins, polysaccharides, nucleic acids) in dilute solution and the assembly of macromolecules into cellular structure. The reason for grouping the two steps together is our lack of sufficient information to carry out the more detailed computation.

Lacking sufficient experimental data, we treat biomass as one gigantic macromolecule and consider the free-energy changes in its synthesis. The general reaction scheme we are considering is a series of polymerization

TABLE 12-8
Heats of Formation and Entropies of the Intermediate System

Case	Heat of formation (kJ/g)	Entropy (J/g/K)
I	−6.506	$11.87 + 0.0682 \ln V$
II	−6.506	$14.98 + 0.0682 \ln V$
III	−6.506	$11.16 + 0.0682 \ln V$
IV	−12.458	$3.87 + 0.0682 \ln V$

steps of the following type:

$$HM_1OH + HM_2OH \rightarrow HM_1 - M_2OH + H_2O \qquad (12\text{-}14)$$

where HM_1OH and HM_2OH represent generalized monomers. The overall process is

$$\sum n_i(HM_iOH) \rightarrow \text{Biomass} + \left(\sum n_i\right)H_2O \qquad (12\text{-}15)$$

The entropy change in this type of process can be estimated by using methods of polymer statistics. If we can then assign an average value to the ΔH for polymer bond formation, we can examine the overall free-energy changes in biomass synthesis.

Polymer statistics depends on enumerating states of a system and computing entropies based on the enumeration. Consider a system in which all states are isoenergetic and therefore all states have equal *a priori* probability. Starting with eq. (8-16) for entropy we can then write

$$S = -k \sum_{i=1}^{\mathscr{W}} f_i \ln f_i = -k\mathscr{W}\left(\frac{1}{\mathscr{W}} \ln \frac{1}{\mathscr{W}}\right) = k \ln \mathscr{W} \qquad (12\text{-}16)$$

This is the Boltzmann formula for the entropy of a microcanonical ensemble in which each member has the same energy. In the system of interest we calculate the number of possible configurations of the monomer mixture and of the cell. Equation (12-16) then permits us to calculate entropies.

The model system for computing entropy change starts with a volume V in which there are $\sum B_i$ dissolved monomer molecules of average volume $\mathfrak{B}$. We divide the large volume into $N = V/\mathfrak{B}$ compartments or subvolumes In the monomer state the molecules may be distributed among the subvolumes in the following number of ways:

$$\frac{N!}{\prod_i B_i!(N - B)!} \qquad (12\text{-}17)$$

where $B = \sum B_i$ and B_i is the number of monomer molecules of the ith species. If there are x possible spatial orientations for each monomer, then the total number of configurations $\mathscr{W}'$ is given by

$$\mathscr{W}' = \frac{N!x^B}{\prod_i B_i!(N - B)!} \qquad (12\text{-}18)$$

$\mathscr{W}'$ is the thermodynamic probability in the Boltzmann sense. If all the molecules are condensed into biomass, the number of possible orientations is reduced to Ny^B, where y is the number of possible configurations of the monomer in the final structure. N gives the number of possible locations of the terminal monomer in the giant macromolecule in the solution volume.

This assumes that the structure of the nonaqueous portion of the biomass is completely fixed. Each monomer is assumed to occupy a fixed place and is capable only of internal degrees of freedom such as vibrations and rotations and is not capable of independent translation. While this involves a slight degree of overspecification, it does reflect the very high degree of organization which is characteristic of all biological systems. Under these assumptions the change in entropy in going from the monomers to the biological structure is given by

$$\Delta S = k \ln N y^B - k \ln \left[\frac{N! x^B}{\prod_i B_i!(N - B)!} \right] \tag{12-19}$$

Applying Stirling's approximation (Appendix IV), we get

$$\Delta S \simeq -kB \ln \left(\frac{Nxe}{y} \right) + k \sum B_i \ln B_i \tag{12-20}$$

For each species of molecule we can define a mole fraction in the biomass as

$$\mathfrak{N}_i = \frac{B_i}{B} \tag{12-21}$$

Recasting the equation in terms of $\mathfrak{N}_i$, we get

$$\Delta S = -kB \left[\ln \left[\left(\frac{N}{B} \right) \left(\frac{x}{y} \right) e \right] - \sum \mathfrak{N}_i \ln \mathfrak{N}_i \right] \tag{12-22}$$

The ratio N/B is the number of sites associated with one molecule of monomer in solution. Since $N = V/\mathfrak{B}$, where $\mathfrak{B}$ is the average volume per molecule, then

$$\frac{N}{B} = \frac{V}{B\mathfrak{B}} = \frac{V}{\sum B_i \mathfrak{B}_i} \tag{12-23}$$

where $\mathfrak{B}_i$ is the partial molecular volume of the ith species. The ratio N/B can then be computed from experimental data and is equal to $V/0.000822$, where V is given in liters.

The quantity x/y is the most difficult part of eq. (12-22) to estimate. It is the ratio of the number of possible orientations of the free monomer compared with the number of orientations in the final biological structure. The change of entropy should be approximated by that calculated for the loss of one degree of rotational freedom since the monomer is much less free to rotate in the polymer than it is in solution. Thus, using the theory of the rigid rotator, we get for one mole

$$\Delta S = R \ln \left(\frac{x}{y} \right) \simeq R \ln \left[\frac{8\pi^2 \mathscr{I} kT}{h^2} \right] \tag{12-24}$$

where $\mathscr{I}$ is the moment of inertia of the molecule. Equation (12-24) can be derived from eq. (8-51) if we note that the rotational entropy per mole before macromolecule formation is $R \ln \mathfrak{z}$ (where $\mathfrak{z}$ is the partition function for rotation) and after condensation is $R \ln 1$ or zero. For an average molecule of the size indicated above, the entropy contribution from restricted rotation can be estimated at $8.05R$ entropy units per mole at 298.15 K.

Table 12-9 sums up the various entropy contributions for the choice of monomers indicated previously. These results substituted in eq. (12-22) provide our first estimate of the entropy of formation of biomass. This first estimate is too large since it neglects the positive contribution due to the production of 82.2×10^{-4} moles of water in the process. The molar entropy of water at 298.15 K is 69.96. However, inclusion of this entire quantity ignores the hydration of the biological structures which reduces the entropy of the water. A compromise will be made by using 54.39 as the positive contribution of the water. This figure is based on -15.56 entropy units as an average entropy value for hydration of one mole of water per 100 g of protein, which is the appropriate order of magnitude for the hydration correction.

Having calculated the entropy change involved in structure formation we must now examine the enthalpy change. Four types of bonds are principally involved in the formation of biomass from monomeric units: (a) peptide bonds, (b) glucosidic linkages, (c) esters of fatty acids and glycerol, and (d) sugar phosphate bonds. The reactions may be represented as shown in table 12-10. Using the concept of bond energies we would expect the enthalpies of formation of the four types of bonds to be (a) 19.7, (b) 0.0, (c) 0.0, and (d) 0.0 kJ/mole. More experimental evidence exists on these energies.

TABLE 12-9
Entropy Changes in Macromolecule Formation

Term	Value
$\ln \frac{N}{B} = \ln \frac{V}{\sum B_i \mathfrak{B}_i}$	$\ln V + 7.128$ (V in liters)
$\ln \frac{x}{y}$	8.05
$\ln e$	1.000
$-\sum \mathfrak{N}_i \ln \mathfrak{N}_i$	3.201
Sum	$19.379 + \ln V$
$R \times$ sum $-$ water correction	$8.314 \ln V + 106.72$
$\Delta S = n \times$ molar entropy	$0.8774 + 0.0682 \ln V$

TABLE 12-10
Bond Changes in Polymer Formation

		Bonds broken	Bonds formed
(*a*)	$H_2N-CR_1H-C(=O)-OH + H-NH-CR_2H-C(=O)-OH \longrightarrow H_2N-CR_1H-C(=O)-NH-CR_2H-C(=O)-OH + H_2O$	C—O H—N	O—H C—N
(*b*)	glucose (ring, CH_2OH) + glucose (ring, CH_2OH) $\longrightarrow$ disaccharide (rings linked by —O—) $+ H_2O$	O—H C—O	C—O O—H
(*c*)	$H_2C(OH)-CH(OH)-CH_2(OH) + H-O-C(=O)-(CH_2)_nCH_3 \longrightarrow H_2C(-O-C(=O)-(CH_2)_nCH_3)-CH(OH)-CH_2(OH) + H_2O$	C—O O—H	C—O O—H
(*d*)	Base—Sugar(—OH)—O—P(=O)(OH)—OH + Base—Sugar(—OH)—O—P(=O)(OH)—OH $\longrightarrow$ Base—Sugar—O—P(=O)(OH)—OH with Sugar—O—P(=O)(OH)—O—Sugar(—OH)—Base $+ H_2O$	P—O O—H	O—H P—O

A summary of the data on peptide bonds is given in table 12-11. Values are available for the free energy of hydrolysis of a number of sugar phosphate bonds as given in table 12-12.

TABLE 12-11
Enthalpy Changes in the Hydrolysis of Peptide Bonds

Peptide	Bond hydrolyzed	ΔH (J/mole)
Benzoyl-L-tyrosylglycinamide	Tyrosine–glycine	-6485 ± 600
Benzoyl-L-tyrosylglycine	Tyrosine–glycine	-5585 ± 375
Carbobenzoxyglycyl-L-leucine	Glycine–leucine	-8830 ± 225
Poly-L-lysine	Lysine–lysine	-5190
Carbobenzoxyglycyl-L-phenylalanine	Glycine–phenylalanine	-10670 ± 225
L-Tyrosylglycinamide	Tyrosine–glycine	-5440 ± 600

TABLE 12-12
Gibbs Free Energy of Hydrolysis of Sugar–Phosphate Bonds

Compound	ΔG_0 (kJ/mole)	Compound	ΔG_0 (kJ/mole)
Glucose-6-phosphate	−12.6	Fructose-1-phosphate	−11.7
Galactose-6-phosphate	−12.6	Mannose-6-phosphate	−11.3
Fructose-6-phosphate	−12.6	3-Phosphoglyceric acid	−12.6
Glucose-1-phosphate	−20.5	2-Phosphoglyceric acid	−17.2

Krebs and Kornberg (see Bibliography) quote the following values:

> The free energy of hydrolysis of the glucosidic bonds of starch or glycogen is about 18.0 kJ, of the peptide bond of the order of 12.6 kJ, and of an ester of the order of 10.5 kJ, per mole.

These free energies contain both enthalpy and entropy terms so that the enthalpy changes can be expected to be of a smaller magnitude than the Gibbs free-energy changes. Considering these data we see that, in most biological polymers, ΔH for synthesis probably ranges between 0–12,500 J/mole. We will take an assumed figure of 8300 J/mole as a basis of calculation. At this point, it is well to note that the calculations cannot, at present, be at all exact but are nonetheless significant in establishing the magnitude of quantities which are important in studying energy exchanges in the biosphere. These approximations must eventually be replaced by more exact data based on calorimetric measurements on biological material. The value of 8300 J/mole of monomer leads to a ΔH for biomass formation of 0.0686 kJ/g. The final free-energy changes in the formation of biomass are shown in table 12-13.

In table 12-13 two points require elaboration. In Case II the entropy change is positive and in Case IV the Gibbs free-energy change is negative. The first situation arises from the fact that we are considering an overall process which includes considerable gas production as well as biomass formation. In Case II large amounts of oxygen gas are made from dissolved nitrate and water. These are highly exentropic processes and lead to an overall entropy increase. Correspondingly, the enthalpy change is larger for this case, since the reactants are in a highly oxidized state which, in general, corresponds to large negative free energy of formation. However, if we were just to consider the entropy change in forming biomass, independently of the large contribution from the oxygen gas, then we would get a substantial negative value. In Case IV it is apparent that a biomass, CO_2, water mixture is more thermodynamically likely than a solution of glucose, ammonia, sulfate, and phosphate. While this may seem strange at first, it accords with the experimental result that if such a solution were isolated in contact with an isothermal reservoir and seeded with an anaerobic bacterium, it would be converted into biomass with a net increase of entropy of the universe.

TABLE 12-13

Free Energy of Formation of One Gram of Biomass

Case	ΔH (kJ/g)	ΔS (J/g/K)	ΔG ($T = 298.15$, $V = 1$ liter) (kJ/g)
I	22.17	$-1.761-0.047 \ln V$	22.69
II	22.76	$0.900-0.083 \ln V$	22.49
III	18.92	$-1.933-0.083 \ln V$	19.50
IV	−1.37	$-0.214-0.146 \ln V$	−1.10

We may now examine the problem of the spontaneous formation of a living system by a fluctuation in an equilibrium ensemble. The maximum probability will be

$$\mathscr{P}_L = e \frac{-\Delta G_{\text{cell}}}{kT} \tag{12-25}$$

For Case I, starting with N_2, CO_2, H_2O, H_3PO_4, H_2SO_4 in one liter of solution, ΔG_g for one gram of material is

$$\Delta G = \Delta H - T\Delta S = 22.70 \quad \text{kJ/g} \tag{12-26}$$

For a cell of mass m, the free energy change is $m\,\Delta G_g$. Thus, in table 12-14 we examine $\mathscr{P}_{L\,\max}$ as a function of the size of the biological structure.

We may now, however, proceed to ask a more realistic question. Suppose that we are given a one-liter vat containing all the monomer units necessary to make a living cell; what is the probability that a cell will form in an equilibrium ensemble? We proceed as in the previous calculations, only the ΔH and ΔS values are considerably smaller and are equal to 0.0686 kJ/g and 0.8774 entropy units. ΔG is thus 326 J/gm. The probabilities are then given in table 12-15.

We have examined, insofar as the data have permitted, the enthalpy, entropy, and Gibbs free-energy changes in biological transformations. Although inadequacies in available data as well as uncertainties in our knowledge of macromolecular chemistry do not allow us to carry out the calculations in a completely satisfactory manner, nonetheless, the equipment is at hand with which to thoroughly study the energetic and entropic bookkeeping in living cells and organisms. This chapter can only represent an early attempt at this type of detailed analysis of bioenergetics. In the following pages we will examine certain features of ecology and biochemistry in terms of the thermochemical framework that we have established here.

TABLE 12-14
Probability of a Cell Arising as a Spontaneous Fluctuation in an Equilibrium System[a]

Cell mass (g)	Maximum probability of forming a living cell
10^{-10}	$10^{-2.4\times10^{14}}$
10^{-11}	$10^{-2.4\times10^{13}}$
10^{-12}	$10^{-2.4\times10^{12}}$
10^{-13}	$10^{-2.4\times10^{11}}$
10^{-14}	$10^{-2.4\times10^{10}}$

[a] Calculated from thermodynamic data for varying size systems.

TABLE 12-15
Probability of a Cell Arising Spontaneously in an Ocean of Monomer Units

Cell mass (g)	Probability of forming
10^{-10}	$10^{-3.4\times10^{12}}$
10^{-11}	$10^{-3.4\times10^{11}}$
10^{-12}	$10^{-3.4\times10^{10}}$
10^{-13}	$10^{-3.4\times10^{9}}$
10^{-14}	$10^{-3.4\times10^{8}}$

TRANSITION

Solar radiation impinging on the earth has a characteristic spectral distribution and flux density which are a predominant influence on the nature of the global ecosystem. A fraction of this energy is converted to a biological form by photosynthetic reactions in the primary producers. This energy then flows through the trophic network with much of it being given up as heat and a small fraction being stored in various forms. From a knowledge of the biochemical composition of living materials the free energy of formation may be calculated. It is thus possible from first principles to work out the major thermodynamic parameters of living systems. For further ecological analysis some familarity with chemical kinetics is required and we go on to review the introductory principles of that science.

BIBLIOGRAPHY

Benson, S. W., "Thermochemical Kinetics." Wiley, New York, 1968.

A detailed discussion of methods for the estimation of thermochemical data and rate parameters.

Gates, D. M., "Energy Exchange in the Biosphere." Harper and Row, New York, 1962.

This is a monograph covering the macroscopic aspects of energy exchange between organisms and their surroundings.

Gorski, F., "Plant Growth and Entropy Production." Zaklad Fizjologii Roslin Pan, Krakow, Poland, 1966.

This monograph extensively reviews the entropy changes in the growth of plants.

Kormondy, E. J., "Concepts of Ecology." Prentice-Hall, Englewood Cliffs, New Jersey, 1969.

An introduction to ecology that develops the trophic energy theme.

Krebs, H. A., and Kornberg, H. L., "Energy Transformations in Living Matter." Springer, New York, 1957.

A detailed review of the relation of intermediary metabolism to bioenergetics.

Lindeman, R. L., *Ecology* **23**, 399 (1942).

This is the paper that sets forth the trophic energy view of ecology.

Morowitz, H. J., "Energy Flow in Biology." Academic Press, New York, 1968.

Many of the ideas just discussed are developed more fully in this book.

13 Kinetics and Cycles

A. CHEMICAL KINETICS

While traditional thermodynamics lends considerable insight into various stable systems, it has little to say about the process by which a system moves from one state to another. Such problems have usually been dealt with by mechanics, hydrodynamics, chemical kinetics, and electrodynamics. There remains a thermal aspect to all of these treatments because of viscosity, friction, and explicit temperature effects. No branch of molecular physics escapes an involvement with thermal factors and efforts have been made to extend various disciplines into a more generalized thermodynamics. We will defer considerations of such topics until Chapter 16, but will here take up the more limited problem of the rates at which chemical reactions take place and the generalized basis of chemical cycles in nonequilibrium chemical systems such as those encountered in ecology.

We start by considering three general classes of chemical reactions:

$$\begin{gathered} A \rightleftharpoons B \\ A + C \rightleftharpoons D \\ A + C \rightleftharpoons E + F \end{gathered} \tag{13-1}$$

A large body of experimental data has grown up on transitions of these types and many follow "mass action kinetics;" that is (as an empirical generalization), the rates of the reactions can be written, respectively, as

$$\frac{d[\mathrm{A}]}{dt} = -k_1[\mathrm{A}] + k_2[\mathrm{B}]$$

$$\frac{d[\mathrm{A}]}{dt} = -k_3[\mathrm{A}][\mathrm{C}] + k_4[\mathrm{D}] \tag{13-2}$$

$$\frac{d[\mathrm{A}]}{dt} = -k_5[\mathrm{A}][\mathrm{C}] + k_6[\mathrm{E}][\mathrm{F}]$$

The brackets indicate concentrations in the classical version of the theory, although the problem may be recast into activities. The k's then uniformly have the dimensions of $(\text{time})^{-1}$. For subsequent developments in this chapter, we assume that we are in the domain where activities can be set equal to concentrations (see Chapter 10). By introducing short-lived interaction intermediates, it is possible to reduce most complex chemical reactions into sequences of the kind shown in eq. (13-1).

The k's are called rate parameters or rate constants; that is, they do not depend explicitly on the concentrations, although in general they are functions of the temperature and other environmental variables. These rate parameters as presented in eq. (13-2) are purely experimental quantities and are not derived from theory. Such quantities usually have an exponential dependence on temperature, which suggests via the Maxwell–Boltzmann distribution, a more deeply thermodynamic basis. These ideas were put forth by Arrhenius in 1889 when he formulated the following equation (empirical generalization).

$$k = \mathfrak{G}e^{-E/RT} \tag{13-3}$$

$\mathfrak{G}$ is the collision number or frequency factor and E is known as the energy of activation. The assumption behind eq. (13-3) is that there is an activation barrier, to the reaction, of magnitude E; $\mathfrak{G}$ measures the collision rate and $e^{-E/RT}$ is the probability that a collision will be sufficiently energetic to get over the barrier. The ideas embodied in eq. (13-3) have been extended in the theory of absolute reaction rates which provides a useful although theoretically incomplete and sometimes troublesome approach to understanding rates. The problem at hand is a dynamical one, and thermodynamic approaches can offer guidance but not a complete solution.

B. ABSOLUTE REACTION RATE THEORY

The assumptions of the theory have been set forth by its founders and expositors (see Glasstone *et al.* in the Bibliography) as follows:

> A chemical reaction or other rate process is characterized by an initial configuration which passes over by continuous change of coordinates into the final configuration. There is, however, always some intermediate configuration which is critical for the process, in the sense that in this state there is a high probability that the reaction will continue to completion. This critical configuration is called the 'activated complex' (transition state) of the reaction. The activated complex is to be regarded as an ordinary molecule, possessing all the usual thermodynamic properties, with the exception that motion in one direction, i.e,, along the reaction coordinate, leads to decomposition at a definite rate.†

The first concept that must be elaborated is that of the reaction coordinate. Consider the following reactions:

$$\begin{gathered} Br\cdot + H_2 \rightarrow HBr + H\cdot \\ \text{cis isomer} \rightarrow \text{trans isomer} \end{gathered} \tag{13-4}$$

In the first case, the transition has the highest probability if the bromine radical approaches along the H_2 molecular axis. We can choose this as the reaction coordinate and define the distance between the radical and the nearest atom as the distance measure. For the cis–trans isomerization we can define an angle of rotation about the bond connecting the two parts of the molecule and allow this to be the relevant coordinate. Figure 13-1 illustrates the energy diagrams corresponding to eq. (13-4).

In more complicated cases, it is theoretically possible to analyze the potential surface and find the most likely trajectory for the reaction. This trajectory is designated the reaction coordinate and the state of the system

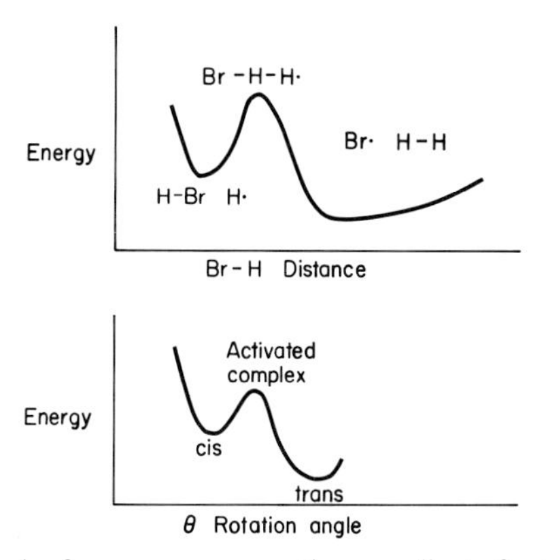

FIGURE 13-1. Graph of energy versus reaction coordinate for two simple reactions.

† From "The Theory of Rate Processes," pp. 10–11, by Glasstone *et al.* Copyright © 1941 by McGraw-Hill, New York. Used with permission of McGraw-Hill Book Company.

at the top of the potential barrier along this coordinate is the activated complex. Actually figuring out the trajectory and calculating the nature of the potential surface is possible only for reactions of molecules containing a small number of atoms. In other instances, the reaction coordinate and activated complex are still useful constructs. We show this generalization in fig. 13-2.

We represent the activated complex by a shallow potential well to indicate that it can be regarded as an actual state of the system of finite lifetime. The width of the well is δ and its height may be represented as about $\frac{1}{2}kT$, the mean thermal energy per degree of freedom. The complex at the top of the barrier may be visualized as a harmonic oscillator with the reaction coordinate being a degree of vibrational freedom. The complex can be assumed to have an effective reduced mass m with respect to the vibration and a force constant of κ.

Now consider a reaction of the form

$$\begin{aligned} \text{Reactants} &\rightleftharpoons \text{Activated Complex} \rightarrow \text{Products} \\ \nu_A \mathrm{A} + \nu_B \mathrm{B} \cdots &\rightleftharpoons M^{\dagger} \rightarrow \text{Products} \end{aligned} \tag{13-5}$$

The rate of the forward reaction will be given by

$$\begin{aligned} \text{Rate} &= (\text{Concentration of activated complexes}) \\ &\quad \times (\text{Rate of crossing the barrier}) \\ &= C^{\ddagger} \times (\text{Rate of crossing}) \end{aligned} \tag{13-6}$$

We use the double dagger notation (‡) to indicate parameters associated with the activated complex. *We assume a partial equilibrium; the reactants are in equilibrium with the activated complex* (postulate). Thus we may write

$$\frac{C^{\ddagger}}{[\mathrm{A}]^{\nu_A}[\mathrm{B}]^{\nu_B}\cdots} = K^{\ddagger} = e^{\frac{-\Delta G^{\ddagger}}{RT}} \tag{13-7}$$

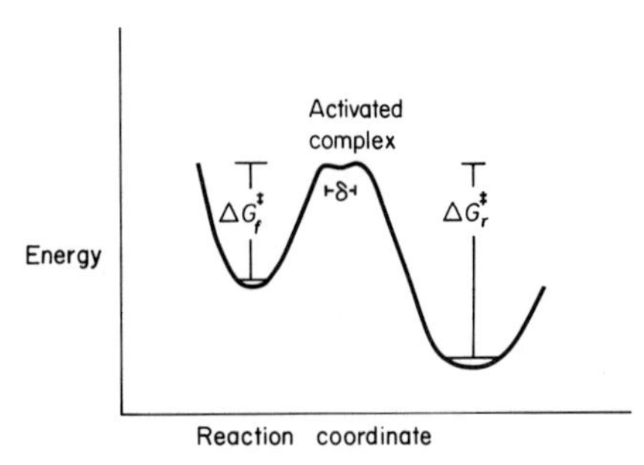

FIGURE 13-2. Model of a chemical reaction indicating that the activated complex is a very shallow energy well ($\sim\frac{1}{2}kT$) at the top of a barrier.

where $\Delta G^{\ddagger}$ is the Gibbs free-energy difference between the activated complex and the reactants.

To obtain the rate of crossing we divide the mean velocity of crossing by the distance to be traversed. *We assume a crossing will involve a single oscillation so that the distance involved is the order of* 2δ, *the length transversed in the time of a single vibrational period*, Ξ (postulate).

$$\text{Rate of barrier crossing} = \frac{\bar{v}}{2\delta} = \frac{1}{\Xi} \tag{13-8}$$

The second equality of eq. (13-8) comes from considering that the mean velocity (speed) is defined as the distance traversed per cycle, 2δ, divided by the period of an oscillation Ξ. We now introduce two results, one from the classical theory of harmonic motion and the second from the quantum mechanical theory of oscillators, in order to evaluate Ξ. First note that the period of a classical harmonic oscillator is given by

$$\Xi = 2\pi\sqrt{\frac{m}{\kappa}} \tag{13-9}$$

The ground state energy of an oscillator from quantum theory is [see eq. (8-45)]

$$\varepsilon_0 = \frac{h}{4\pi}\sqrt{\frac{\kappa}{m}} = \frac{h}{2\Xi} \tag{13-10}$$

The right-hand equality comes from introducing (13-9) into (13-10). Noting our assumption that the depth of the well is $\frac{1}{2}kT$ we get

$$\frac{1}{\Xi} = \frac{2\varepsilon_0}{h} = \frac{2(\frac{1}{2}kT)}{h} = \frac{kT}{h} \tag{13-11}$$

Comparison with eq. (13-8) shows that there is a barrier crossing rate independent of δ, κ, and m, and as is seen from eq. (13-11) is dependent only on the barrier height. Since we assume the same barrier height ($\frac{1}{2}kT$) for the shallow well at the top of barriers for all reactions, the crossing rate is generalized to a universal function of temperature alone.

If we now go back to eq. (13-6) and substitute (13-7) and (13-11), we get

$$\text{Rate} = \frac{kT}{h}\left(e\frac{-\Delta G^{\ddagger}}{RT}\right)[A]^{\nu_A}[B]^{\nu_B}\cdots \tag{13-12}$$

From a consideration of eq. (13-2), we get

$$\text{Rate} = k_{\text{phenomenological}}[A]^{\nu_A}[B]^{\nu_B}\cdots \tag{13-13}$$

Hence

$$k_{\text{phenomenological}} = \frac{kT}{h} e^{-\Delta G^{\ddagger}/RT} \tag{13-14}$$

This result is the fundamental expression of the theory of absolute reaction rates. For very simple cases we can formulate potential surfaces and attempt to calculate $\Delta G^{\ddagger}$. For more complicated transitions $\Delta G^{\ddagger}$ is an empirical parameter which is a useful bookkeeping device in dealing with rate data as we shall see in the next chapter.

Returning to fig. 13-2, we next consider a reversible reaction. To be specific, let it be

$$A + C \rightleftharpoons E + F \tag{13-15}$$

The forward and reverse reaction rates may be written:

$$\begin{aligned} \text{Forward rate} &= k_f[A][C] \\ \text{Reverse rate} &= k_r[E][F] \end{aligned} \tag{13-16}$$

At equilibrium the forward and reverse reactions occur at equal rates. This principle of detailed balance will later be expanded upon when we deal with Onsager's approach to irreversible thermodynamics. As a consequence of (13-16) and detailed balance at equilibrium we note

$$k_f[A]_{eq}[C]_{eq} = k_r[E]_{eq}[F]_{eq} \tag{13-17}$$

$$K = \frac{[E]_{eq}[F]_{eq}}{[A]_{eq}[C]_{eq}} = \frac{k_f}{k_r} = \frac{(kT/h)e^{-\Delta G_f^{\ddagger}/RT}}{(kT/h)e^{-\Delta G_r^{\ddagger}/RT}} \tag{13-18}$$

The first equality in (13-18) comes from the definition of the equilibrium constant, the second comes from eq. (13-17) (detailed balance), and the third from eq. (13-14). By reference to fig. 13-2, we note that we can rewrite (13-18) as

$$K = \exp\left(\frac{-\Delta G_f^{\ddagger} + \Delta G_r^{\ddagger}}{RT}\right) = \exp -\left(\frac{G_{\text{products}} - G_{\text{reactants}}}{RT}\right) \tag{13-19}$$

This result, identical to the one derived from thermodynamics alone (eq. (10-55), shows the consistency of this kinetic approach with previous developments.

Detailed balance allows us to formulate relations among rate parameters. This may be most easily seen in the isomerization reactions shown in fig. 13-3. At equilibrium, detailed balance demands that

$$k_1[A] = k_2[B], \qquad k_3[B] = k_4[C], \qquad k_5[C] = k_6[A] \tag{13-20}$$

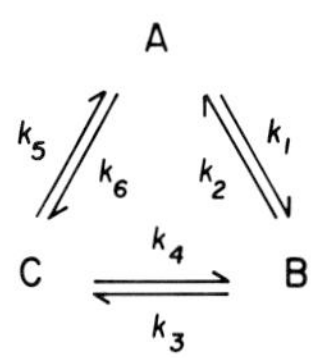

FIGURE 13-3. Reaction sequence of three isomeric compounds, A, B, and C.

from which

$$k_1k_3k_5 = k_2k_4k_6 \tag{13-21}$$

More general relations may be developed for complex reaction networks. We will derive two of these that are of special importance in biological nutrient flow.

Two essential aspects of global ecology are the existence of major material cycles and the presence of a pool of thermodynamic ground-state molecules as reactants in the cycle. We wish to show that these features are a necessary consequence of the physics and chemistry of nonequilibrium systems undergoing electronic excitations by an energy flux. The overall behavior of ecosystems is thus, in a sense, independent of the details of the interactions within the system.

C. *REACTION NETWORKS*

Our discussion will necessitate first establishing the outlines of a general formalism for dealing with networks of organic reactions. We will then derive a cycling theorem which says that steady-state organic networks maintained away from equilibrium by the input of electronic excitation will be characterized by flows of matter around closed loops in the network. We will then focus on the special role of the thermodynamic ground state as an attractor in directing the flow properties of the network.

In recent years a number of workers have attempted to establish organic chemistry on a more axiomatic foundation, or to provide a set of algorithms to analyze synthetic or degradative pathways in a complex reaction network. A feature of these studies has been the necessity of representing an organic molecule as a linear array of symbols. These programs have indicated that any organic molecule whose structure is known can be so represented. As a result, a molecule can be represented as a point in an abstract n-dimensional space (an n-tuple). Given this representation, we can now form a graph from the set of points by using lines to connect valid subsets of reactants and products. Rather than the edges of conventional graph theory, we need a more complex line structure to represent chemical reactivity. By allowing short-lived intermediates as points in our space (part of the n-tuple could be the energy state of the molecule) and dividing reaction sequences into unit steps, we can represent chemical reactions by combinations of the operations shown in fig. 13-4. These are the reactions shown in eq. (13-1). The *points* represent molecular descriptions and the *lines* represent reaction pathways. Whether or not reaction lines can be drawn between points depends on conservation rules as well as on detailed quantum mechanical considerations. While such relations are not generally available from first principles, synthetic organic chemists have developed a high degree of predictive ability in studying such problems. In any case, the space of organic molecules is clearly representable by a graph as outlined above. The graph can be made directed by assigning arrows in the direction of decreasing standard free energy of the reactions (ΔG_0). Network theory representations of systems of chemical reactions have been formulated for a number of cases.

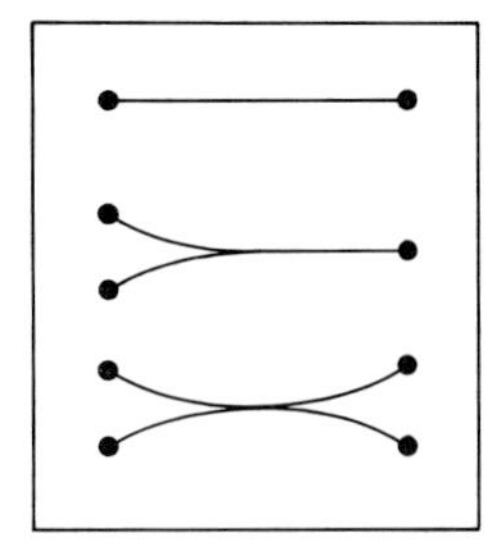

FIGURE 13-4. Graph theory representation of reaction types. Shown are isomerization, condensation, and condensation splitting.

The input of electronic energy can now be represented by a charging operation in parallel with a reaction line in the graph, as is shown in fig. 13-5. Consider next a model system closed to the flow of matter and in contact with an infinite isothermal reservoir. For definiteness, assume that the

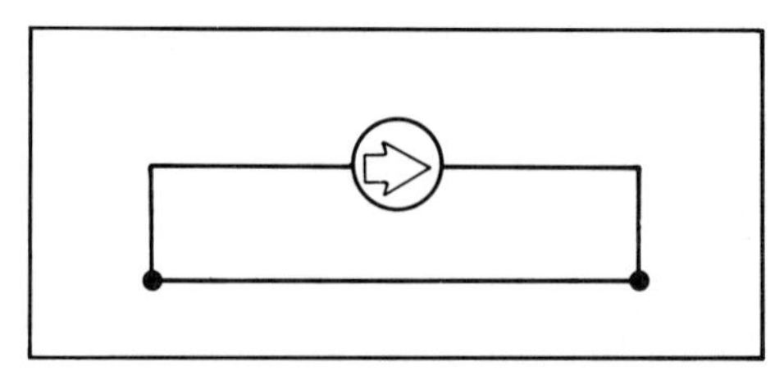

FIGURE 13-5. Representation of an isomerization reaction driven by an external energy input.

system contains a collection of molecules made up of the atoms C, H, N, O, P, and S. For homogeneous systems, the complete phenomenological description consists of specifying all the concentrations and reaction rates. In terms of the graph described below, we now think of that graph as a network, assigning concentration values to the points (which roughly correspond to a voltage assignment in electrical network theory) and flow values to the lines (which correspond to currents in electrical network theory). The input of electronic energy is analogous to the introduction of voltage sources in the electrical case.

D. CYCLIC FLOWS

Next, consider a constant flux of excitation energy into the model system. This can be in the form of photons, ionizing radiation, very high-temperature thermal sources, spark discharge, or any other source that causes electronic transitions from lower-lying to higher-lying energy states. Under constant energy flux, new values of concentrations and flows emerge. The system will age and form a steady state where the time derivatives of the concentrations vanish. For this condition, Kirchhoff's current law holds at every point in the chemical network. This balance requires that the flows of material generated by the electronic transitions return to the points in the network where they originated. Thus, the steady state is of necessity characterized by cyclic flows of material around loops in the reaction network. As the material flows around the cycles, the input energy flows as heat into the isothermal reservoir. Thus, material cycling is seen to be a very general feature of chemical networks kept at a far-from-equilibrium steady state by the constant influx of energy in a form capable of electronic excitation. The relation of the model system to a planetary surface is clear. Such a surface is closed to the flow of matter and is exposed to a time-averaged constant flux of solar radiation. Outer space serves as an isothermal reservoir for the flow of heat, which is transferred in the form of infrared radiation. Global ecology is thus an example of a general principle. Cycling is not uniquely biological; rather the biological example is a special case of a more general principle.

To introduce the special role of low-lying energy levels consider the previous model system after it has come to a steady state in an energy flux, and then remove the energy source, leaving it in contact with the isothermal reservoir. The system will decay to equilibrium which, in the case of a biological atomic composition, will be characterized by CO_2, H_2O, and N_2. Eventually the entire system will consist almost entirely of these energetically lowest-lying atomic configurations. The equilibrium state is thus a powerful molecular space attractor in the topological sense.

The attractor nature exhibited by the equilibrium distribution is a property of the network, not just a property of the equilibrium state. Therefore, during the dynamic states, the low-lying configurations are also acting as attractors, directing flows in the network toward those points. Dynamic states of the system in which low-lying molecules are not pumped up to higher states will tend to disappear as the high-energy material drains into the sinks established by the attractors. For steady states to be maintained in the network, the energetically pumped cycles must involve the low-lying states. Thus, in the biosphere, there is the necessity that photosynthetic processes operate on molecules such as water and carbon dioxide. Again, this is not a uniquely biological result but follows from the analysis of chemical networks.

We have thus demonstrated that the previously stated features of global ecology have their origin in very general properties of networks of reacting molecules.

A somewhat more general consideration of cycles can be approached by going back to fig. 13-3 and eq. (13-20) and postulating that the reaction takes place in a closed system. We may now introduce the condition that the total amount of material is constant. Setting n_T equal to the total mole number we get

$$[A] + [B] + [C] = n_T \tag{13-22}$$

and then solve for the concentrations

$$\begin{aligned} A_{eq} &= \frac{k_2 k_4 n_T}{k'} \\ B_{eq} &= \frac{k_1 k_4 n_T}{k'} \\ C_{eq} &= \frac{k_1 k_3 n_T}{k'} \end{aligned} \tag{13-23}$$

$$k' = k_2 k_4 + k_1 k_4 + k_1 k_3 \tag{13-24}$$

We could have also determined the equilibrium concentrations directly from thermodynamics by noting

$$\begin{aligned} A_{eq} &= \frac{n_T}{1 + \exp(-\Delta G_{AB}/RT) + \exp(-\Delta G_{AC}/RT)} \\ B_{eq} &= \frac{n_T \exp(-\Delta G_{AB}/RT)}{1 + \exp(-\Delta G_{AB}/RT) + \exp(-\Delta G_{AC}/RT)} \\ C_{eq} &= \frac{n_T \exp(-\Delta G_{AC}/RT)}{1 + \exp(-\Delta G_{AB}/RT) + \exp(-\Delta G_{AC}/RT)} \end{aligned} \tag{13-25}$$

A is assumed to be at the lowest free energy and is chosen as the reference level for computing ΔG's. These three eqs. (13-25) are the only requirements of thermodynamics. The rate-balance equations are a separate condition of detailed balance. Thus, from equilibrium constraints alone we could have the same final concentration by maintaining a net flow of material around the system. The steady-state condition would then require that

$$k_1[\mathrm{A}] - k_2[\mathrm{B}] = k_3[\mathrm{B}] - k_4[\mathrm{C}] = k_5[C] - k_6[\mathrm{A}] = J \quad (13\text{-}26)$$

where J is the flow, the rate at which the material is cycling around the system. It is precisely this flux of material around a closed loop that we have designated as a cycle. It is prohibited at equilibrium by detailed balance, but might obtain in other steady states.

To construct a formal theory, we first consider the particular case where the energy flows into the system as electromagnetic energy and flows out as heat. We will then generalize this result to other energy flows, including the case where energy flux is linked to the transition of chemical species from a high to a low chemical potential.

We postulate a canonical ensemble of systems at equilibrium at constant mass and temperature. The possible states are designated by the subscript i and each characterized by energy ε_i. The equilibrium description consists of a set of numbers f_i designating the fraction of ensemble members in the ith state. Any individual member is not static and is constantly changing state. Designate by t_{ij} the transition probability that a system in state i will change to state j in unit time. Then the principle of dynamic reversibility, or detailed balance, requires that at equilibrium

$$f_i t_{ij} = f_j t_{ji} \quad (13\text{-}27)$$

Next, consider the same ensemble which is in contact with an infinite isothermal reservoir but is irradiated with a constant flux of electromagnetic radiation such that there is a net absorption of radiation. The steady state will be characterized by a flow of heat to the reservoir equal to the flow of electromagnetic energy into the system.

When the ensemble finally achieves the steady state, it will be characterized by a new set of occupation numbers and transition probabilities f_i' and t_{ij}'. The new transition probabilities will include the initial type of random thermal transitions as well as radiation-induced transitions. We shall now show that there are some pairs of states for i and j for which the following relation cannot hold:

$$f_i' t_{ij}' = f_j' t_{ji}' \quad (13\text{-}28)$$

First, assume eq. (13-28) holds for all pairs of states. This means that, for every transition involving the absorption of radiation, a reverse transition will exist in which the system will radiate a photon. There will then be no net absorption of electromagnetic energy. This contradicts our assumption that the system absorbs radiant energy; hence, we must conclude that there are pairs of states for which the relation does not hold. In the steady state all the f_i' terms are time independent. Therefore, we note that

$$\frac{df_i'}{dt} = 0 = \sum_{ij} (f_i' t_{ij}' - f_j' t_{ji}^i) \qquad (13\text{-}29)$$

For some f_i' terms we have established that $f_i' t_{ij}' - f_j' t_{ji}'$ must be nonzero. Other pairs must then also be nonzero. That is, members of the ensemble must leave certain states by one path and return by other paths. Formally, this constitutes a cycle, and leads to a statement that we designate as the cycling theorem.

Theorem. *In steady-state systems, the flow of energy through the system from a source to a sink will lead to at least one cycle in the system.*

The above example is a special case, but it can be generalized by the realization that the flow of energy into the system, by any means, will always by accompanied by a transition $i \to j$ where $\varepsilon_j > \varepsilon_i$.

For the transitions involved in energy flow into the system, detailed balance cannot obtain because it is contrary to the net flowthrough of energy. The breakdown of detailed balance at any point leads to cycling in the steady state. The same argument could have been applied to those transitions whereby energy leaves the system and flows into the sink.

The flow of energy can be accompanied by the flow of matter. Thus, the steady-state flow of compounds of high chemical potential into a system and low chemical potential out of the system will also lead to cycling. If the induced states j represent a different distribution of chemical compounds than the i states, the cycling will represent a chemical flow of matter around the systems.

E. MODEL SYSTEMS WITH CYCLING BEHAVIOR

The theorem as derived is very general and refers to states of the system as quantum-mechanical states. There are no restrictions as to the complexity of the system. Thus, if our theorem were applied to a simple case such as pure argon gas, the cycling would be between various excited states of argon atoms. In order for the theorem to be biochemically interesting, the states must be chemically distinguishable. If the incoming photons cause electronic transitions leading to chemical rearrangements in the appropriate system of atoms, this condition will be met. Two simple examples are now presented.

Example 1. Consider a closed isothermal vat containing compounds A, B, and C with the isomerization reaction scheme as previously used. At equilibrium, eqs. (13-20) are applicable.

If we now irradiate the vat with monochromatic radiation of frequency ν such that the following photochemical reaction occurs,

$$A + h\nu \rightarrow B \tag{13-30}$$

then a cycle will take place and, in the steady state, there will be a net flow of material around the cycle from A to B to C to A.

Example 2. Next we postulate the following reaction scheme:

$$\begin{aligned} A + D &\rightleftharpoons B + E \\ B &\rightleftharpoons C \\ C &\rightleftharpoons A \end{aligned} \tag{13-31}$$

Consider the vat from Example 1 in contact, through a semipermeable membrane, with a reservoir of substances D and E at constant chemical potentials where $\mu_D > \mu_E$. The membrane is taken to be impermeable to A, B, and C. Material will now cycle around the system from A to B to C to A.

In the steady state, there will be a constant flow D into the system and a flow of E out of the system. A net amount of enthalpy, $\Delta H = H_E - H_D$, will be transferred to the intermediate system per mole of reaction. Since energy cannot accumulate in the steady state, there is a net flow of heat, $-\Delta H$, to the isothermal reservoir. Thus, high-energy compounds flow in and low-energy compounds and heat flow out; energy flow is maintained and cycles occur in the intermediate system.

Biology, as already noted, is characterized by cycles in systems which are approaching steady states of intermediate stability. The theorem discussed earlier indicates that such cycles are a general consequence of the physics of systems undergoing energy flux and need not be introduced as special biological phenomena.

It will be instructive at this point to examine Example 1 in some detail. First, this treatment will point up the difficulties that will be encountered in a brute force kinetic analysis of chemical networks of any degree of complexity. Second, a number of features of energy flow organization will emerge from this case. Third, this system represents the most primitive type of photosynthetic cycle and will be useful in making contact between theory and actual problems of terrestrial energy balance.

We start by representing our system on an energy level diagram and an energy plot as used in absolute reaction rate theory. This is shown in fig. 13-6. Next, consider an actual physical setup shown in fig. 13-7 consisting of a

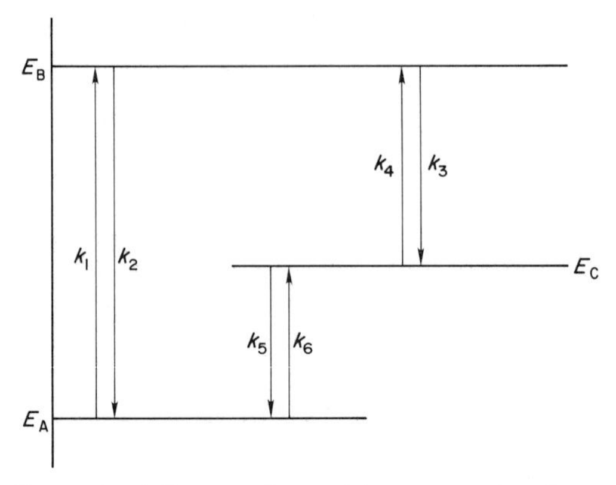

FIGURE 13-6. Energy level diagram of a model system used to demonstrate some of the fundamental features of a photosynthetic cycle. A, B, and C are interconvertible isomers.

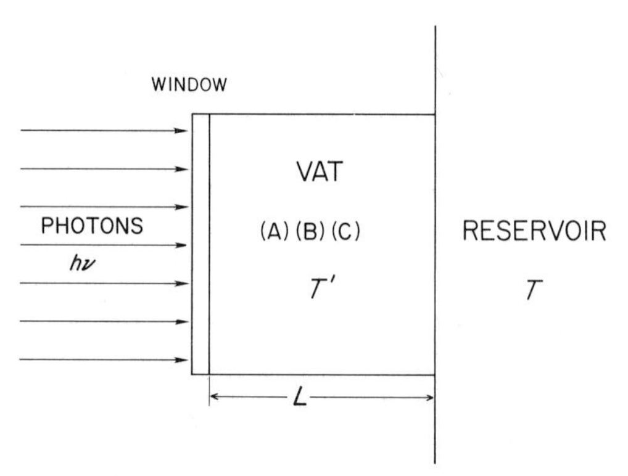

FIGURE 13-7. Diagram of the system used to demonstrate photosynthetic cycles. Electromagnetic energy enters through the window, photochemical reactions take place in the vat, and heat flows to the reservoir.

rectangular vat of length L in contact with an infinite isothermal reservoir of temperature T. One side of the vat is a window through which a flux of photons per unit area, J_p, is maintained. The wavelength of the photons is given by

$$h\nu = E_B - E_A \tag{13-32}$$

The absorption coefficient γ_A is finite, but γ_B and γ_C are zero; that is, no photon energy is absorbed by compounds B and C. Uniform mixing is maintained. The following equations describe the behavior of the system:

$$k_1[\mathrm{A}] + \frac{J_p}{L}[1 - \exp(-\gamma_{\mathrm{A}} L[\mathrm{A}])] - k_2[\mathrm{B}] = J_{\mathrm{AB}} \tag{13-33}$$

$$k_3[\mathrm{B}] - k_4[\mathrm{C}] = J_{\mathrm{BC}} \tag{13-34}$$

$$k_5[\mathrm{C}] - k_6[\mathrm{A}] = J_{\mathrm{CA}} \tag{13-35}$$

where $(J_p/L)[1 - \exp(-\gamma_{\mathrm{A}} L[\mathrm{A}])]$ is the number of photons absorbed per unit volume by the Beer–Lambert law and the J_{ij} terms are flows of material from isomer i to isomer j per unit volume. The quantum efficiency is taken as having a value of one. We also note that the flow of heat to the reservoir is

$$\mathfrak{T}(T' - T) = J_q \tag{13-36}$$

where T' is the steady-state temperature of the system, $\mathfrak{T}$ is the thermal transfer coefficient, and J_q is the energy flux per unit area. The total amount of material in the system is n_T. The steady state is characterized by a balanced flow around the network:

$$J_{\mathrm{AB}} = J_{\mathrm{BC}} = J_{\mathrm{CA}} \tag{13-37}$$

In addition an energy balance must obtain:

$$h\nu J_p[1 - \exp(-\gamma_{\mathrm{A}} L[\mathrm{A}])] = \mathfrak{T}(T' - T) \tag{13-38}$$

For the equilibrium condition, J_p is zero and J_{AB}, J_{BC}, and J_{CA} are all zero. This leads to eqs. (13-23)–(13-25).

In general, the existence of cycles means that the reaction rate constants are not all independent, since the reaction rates are functionally related to the equilibrium constants and the free-energy changes around a closed loop must add up to zero. The steady-state equations in the photon flux case become

$$k_1[\mathrm{A}] - (k_2 + k_3)[\mathrm{B}] + k_4[\mathrm{C}] = -\frac{J_p}{L}[1 - \exp(-\gamma_{\mathrm{A}} L[\mathrm{A}])] \tag{13-39}$$

$$k_6[\mathrm{A}] + k_3[\mathrm{B}] - (k_4 + k_5)[\mathrm{C}] = 0 \tag{13-40}$$

which along with eqs. (13-21) and (13-22) completely define the problem. Because of the exponentials, the set of equations cannot generally be solved analytically. In this regard we might note that the k's are themselves exponential functions of T. Two special cases can, however, be handled: (1) where $\gamma_{\mathrm{A}} L[\mathrm{A}] \gg 1$ and $J_p[1 - \exp(-\gamma_{\mathrm{A}} L[\mathrm{A}])] = J_p$—that is, when all the incoming radiant energy is absorbed; and (2) the case where $\gamma_A L[A] \ll 1$ and $J_p \gamma_A L[A]$ is absorbed.

For the first condition, we can solve directly for T', which becomes $T + (h\nu J_p/(\mathfrak{T})$. This value may then be substituted into the equations for the rate constants. We may then employ determinants and write down the solutions to eqs. (13-22), (13-39), and (13-40) in the form

$$[A] = \frac{\begin{vmatrix} -\dfrac{J_p}{L} & -(k_2 + k_3) & k_4 \\ 0 & k_3 & -(k_4 + k_5) \\ n_T & 1 & 1 \end{vmatrix}}{\begin{vmatrix} k_1 & -(k_2 + k_3) & k_4 \\ k_6 & k_3 & -(k_4 + k_5) \\ 1 & 1 & 1 \end{vmatrix}} \tag{13-41}$$

with related expressions for [B] and [C]. We can immediately write down the full solutions by evaluating the determinants. It can be shown that as J_p goes to zero, [A], [B], and [C] assume the equilibrium values shown in eqs. (13-23)–(13-25). Note that, as J_p increases, [A] decreases, [B] increases, and [C] may either increase or decrease, depending on whether $(k_6 - k_3)$ is positive or negative.

The flux of material around the cycle is

$$J_{AB} = \frac{J_p(k_3k_5 + k_3k_6 + k_4k_6)}{L(k_2k_4 + k_2k_5 + k_3k_5 + k_1k_4 + k_1k_5 + k_4k_6 + k_1k_3 + k_2k_6 + k_3k_6)} \tag{13-42}$$

a quantity directly proportional to J_p.

A series of experiments carried out by Groth and co-workers gives experimental evidence for the existence of cycles and shows how a relatively simple system exhibits behavior surprisingly like some features of the terrestrial ecosystem. They first investigated the primary photochemical dissociation of CO_2 and H_2O by light (mostly of wavelength 147.0 and 129.5 nm) from a xenon lamp and established the following primary reactions:

$$CO_2 + h\nu \rightarrow CO + O, \qquad H_2O + h\nu \rightarrow OH + H$$

When mixtures of H_2O and CO_2 were irradiated, formaldehyde and free oxygen were formed for which they proposed the reactions

$$H + CO \rightarrow HCO, \qquad 2HCO \rightarrow H_2CO + CO, \qquad O + O \rightarrow O_2$$

Aldehydes and oxygen will spontaneously react to yield carbon dioxide and water. The overall reaction scheme can then be represented as

$$CO_2 + H_2O \underset{}{\overset{h\nu}{\rightleftarrows}} \text{aldehydes and } O_2$$

This is, of course, a cycle in terms of our previous analysis. It is indeed a good physical representation of what is embodied in the cycling theorem. The model presented is an obvious oversimplification since more complex reaction products will form after prolonged irradiation.

What is perhaps most fascinating is that, in a very simple laboratory system, there emerge many of the significant features of biological photosynthesis and the ecological carbon cycle. That is, CO_2 is being photosynthetically fixed into a simple carbohydrate (formaldehyde) which is subsequently oxidized, yielding energy and returning the carbon to CO_2. Lastly, we note that this is an example of the energy flow principle; the flow of energy through the CO_2–H_2O system is leading to a degree of chemical organization, indicated qualitatively by the existence of more complex molecular species such as the aldehydes.

TRANSITION

Chemical kinetics usually starts with a set of differential equations describing rates of reactions in terms of concentrations or activities. These equations contain quantities called rate parameters. The evaluation of these parameters from first principles is called absolute reaction rate theory. While a satisfactory solution to this problem is not available, an approximate theory follows from considering the reaction pathway as a vibrational mode and postulating the existence and properties of an activated complex as an intermediate between reactants and products. Complex series of chemical reactions can be treated by reaction networks. In the steady state, cyclic flows occur in these networks and a series of theorems may be formulated about the cycles. These cyclic processes, in energy-driven steady states, show many similarities with global ecological processes. The existence of this theory provides a background for some of the more detailed ecological energetics that we will consider next.

BIBLIOGRAPHY

Benson, S. W. "Thermochemical Kinetics." Wiley, New York 1968.

This book covers methods of handling kinetic data and estimating thermochemical data and rate parameters.

Daniels, F. (ed.), "Photochemistry in the Liquid and Solid States." Wiley, New York, 1960.

This book contains a number of research articles including the studies by W. Groth discussed in this chapter.

Glasstone, S., Laidler, K. J., and Eyring, H. "The Theory of Rate Processes." McGraw-Hill, New York, 1941.

This monograph is the first book on this general approach. It explores the theory in great detail.

Onsager, L., *Phys. Rev.* **37**, 405 (1931); **38**, 2265 (1931).

These two classic papers discuss the principle of detailed balance, the underlying theory, and some of the consequences.

14 Ecological Energetics

A. BIOCHEMICAL OUTLINES

In Chapter 12 we reviewed data and established computational mechanisms which allow us to deal with the concept of free energy of formation of living material relative to some standard state of the starting material. In this chapter it is our purpose to apply this concept to certain problems in cell division, predator–prey relations, and trophic ecology. Our aim is to combine the insights of thermodynamics with those of biochemistry and trophic ecology in order to examine the relationships between energy flows and biological processes.

In order to proceed as outlined, we must first examine certain aspects of biochemistry which will underlie our discussion. This examination is based in large part on the monograph of Krebs and Kornberg (see Bibliography) which reviews in detail the classical biochemistry of energy transformations in living matter.

The first principle that we are concerned with is the ubiquity of energy mechanisms in biology. There is a general scheme, parts of which are common to all organisms, which is followed in the combustion of foodstuffs to produce energy that is utilizable in biological processes. In the overwhelming majority of cases, the energy produced in the breakdown of biomass is stored as chemical energy in the pyrophosphate bonds of adenosine triphosphate (ATP). In the words of Krebs and Kornberg,

> The first major stage, then, of the energy transformations in living matter culminates in the synthesis of pyrophosphate bonds of ATP, at

the expense of the free energy of the degradation of foodstuffs. The overall thermodynamic efficiency of this process is estimated at 60–70%. [Current estimates are considerably lower.] This is high when compared with the efficiency of man-made machines depending on the burning of a fuel as a source of energy. The greater efficiency is possible because living matter is not a heat engine, but a 'chemical' engine organized in a special manner.

The transformation of the potential chemical energy of biomass to a utilizable form occurs in three phases, as outlined in table 14-1. In the first phase, digestive enzymes hydrolyze the macromolecules down to their monomeric constituents. In the second phase, the released monomers are partially oxidized to one of three substances: acetic acid in the form of acetyl coenzyme A, α-ketoglutarate, and oxaloacetate. These three compounds all funnel into a common reaction pathway. the tricarboxylic acid cycle (fig. 14-1). In a complete cycle the material is further oxidized to two carbon dioxide molecules and four pairs of hydrogen atoms, which are subsequently oxidized by O_2 with a concomitant release of energy. This last process, oxidative phosphorylation, leads to the production of ATP.

TABLE 14-1
The Three Main Phases of Energy Production from Foodstuffs

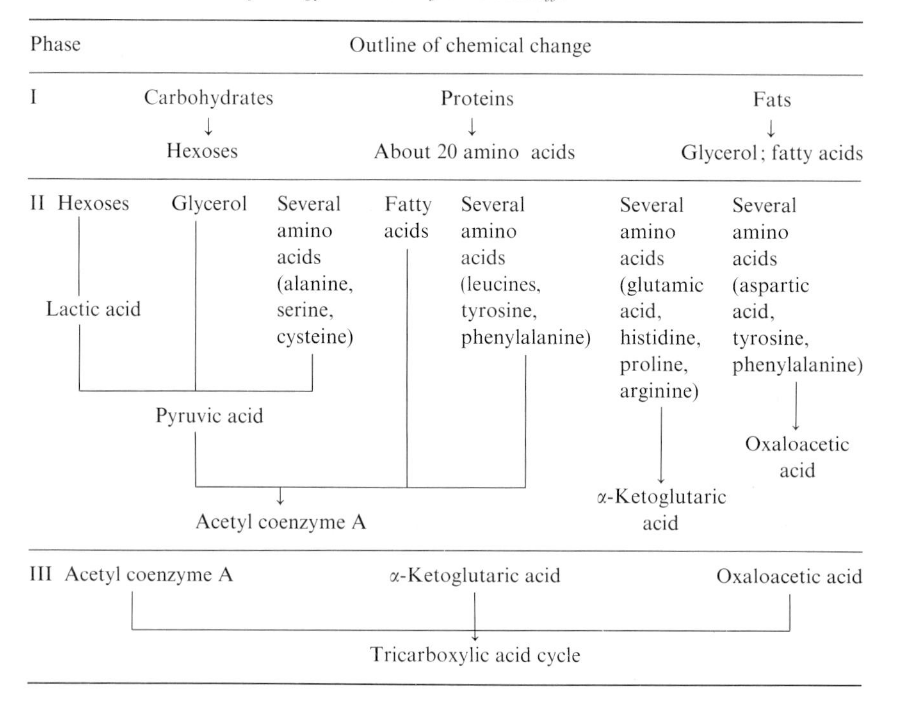

Phase	Outline of chemical change
I	Carbohydrates → Hexoses; Proteins → About 20 amino acids; Fats → Glycerol; fatty acids
II	Hexoses → Lactic acid → Pyruvic acid; Glycerol → Pyruvic acid; Several amino acids (alanine, serine, cysteine) → Pyruvic acid
	Pyruvic acid; Fatty acids; Several amino acids (leucines, tyrosine, phenylalanine) → Acetyl coenzyme A
	Several amino acids (glutamic acid, histidine, proline, arginine) → α-Ketoglutaric acid
	Several amino acids (aspartic acid, tyrosine, phenylalanine) → Oxaloacetic acid
III	Acetyl coenzyme A; α-Ketoglutaric acid; Oxaloacetic acid → Tricarboxylic acid cycle

The universality of the schemes of table 14-1 and fig. 14-1 underlies trophic dynamics. It is the common denominator of food chains. The primary requirement of a predator is that it can extract energy from its prey. The existence of a trophic network requires that common energy-yielding molecules connect all points in the network through which energy must flow. The existence of a small and ubiquitous set of energy-yielding reactions, as well as the ubiquity of macromolecular constituents, guarantees the possibility of this kind of interrelatedness. If this were not so, there could be a series of nonbiochemically interacting ecological systems inhabiting the same niches. Noninteraction is, of course, contrary to experience and imposes a good deal of biochemical overlap as a *sine qua non* of food chains and other aspects of trophic ecology.

A number of other energy-yielding reactions are possible for special fermentations, but they all are related to table 14-1 through the utilization of lactate or pyruvate. In many organisms the pentose-phosphate cycle represents an alternative energy-yielding pathway for glucose. It seems likely that there are other energy pathways yet to be discovered, but it also seems most probable that they will closely relate to present ones and will not alter our concepts of the major energy-yielding pathways.

B. HEATS OF COMBUSTION

We now turn to the question of the actual energy levels of living systems. In Chapter 12 we calculated that our theoretical biomass plus O_2 was 22.17 kJ/g higher in energy and 1.761 J/g/K lower in entropy than the

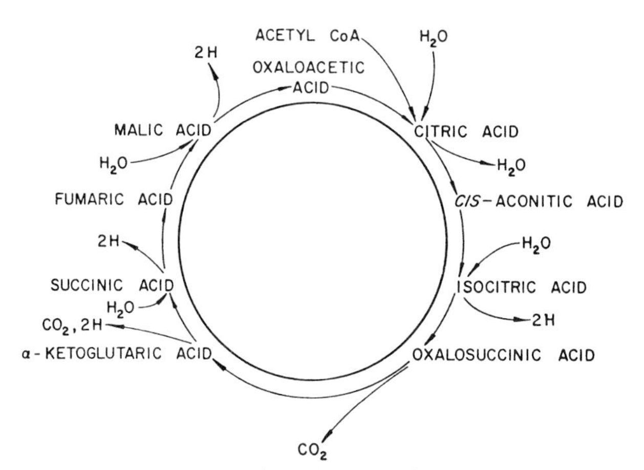

FIGURE 14-1. The tricarboxylic acid cycle. One of the ubiquitous systems for energy conversion in living organisms.

starting materials (from table 12-13, assuming V equals 1 liter.) If we look at the reverse of the synthesis in eq. (12-1), we note that it corresponds to the usual bomb calorimeter oxidation of biomass to CO_2, H_2O, and N_2. Heats of combustion of biological materials should then provide a check of the validity of the enthalpy calculated in Chapter 12.

Experimental data are available on a number of plant and animal systems and some are presented in tables 14-2 and 14-3. The data provide support for the essential validity of the calculations just outlined.

The range of values makes it clear that we need to reformulate the concept of biomass with a more detailed view that reflects the differences between organisms. A study on birds enables us to see clearly the dependence of heats of formation on the gross chemical composition. The heats of combustion of birds having widely different fat contents varies with the migratory season in which they were taken; the results are shown in table 14-4.

Since the heats of formation depend, as would be expected, on the composition, a formal representation can be made in the following way:

$$\Delta H = \Delta H_{\text{fat}} f_{\text{fat}} + \Delta H_{\text{protein}} f_{\text{protein}} + \Delta H_{\text{carbohydrate}} f_{\text{carbohydrate}} \tag{14-1}$$

$$f_{\text{fat}} + f_{\text{protein}} + f_{\text{carbohydrate}} = 1 \tag{14-2}$$

TABLE 14-2
Heats of Combustion of Animal Material[a]

Organism	Species	Heat of combustion (kJ/ash-free g)
Ciliate	*Tetrahymena pyriformis*	−24.84
Hydra	*Hydra littoralis*	−25.24
Green hydra	*Chlorohydra viridissima*	−23.97
Flatworm	*Dugesia tigrina*	−26.30
Terrestrial flatworm	*Bipalium kewense*	−23.78
Aquatic snail	*Succinea ovalis*	−22.65
Brachiopod	*Gottidia pyramidata*	−18.39
Brine shrimp	*Artemia* sp. (nauplii)	−28.18
Cladocera	*Leptodora kindtii*	−23.45
Copepod	*Calanus helgolandicus*	−22.59
Copepod	*Trigriopus californicus*	−23.07
Caddis fly	*Pycnopsyche lepido*	−23.79
Caddis fly	*Pycnopsyche guttifer*	−23.87
Spit bug	*Philenus leucopthalmus*	−29.12
Mite	*Tyroglyphus lintneri*	−24.30
Beetle	*Tenebrio molitor*	−26.41
Guppie	*Lebistes reticulatus*	−24.36

[a] From Slobodkin and Richmond, *Nature* **191**, 299 (1961).

TABLE 14-3
Energy Values in an Andropogon virginicus Old-Field Community in Georgia[a]

Component	Energy value (kJ/ash-free g)
Green grass	−18.29
Standing dead vegetation	−17.94
Litter	−17.31
Roots	−17.43
Green herbs	−17.94
Average	−17.78

[a] From Golley, *Ecology* **42**, 581 (1961).

TABLE 14-4
Heats of Combustion of Migratory and Nonmigratory Birds[a]

Sample	Ash-free material (kJ/g)	Fat ratio (% dry weight as fat)
Fall birds	−33.81	71.7
Spring birds	−29.46	44.1
Nonmigrants	−26.19	21.2
Extracted bird fat	−37.78	100.0
Fat extracted: fall birds	−22.87	0.0
spring birds	−22.64	0.0
nonmigrants	−22.76	0.0

[a] From Odum, Marshall, and Marples, *Ecology* **46**, 901 (1965).

where ΔH represents the individual enthalpies of formation and f is the fraction of the dry biomass of a particular kind. Similar linear equations can be written for the entropies of formation.

Heats of combustion of biological materials can be calculated from the data in Chapter 12 and are, in many cases, available experimentally. Selected data are presented in table 14-5.

We can calculate the heat of combustion of a model protein system. This yields 24.35 kJ/g dry weight compared with the range of values 20.92–24.27 presented in table 14-5. The heat of combustion of pure glucose is 15.64 kJ/g, which is a characteristic value for carbohydrates. A maximum heat of combustion for fats is calculated for glycerol tristearate and found to be 46.73 kJ/g. If one of the stearates is replaced by a phosphate, the calculated

TABLE 14-5
Heat of Combustion of Components of Biomass

Material	ΔH protein (kJ/g)	ΔH fat (kJ/g)	ΔH carbohydrate (kJ/g)
Eggs	−24.06	−39.75	−15.69
Gelatin	−22.05	−39.75	—
Glycogen	—	—	−17.53
Meat, fish	−23.64	39.75	—
Milk	−23.64	−38.70	−16.53
Fruits	−21.76	−38.91	−16.74
Grain	−24.27	−38.91	−17.57
Sucrose	—	—	−16.53
Glucose	—	—	−15.69
Mushroom	−20.92	−38.91	−17.15
Yeast	−20.92	−38.91	−17.57

value is 38.62 kJ/g. The values, based on the chemistry of pure compounds, are thus in reasonable agreement with the combustion experiments on biological material. If we refer all of our data to N_2, CO_2, H_2O, H_2SO_4, and H_3PO_4 as the ground state, then the heats of combustion will be the same as the heats of formation. From an examination of table 12-13 it is clear that, for the overall process, the ΔH term (enthalpy) is the dominant one in ΔG (Gibbs free energy). For the present overall examination of trophic energy relations, we will consider ΔG as approximated by ΔH. The procedure of approximating the Gibbs free energy by the enthalpy is, in general, a risky one and care must be taken not to permanently forget the $T\Delta S$ term. In much of the literature of ecological energetics, the heat of combustion is taken as the sole thermodynamic parameter necessary to characterize a system. Thus, eqs. (14-1) and (14-2) establish a measure of energy of formation as a function of overall composition. From the data presented we can assume the following values:

$$\Delta H_{\text{protein}} = -23.01 \quad \text{kJ/g}$$
$$\Delta H_{\text{fat}} = -38.91 \quad \text{kJ/g}$$
$$\Delta H_{\text{carbohydrate}} = -17.15 \quad \text{kJ/g}$$

It can be seen from tables 14-3 to 14-5 that the primary producers have a lower free energy of formation than the upper trophic levels. Therefore, there are strong energy constraints in moving material up the trophic pyramid and, of necessity, much of the material in the primary trophic level must be degraded to raise the rest of the material to an appropriate free energy. Thus, there are important energetic considerations to trophic efficiencies.

It is useful to represent graphically the free energy as a function of composition of biomass. To do this, we construct a graph in which the abscissa is protein fraction (f_{protein}) and the ordinate is fat fraction (f_{fat}). Since the overall system is represented by eq. (14-2) a point on this graph uniquely represents a possible composition. The free energy of formation is given by

$$\Delta G = -23.01 f_{\text{protein}} - 38.81 f_{\text{fat}} - 17.15(1 - f_{\text{protein}} - f_{\text{fat}}) \qquad (14\text{-}3)$$

Isoenergetic loci are straight lines in this representation; fig. 14-2 is a graph of this type. The primary trophic level occupies a small energy triangle in the lower left-hand corner of the graph. Most animals have relatively little carbohydrate and lie on the graph in a zone along the zero carbohydrate line. This emphasizes the major jump in moving material from primary producers into animal biomass.

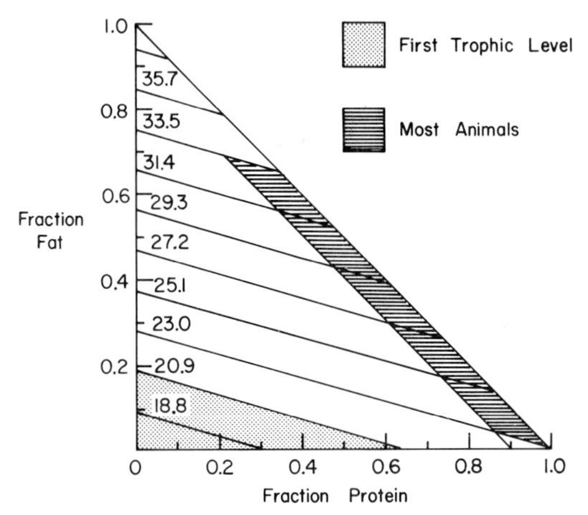

FIGURE 14-2. Relation between macromolecular composition and heat of combustion of living organisms. First trophic level is represented in the lower left-hand corner of the graph. Most animal species have values in the lined area.

C. TROPHIC LEVELS

We are now in a position to examine trophic transformations from the point of view of thermodynamics and to see where the analysis in these terms limits or better defines the problem. Figure 14-3 is a flow chart representing the major processes in the transfer of biomass between trophic levels. We have specifically included microorganisms in our scheme, since the digestive tract of almost all metazoa contains microorganisms which play a significant role in the metabolism and energy transformations of the organism. Since the microorganisms live inside the host, it is easier to consider

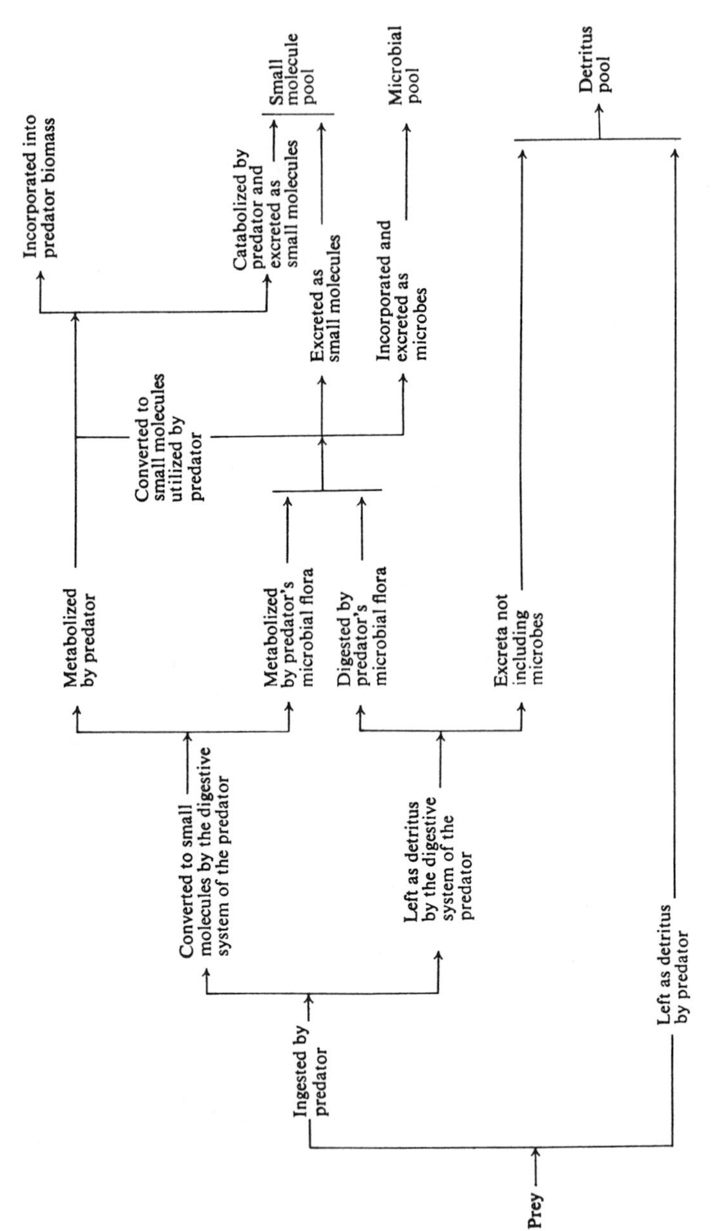

FIGURE 14-3. Flow chart showing the biomass conversions between predator, prey, microorganisms, and detritus.

them in a general scheme than to try to formulate two independent interacting energy budgets. Three rules govern each step in the flow diagram of fig. 14-3:

(1) Mass is conserved. Specifically the number of atoms of each type is conserved.
(2) Energy is conserved.
(3) Entropy increases.

The ingestion step establishes the amount of matter and energy entering the system. The ingested material may be either parts of an organism from a lower trophic level or detritus. ("Detritus" is used here to mean products of living organisms, including their remains after death.) The energy entering the system must appear either as heat, as work in the most general sense, or as chemical potential energy of biomass or products.

In order to enter the metabolic processes of either the predator or the predator's microbes, the ingested material must first be split into relatively small molecules. This may be stated as a general rule of ecology: *The utilization of prey biomass by a predator involves its degradation into its monomeric constituents* (empirical generalization). This is a key point of contact between biochemistry, ecology, and thermodynamics. There are two reasons for the degradation to monomers. First, the ubiquitous energy utilization mechanism already discussed employs small molecules. Second, the synthesis of macromolecules in living systems always proceeds through monomers. Since many macromolecules are information-specific and genetically determined, it is not possible to utilize them directly; they must be degraded to common components. This process involves a heat release of about 68.6 J/g and an entropy increase of 0.877 J/g/K, corresponding to a decrease in Gibbs free energy of 330 J/g. There are no processes by which to couple this free energy release to do useful work, although it may contribute to temperature maintenance in the predator. The enthalpy change is eventually transferred to heat in the environment.

One important exception to the breakdown and utilization rule should be noted. This is the phenomenon of genetic transformation by deoxyribonucleic acid (DNA). From experiments on bacterial transformation, it seems clear that large pieces of DNA are taken into the cell intact and whole cistrons or groups of cistrons are then incorporated into the cell's genome. This process generally requires close homology between the donor and receptor DNA. The total mass of material transferred in this way in natural habitats is probably negligibly small so that, while transformation may be of great genetic and evolutionary significance, it is a very minor pathway in trophic ecology.

D. ENERGY UTILIZATION

The small molecules produced by the digestive enzymes of the predator or its bacteria now enter the usual pathways of intermediary metabolism where they either are degraded to yield energy and are subsequently excreted, or are incorporated into predator macromolecules. The energy produced is used for three basic processes: maintenance, synthesis, and mechanical energy.

1. Maintenance. As already noted, a nonequilibrium system isolated from energy sources slowly decays toward equilibrium. Although thermodynamics does not provide a time scale, such a system moves toward more disordered states, and a constant expenditure of work is required to restore the system to its proper state. Two examples will illustrate this point. In a living organism the macromolecules, such as proteins, are relatively unstable and are constantly undergoing thermal denaturation. In order to renature the proteins, they must be digested to amino acids and resynthesized. This process takes energy and, consequently, work must constantly be done to maintain the "static" molecular structure of a living organism. For a second example, suppose the functioning state of an organism requires a concentration difference of ions across a biological membrane. The concentration gradient is originally produced by the active transport of ions. In general, there will be a small leakage of ions back across the membrane from the high chemical potential side to the low. In order to maintain the state of the system, work must constantly be done to pump ions back to the high side. If the leak rate is J_i moles of ion per second and the chemical potentials are μ_H and μ_L on the high and low sides, respectively, then the energy must be supplied at a rate exceeding $J_i(\mu_H - \mu_L)$ in order to maintain the system. Maintenance energy must thus be used to combat the spontaneous entropy production which characterizes nonequilibrium systems. In addition, simply maintaining an aqueous system such as a terrestrial organism requires work to combat the entropy-generating processes of evaporation and diffusion.

Any attempts to analyze maintenance energy from a physical point of view rest heavily on the physical and chemical details of the system being studied. One aspect, however, the thermal denaturation of macromolecules, appears to be susceptible to some generalizations. Among the quantitatively most plentiful macromolecules, proteins appear to be most sensitive to thermal effects. A denatured protein must be completely degraded to amino acids by the organism before reuse. This constant breakdown of proteins leads to substantial amino acid turnover in mammalian systems. A 70-kg adult man synthesizes and degrades about 70 g of protein nitrogen per day. Since the total nitrogen content of such an individual is about 900 g, this indicates a protein turnover of about 7.7% per day. All of this protein turnover does not result from protein denaturation; cell death and autolysis by proteolytic enzymes also contribute to the turnover. Part of the basal metabolism contributes energy to the resynthesis of this protein.

Since proteins are in one sense a fairly uniform set of molecules, all being polymers of the same twenty α-amino acids (more properly nineteen amino acids plus proline which is an imino acid) linked by peptide bonds, one might anticipate common features in the thermal stability. The kinetics of most protein inactivation processes are exponential, following an equation of the form

$$N = N_0 \exp(-k_1 t) \tag{14-4}$$

where N is the amount of active or native protein, N_0 is the original amount of active material, t is time, and k_1 is a rate constant which may be expressed in terms of the activation enthalpy and entropy as discussed in Chapter 13. Consequently, data on thermal inactivation are generally expressed in terms of $\Delta H^{\ddagger}$ and $\Delta S^{\ddagger}$ (table 14-6).

We average the values from known proteins and obtain a $\Delta H^{\ddagger}$ of 343,000 J/mole and a $\Delta S^{\ddagger}$ of 686.6 J/K/mole. Averaging the energy functions is equivalent to taking the geometric mean of the rate values. If we assume that this average holds as an appropriate average for the protein of biomass, we can calculate the necessary replacement rate as a function of temperature. The values are given in table 14-7 and shown in fig. 14-4. Organisms utilizing this class of proteins would have difficulty in homeostasis at temperatures much above 318 K because the recycling of protein would impose a heavy burden on the metabolic system. The exponential character of this type of curve makes it clear that there must be a sharp cutoff in the temperature at which organisms can maintain homeostasis. Figure 14-4 contains a number of representative temperatures of animals which possess some temperature regulation. It appears from these data that animals do regulate at temperatures lower than those at which they are menaced by the high rate of protein denaturation.

There are of course thermophilic organisms, in particular a number of microbial species grow at temperatures up to 338 K. These organisms replicate at a sufficiently rapid rate that 24 h is not an appropriate time scale for considering protein replacement. It has been argued that thermophiles synthesize proteins which are inherently more heat stable than the usual class of proteins from mesophiles. It is not easy to see what features in protein structure would confer this heat resistance. Such a class of proteins would lead to a curve in fig. 14-4 which would be displaced to the right. The curve shape would, however, be the same and our reasoning would

TABLE 14-6
Thermal Inactivation of Proteins and Related Systems in Aqueous Solution

Substance	$\Delta H^\ddagger$ (kJ/mole)	$\Delta S^\ddagger$ (J/mole/K)
Insulin	149.0	99.6
Trypsin	168.2	187.0
Catalase	211.3	339.0
Pepsin	232.6	474.0
Peroxidase (milk)	775.3	1949.7
Enterokinase	176.4	220.9
Trypsin kinase	185.2	241.0
Proteinase (pancreatic)	158.4	169.9
Amylase malt	174.1	218.8
Emulsin	187.9	273.2
Leucosin	352.7	774.0
Hemoglobin	316.3	638.9
Hemolysin (goat)	828.4	2246.8
Vibriolysin	535.6	1364.0
Tetanolysin	722.4	1920.5
Rennen	373.6	870.3
Egg albumin, *p*H 7.7	561.9	1326.7
*p*H 3.4	405.0	936.0
*p*H 1.35	147.3	151.9
Yeast invertase, *p*H 5.7	219.2	354.3
*p*H 5.2	361.5	774.0
*p*H 4.0	461.9	1098.3
*p*H 3.0	311.3	637.6
Inactivation of tobacco mosaic virus	167.4	75.3
Inactivation of T_1 bacteriophage	397.5	866.1
Average	375.4	819.2

TABLE 14-7
Thermal Denaturation of Average Protein (Biomass) as a Function of Temperature

Temperature (K)	Rate (s^{-1})	Percentage replaced in 24 h	Temperature (K)	Rate (s^{-1})	Percentage replaced in 24 h
310	1.25×10^{-7}	1.08	317	2.37×10^{-6}	20.50
311	1.89×10^{-7}	1.63	318	3.52×10^{-6}	28.65
312	3.70×10^{-7}	3.20	319	5.35×10^{-6}	46.20
313	5.12×10^{-7}	4.42	320	7.92×10^{-6}	68.40
314	6.62×10^{-7}	5.71	321	1.35×10^{-5}	116.64
315	1.05×10^{-6}	9.08	322	1.81×10^{-5}	161.51
316	1.50×10^{-6}	13.81	323	2.71×10^{-5}	234.14

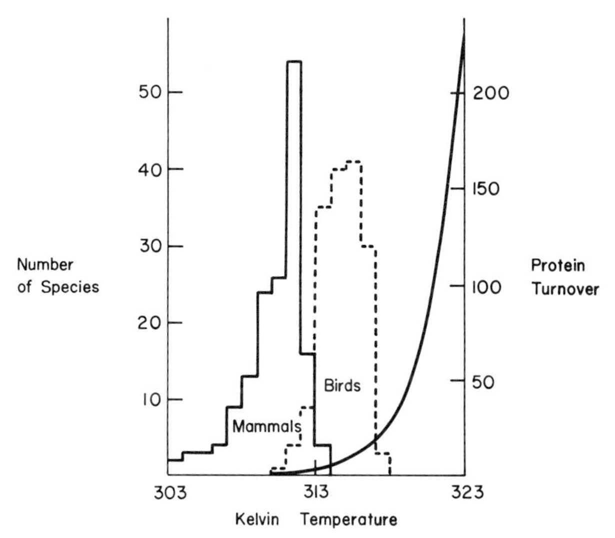

FIGURE 14-4. Graph of temperature homeostasis versus the number of species that regulate at that temperature. Also shown are protein turnover rates as a function of temperature. Most species regulate at temperatures where the thermal decay of proteins is small.

hold that large organisms must exist in a temperature range in which the protein replacement rate can be satisfied without burdening the energy production system.

2. *Synthesis.* Since the breakdown of macromolecules to smaller structures releases energy, energy must be supplied in the reverse process. This is the irreducible minimum loss in the transfer of biomass between trophic levels. We can use our previously developed information to calculate what this loss must be. Assume for the moment that the chemical composition of predator and prey are the same and correspond to that of the biomass in table 12-2. We then will require 330 J/g to synthesize macromolecules. This must come from the oxidation of some of the ingested material. From tables 12-6 and 12-8 and using Case III, we can calculate that the free-energy change on oxidation to CO_2, H_2O, and NH_3 is 18.95 kJ/g. As previously noted in this chapter, such energy can be converted to phosphate-bond energy with an average efficiency of about 65% so that the available energy is 12.14 kJ/g. Thus, 0.023 g of ingested material must be used to obtain the energy for synthesis of a gram of macromolecule. Actually, the coupling of phosphate-bond energy to synthesis is not 100% efficient as would be indicated by this calculation. If we examine the actual synthetic pathways, we note that, in fact, two phosphate bonds are hydrolyzed in the synthesis of each macromolecular linkage.

The free-energy change in going from adenosine triphosphate to adenosine monophosphate and two orthophosphates is 59 kJ/mole. Since 1 g of biomass involves 82.2×10^{-4} moles of polymeric linkages, 481 J/g are required. This necessitates the oxidation of 0.038 g of biomass per gram of material synthesized. The macromolecular synthetic processes are remarkably efficient in utilizing energy for the transformation of biomass from one species to another.

In the previous paragraph we have assumed that the predator and its food had the same molecular composition. Usually they will not be exactly the same and energy will have to be expended in converting the small molecules into their appropriate forms. Frequently the predator biomass may be at a higher free energy than the food and large quantities of nutrient must be oxidized to raise the energy of the associated incorporated material. Referring to tables 14-2 and 14-3, we note that, in general, the first trophic level supplies food about 16.7 kJ/g above ground level, where upper trophic levels are at an energy of about 23 kJ/g or higher. Such a system can, at most, be 77% efficient in transfer between trophic levels. A second restriction enters for such systems. The lowest trophic level is also low in nitrogen so that ingesting sufficient nitrogen for growth can limit the efficiency rather than energy limitations. We note that man is 5.14% nitrogen while alfalfa is 0.825% nitrogen. If we were to convert alfalfa biomass into human biomass at maximum efficiency, we would require (5.14/0.825) or 6.22 g dry weight of alfalfa per gram dry weight of human biomass, which is an efficiency of conversion of 16%. In the adult stage of animals, the problem of nitrogen balance changes and animal ingests excess nitrogen compounds in the process of obtaining sufficient maintenance energy. Since no method of storing nitrogen exists, various pathways of nitrogen excretion via such compounds as uric acid, urea, and amino acids are used. This factor greatly limits the maximum possible efficiency of total mass transfer.

3. Mechanical Energy. Most animals expend some energy in the mechanical motion associated with obtaining food, avoiding predators, locating suitable environments, and procreating. The processes are highly species-specific and will not, at the moment, be subject to any thermodynamic generalizations. It should be noted, however, that this category of energy cannot be sharply distinguished from maintenance energy. Moving into the warm sun to acquire heat is as much a part of the maintenance energy as obtaining the thermal energy by oxidation of carbohydrates.

E. ENERGY FLOW

Although we have discussed at some length the flow of energy as a function of the detailed chemical composition of the organisms involved, a generalization emerges which may simplify the bookkeeping procedures. If heats of combustion refer to the carbon content of biological materials, rather than to the total mass, then the energy available per gram of carbon is approximately a constant, independent of the material. This assumes that the major components are protein, carbohydrate, lignin, and lipid, and is shown in table 14-8. As a consequence, unless the highest degree of accuracy is needed, carbon and energy flows may be used interchangeably. Given the spread of the data, a value of 42 kJ/g will be more than adequate in precision.

It is worth concentrating for a moment on the nature of the constancy of the heat of combustion per unit of carbon. This is a biological generalization and we do not have firm underlying reasons as to why evolutionary processes led to a series of compounds which exist on a more or less isoenergetic surface with respect to carbon. Nonetheless, any constant is intriguing in a field that exhibits such diversity in so many aspects. The range of possible values is quite large. Since CO_2 has a heat combustion of zero, pure carbon in its various forms shows values around 33.5 kJ/g and methane is 73.6 kJ/g carbon. Almost all biological material, however, falls in the range of 37–46 kJ/g of carbon. One possible suggestion for the uniformity in heats of combustion comes from the fact that most of the carbon in biological macromolecules is involved in C–C bonds (245.2 kJ/mole), C–N bonds (203.3 kJ/mole), and C–H bonds (365.3 kJ/mole). The oxidation products CO_2, H_2O, and N_2 are such that a fairly uniform heat of oxidation results. This idea needs to be explored in greater detail.

We can now focus on the limitations of the energy flow model in ecology. While it is a useful bookkeeping device, it does not provide us with additional information on the structure and stability of ecosystems. The failure of thermodynamic theories to provide structural information is a general feature of that science. Detail always enters through the equation of state or some accessory phenomenological equation. Given these experimental results, classical thermodynamics is able to impose limitations on the state of the system. Traditional thermodynamics is also able to give information about stability considerations of the various free-energy minima. We have

TABLE 14-8
Available Energy per Unit of Carbon

Material	ΔH (kJ/g)	Carbon content	ΔH (kJ/g carbon)
Protein	−23.01	51.4%	−44.77
Fat	−38.91	87.3%	−44.56
Carbohydrate	−17.15	44.4%	−38.63
Lignin	−26.28	63.6%	−41.30

at the moment no such general principle for far-from-equilibrium flow systems, so that we are not able to go directly from an energy flow model of ecology to a consideration of the response to perturbations. We can use the compartmentation model of fig. 12-2 or 12-3 and develop ideas of stability from intercompartmental flow relations, but these arise from the kinetics rather than from any special feature of the energetics.

TRANSITION

The pathways of biochemistry provide a unifying framework for the flow of matter and energy between trophic levels. The measured heats of combustion of organisms are in agreement with the thermodynamic parameters calculated in Chapter 12. Trophic energy transfer can thus be viewed in a general framework. Energy utilization is subject to more detailed analysis in terms of maintenance, synthesis, and expenditures of mechanical energy. We are thus able to make contact between thermodynamics and the study of energy flow in ecology.

So far we have dealt with rather general considerations of radiant and chemical energy. Other terms in the internal energy such as charge transfer, surface energy, and osmotic work are of major importance in biology. In the next chapter we present some approaches used to deal with these topics.

BILIOGRAPHY

Krebs, H. A., and Kornberg, H. L., "Energy Transformations in Living Matter." Springer, New York, 1957.

A detailed review of the relation of intermediary metabolism to bioenergetics.

Lehninger, A. L., "Biochemistry." Worth Publ., New York, 1975.

This very extensive textbook of biochemistry details the metabolic pathways involved in bioenergetics and discusses various aspects of the subject.

Morowitz, H. J., "Energy Flow in Biology," Academic Press, New York, 1968.

Much of this chapter comes from Chapter IV of this work.

Slobodkin, L. B., "Growth and Regulation of Animal Populations." Holt, New York, 1961.

Chapter 12 discusses the efficiency of predator-prey energy conversions.

Watt, B. K., and Merrill, A. L., "Composition of Foods." U.S. Dept. of Agriculture, Washington, D.C., 1963.

Contains extensive data on heats of combustion of a wide variety of biological materials.

15 *Applications of the Gibbs Free Energy*

The concepts of the Gibbs free energy, chemical potential, and activity provide a framework for the solution of a number of problems of biological and biochemical interest. We will choose a few selected examples to indicate the methods of approach and the kinds of problems that can be handled.

A. ELECTROCHEMISTRY

A wide variety of chemical reactions can be carried out in such a manner so as to link the reaction to the flow of electricity in an external circuit. The setup is shown in fig. 15-1. The potentiometer acts as a variable voltage source and provides a voltage equal to that across the electrodes of the reaction vessels, so that there is no actual flow of current through the circuit at equilibrium. A slight increase of potentiometer voltage would cause current to flow in one direction through the reaction vessel and a slight decrease would cause current to flow in the other direction. Similarly, the linked chemical reaction going from reactants to products with the current flow in one direction will go from products to reactants if the current is reversed. This is very close to an idealized reversible situation and the ability to measure voltage with a high degree of precision makes this a choice method of obtaining thermodynamic data. The appropriate form of the free-energy function for this situation [see Chapter 10, eq. (10-14)] is

$$dG = -S\,dT + V\,dP + \sum \mu_i\,dn_i + \mathscr{E}\,de \qquad (15\text{-}1)$$

The conditions we wish to apply are constant temperature and constant pressure. As previously indicated, for stoichiometric reactions the dn_i are

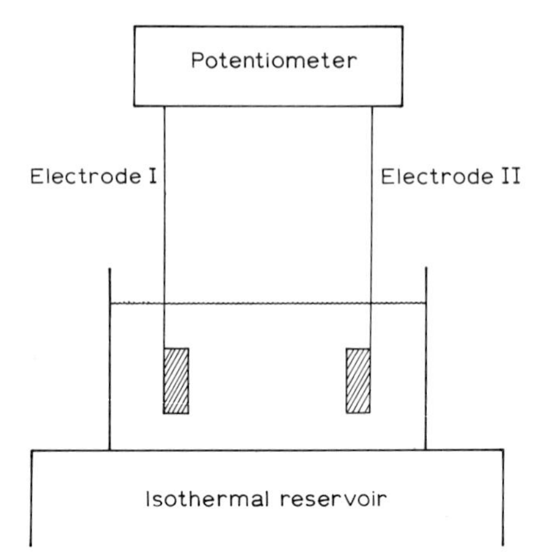

FIGURE 15-1. Experimental setup for studying electrochemical reactions.

all related by

$$dn_i = \nu_i \, d\xi \tag{15-2}$$

where the ν_i are the stoichiometric coefficients and ξ is the degree of advancement of the reaction. As written above, the ν_i are positive for products and negative for reactants. The flow of current is also linked to the degree of advancement, since electrode reactions are part of the process being described. Hence we can write

$$de = \nu_e \mathscr{F} \, d\xi \tag{15-3}$$

where ν_e is the number of moles of electrons flowing in the external circuit per unit advancement and $\mathscr{F}$ is the faraday, the number of coulombs of charge per mole of electrons. Choosing ν_e as positive leads to $\mathscr{E}$ values that increase with the tendency of the reaction to go in the forward direction (see fig. 15-2). Equations (15-2) and (15-3) substituted into eq. (15-1) with the conditions of equilibrium ($dG = 0$) under isothermal ($dT = 0$) and isobaric ($dP = 0$)

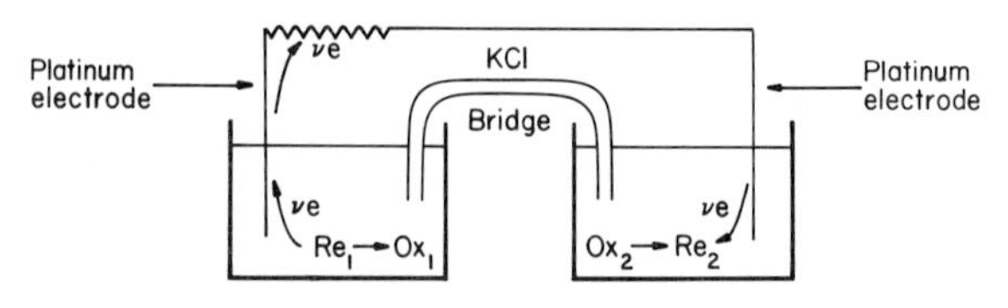

FIGURE 15-2. An oxidation–reduction reaction resolved into two half-cell reactions. For each unit of advancement of the reaction, ν moles of electrons flow through the external circuit.

constraints yields

$$dG = \sum_i (\mu_i \nu_i + \nu_e \mathscr{E} \mathscr{F}) \, d\xi = 0 \tag{15-4}$$

Equation (15-4) can be met only if the coefficient of $d\xi$ vanishes:

$$\sum \mu_i \nu_i + \nu_e \mathscr{E} \mathscr{F} = 0 \tag{15-5}$$

Equation (15-5) is the fundamental equation of electrochemistry. It can be placed in a somewhat different form by utilizing the activity and replacing μ_i by $\mu_{i0} + RT \ln a_i$. This leads to

$$\sum \nu_i \mu_{i0} + RT \ln \prod_i a_i^{\nu_i} = -\mathscr{E} \nu_e \mathscr{F} \tag{15-6}$$

The first term on the left is ΔG_0, the standard free energy of the reaction, and we proceed to define a quantity $\mathscr{E}_0$ which depends only on the reaction and is independent of the reaction conditions:

$$\mathscr{E}_0 = -\frac{\sum \nu_i \mu_{i0}}{\nu_e \mathscr{F}} = -\frac{\Delta G_0}{\nu_e \mathscr{F}} \tag{15-7}$$

We can thus rewrite eq. (15-6):

$$\mathscr{E} = \mathscr{E}_0 - \frac{RT}{\nu_e \mathscr{F}} \ln \prod_i a_i^{\nu_i} \tag{15-8}$$

Let us consider an example to make the preceding theory more definite. For the reaction vessel of fig. (15-1), take a beaker of dilute hydrochloric acid in water. For electrode I, choose a hydrogen electrode, which is an activated piece of platinum against which a stream of hydrogen (H_2) bubbles is directed at 1 atm pressure. Electrode II is a silver wire coated with silver chloride. At electrode I the following reaction occurs:

$$\tfrac{1}{2} H_2(g) \rightleftharpoons H^+ + e^- \tag{15-9}$$

At electrode II the reaction is

$$AgCl(s) + e^- \rightleftharpoons Ag(s) + Cl^- \tag{15-10}$$

The overall chemical reaction is

$$\tfrac{1}{2} H_2(g) + AgCl(s) \rightleftharpoons H^+(c_1) + Cl^-(c_1) + Ag(s) \tag{15-11}$$

In order for the reaction to occur, electrons must flow through an external circuit from the platinum electrode to the silver electrode. If the opposing voltage of the potentiometer is exactly $\mathscr{E}$, no current will flow and the system will be at equilibrium. One mole of electrons flows per unit of reaction, as defined by eq. (15-11), so that ν_e is unity. Using eqs. (15-11) and (15-8) we get

$$\mathscr{E} = \mathscr{E}_0 - \frac{RT}{\mathscr{F}} \ln \frac{a_{H^+} a_{Cl^-} a_{Ag}}{a_{H_2}^{1/2} a_{AgCl}} \tag{15-12}$$

Since the standard state of H_2 is a gas at 1 atm pressure and the standard states of silver and silver chloride are the solids, a_{H_2}, a_{Ag}, and a_{AgCl} are all unity. Equation (14-12) then reduces to the simple form

$$\mathscr{E} = \mathscr{E}_0 - \frac{RT}{\mathscr{F}} \ln a_{H^+} a_{Cl^-} \tag{15-13}$$

If we choose an HCl solution such that a_{H^+} and a_{Cl^-} are unity, then we can directly measure the standard Gibbs free energy of the reaction. Alternatively, if we know $\mathscr{E}_0$, we can use the measurement of $\mathscr{E}$ to determine the activities of the H^+ and Cl^- ions. In thermodynamic terms the pH of a solution can be defined as follows:

$$pH = -\log a_{H^+} \tag{15-14}$$

which allows us to write

$$\frac{\mathscr{F}(\mathscr{E} - \mathscr{E}_0)}{(2.303)RT} + \log a_{Cl} = pH \tag{15-15}$$

where the factor 2.303 converts natural logarithms to common logarithms (base 10). Under the appropriate conditions this type of measurement can be used to measure pH and, indeed, pH meters involve standard electrodes and potentiometer circuits, and are based on this general method of approach. We will develop a more detailed view of pH measurement later in this chapter [eqs. (15-39)–(15-42)].

To sum up the thermodynamic approach to electrochemistry, we note that the essential step was to add the electrical term to the expression for Gibbs free energy. Once this is done the stoichiometric equation for the electrode reactions had to be formulated. This led to eq. (15-8) which embodies the theoretical background. The hard work of electrochemistry consists in appropriate evaluation of activities, which is a major problem of solution chemistry. Indeed, it is often necessary to reconsider the electrodes and electrode reactions used in order to evaluate all the necessary experimental parameters.

B. OXIDATION–REDUCTION

Certain biochemical transformations have been classified as oxidation–reduction reactions. These are of major importance in energy processing and constitute the terminal steps in the oxidation of foodstuffs. The study of the biochemistry and bioenergetics of these processes has provided difficulty and confusion for a number of reasons. First, the historical development of the subject has been plagued by different interpretations and sign conventions. Second, the thermodynamics of electrochemistry has a number of subtle points which must be understood to deal with the biochemical applications. Third, there are aspects of energy transduction in biology which are not completely understood and a number of competing theories exist. At present it is possible, in principle, to clear away the first two causes of confusion so that students can focus attention on the excitement of elucidating the mechanisms of energy transfer in living cells.

Oxidation and reduction have acquired a number of meanings since their introduction in the 1700s and W. Mansfield Clark† has summed up these concepts in the following way:

> Oxidation is the addition of oxygen or the withdrawal of hydrogen or the withdrawal of electrons with or without the withdrawal of protons. Reduction is the withdrawal of oxygen or the addition of hydrogen or the addition of electrons with or without the addition of protons.

Clark gives a number of examples which may be constructively reviewed. The most familiar oxidation is rust formation, the conversion of ferrous iron to ferric iron:

$$2FeO + \tfrac{1}{2}O_2 \rightleftarrows Fe_2O_3 \tag{15-16}$$

By analogy the addition or loss of halogens is also considered an oxidation or a reduction reaction:

$$2FeCl_2 + Cl_2 \leftrightarrows 2FeCl_3 \tag{15-17}$$

The common feature of these two reactions is the conversion of divalent iron to trivalent iron. The change of oxidation state of this metal atom turns out to be of particular importance in biochemistry because of the large number of iron containing structures which are reversibly oxidized and reduced at the iron atom. The change of valence can also be represented as

$$Fe^{++} \quad Fe^{+++} + e \tag{15-18}$$

This reaction, as written, is of a somewhat different character from the first two which can take place in bulk phase. Because of the generation of free electrons in reaction (15-18) it can only take place at an interface, usually a metallic electrode.

The loss and gain of hydrogen atoms is shown in the oxidation of hydroquinone to benzoquinone:

$$HO{-}C_6H_4{-}OH \rightleftarrows H_2 + O{=}C_6H_4{=}O \tag{15-19}$$

† Clark, "Oxidation–Reduction Potentials of Organic Systems," p. 4, Krieger, Huntington, New York, 1972. Copyright © by Robert E. Krieger Publishing Co.

This oxidation involves the removal of hydrogen atoms, i.e., electron–proton pairs. However at high *p*H the following dissociation occurs:

$$\mathrm{HO{-}C_6H_4{-}OH} \rightleftarrows 2\mathrm{H}^+ + {}^-\mathrm{O{-}C_6H_4{-}O}^- \tag{15-20}$$

The oxidation–reduction reaction then becomes

$$^-\mathrm{O{-}C_6H_4{-}O}^- \rightleftarrows 2e + \mathrm{O{=}C_6H_4{=}O} \tag{15-21}$$

The quinone group is very prominent in mitochondrial oxidations in the form of ubiquinone or coenzyme Q,

$$\text{(quinone ring with } C{=}O \text{ at top and bottom, } C{-}CH_3, \; C{-}(CH_2{-}CH{=}\overset{\displaystyle CH_3}{\overset{|}{C}}{-}CH_2)_N H)$$

where for most mammalian mitochondria $N = 10$, while for some microorganisms $N = 6$. Thus, although oxidation–reduction reactions are primarily electron transfers, the electron may be accompanied by a proton in selected cases. Equations (15-19)–(15-21) make it clear as to why certain oxidation–reduction couples are *p*H dependent. There exist reactions in which electrons are partially transferred [eq. (15-16)] but we shall reserve the term oxidation–reduction in our subsequent discussion to those cases where there is a complete transfer of electrons. Thus every oxidation–reduction that takes place in bulk phase requires an electron donor or reductant and an electron receptor or oxidant. We will denote the reductant as Re and the oxidant as Ox. A generalized oxidation–reduction reaction can be written as

$$\mathrm{Re}_1 + \mathrm{Ox}_2 \rightleftarrows \mathrm{Ox}_1 + \mathrm{Re}_2 \tag{15-22}$$

Thus reductant 1 transfers electrons to oxidant 2. Reductant 1 is thus changed to its conjugate, oxidant 1, while oxidant 2 is changed to its conjugate, reductant 2. Consider an example,

$$\mathrm{Zn} + 2\mathrm{H}^+ \rightleftarrows \mathrm{Zn}^{++} + \mathrm{H}_2 \tag{15-23}$$

Reduced metallic zinc is oxidized to zinc ion, while protons are reduced to hydrogen gas. Two electrons are transferred from the zinc atom to the hydrogen ions. Because of an overall requirement for electroneutrality in solution oxidation–reduction reactions always occur pairwise as indicated in eq. (15-22). Such transformations are of the general class of coupled reactions which we often observe in biochemistry:

$$\begin{aligned} &(a) \quad A \rightleftarrows B + C \\ &(b) \quad C + E \rightleftarrows D \\ &(c) \quad A + E \rightleftarrows B + D \end{aligned} \tag{15-24}$$

C is the common intermediate which couples the two reactions. In the case in point the common intermediate is the electron which has no independent existence in solution, but the formalism of eq. (15-24) remains intact and we can write the equilibrium equation of (15-24c) as

$$\frac{a_B a_D}{a_A a_E} = e^{-\Delta G_0/RT} \tag{15-25}$$

For oxidation–reduction reactions we can conceptually split up eq. (15-22) into two processes:

$$\begin{aligned} \mathrm{Re}_1 &\rightleftarrows \mathrm{Ox}_1 + \nu_e \quad \text{electrons} \\ \mathrm{Re}_2 &\rightleftarrows \mathrm{Ox}_2 + \nu_e \quad \text{electrons} \end{aligned} \tag{15-26}$$

We can now go one step further and experimentally split up the two reactions in the appropriate electrolytic cells as is shown in fig. 15-2.

Here the electrons flow through an external circuit from reductant 1 to oxidant 2. To maintain electroneutrality there must be a corresponding flow of ions through the salt bridge. We may now subject the situation in fig. 15-2 to a thermodynamic analysis if we place an opposing source of electrical potential in the external circuit so that the flow of current is infinitesimal (figs. 15-1 and 15-3). The reaction now takes place at equilibrium and we can recall eq. (15-1) for the change of Gibbs free energy of a system with a reversible flow of external current:

$$dG = -S\,dT + V\,dP + \sum \mu_i\,dn_i + \mathscr{E}\,de = 0 \tag{15-27}$$

At constant temperature and pressure the reaction in fig. 15-1 yields

$$(\mu_{\mathrm{Ox}_1} + \mu_{\mathrm{Re}_2} - \mu_{\mathrm{Re}_1} - \mu_{\mathrm{Ox}_2})\,d\xi + \mathscr{E}\nu_e\mathscr{F}\,d\xi = 0 \tag{15-28}$$

$\mathscr{E}$ is by convention the potential of the right-hand electrode minus the potential of the left-hand electrode. $\mathscr{F}$ is the faraday. Using $RT \ln a_i = \mu_i - \mu_{i0}$, eq. (15-28) may be rewritten

$$\Delta G_0 + RT \ln \frac{a_{\mathrm{Ox}_1} a_{\mathrm{Re}_2}}{a_{\mathrm{Ox}_2} a_{\mathrm{Re}_1}} + \nu_e \mathscr{F}\mathscr{E} = 0 \tag{15-29}$$

$\mathscr{E}$ is a quantity that can be measured with great precision, thus opening a whole range of thermodynamically related measurements of biochemical importance. We consider the circuit of a very simple potentiometer which is shown in fig. 15-3. A current from a storage battery causes a potential

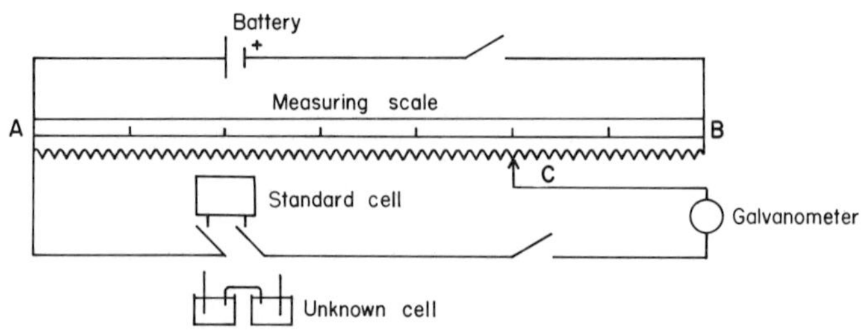

FIGURE 15-3. Potentiometer—see text for description.

drop along the linear slide wire AB. When the standard cell ($\mathscr{E}_s$) is placed in the circuit, the point C is moved until no current flow occurs in the galvanometer and AC_1 is measured. The unknown cell is now substituted for the standard cell and once again the galvanometer contact is moved until no current again flows and the distance AC_2 is now measured. The unknown voltage is thus given by $(AC_2/AC_1)\mathscr{E}_s$.

If the galvanometer contact is moved slightly to the left a tiny current will flow and electrons will move from the left half-cell to the right half-cell driving the reaction:

$$\mathrm{Re}_1 + \mathrm{Ox}_2 \rightarrow \mathrm{Ox}_1 + \mathrm{Re}_2 \tag{15-30}$$

If the galvanometer contact is moved slightly to the right of the balance point a small current flows in the opposite direction and the reaction proceeds:

$$\mathrm{Ox}_1 + \mathrm{Re}_2 \rightarrow \mathrm{Re}_1 + \mathrm{Ox}_2 \tag{15-31}$$

This is as close to a true reversible reaction as one can get and is one of the reasons redox methods yield such useful thermodynamic data. Note that all electrochemical measurements involve the potential difference between the electrodes of two half-cells. There is no such thing as an absolute electrode potential, just a potential difference. To have a common reference point a standard is chosen for the left-hand half-cell and all measurements are referred to that standard. The rather cumbersome standard chosen is the hydrogen half-cell in which the following reaction takes place:

$$\underset{\text{Reductant 1}}{\tfrac{1}{2}\mathrm{H}_2} \rightleftarrows \underset{\text{Oxidant 1}}{\mathrm{H}^+} + e \tag{15-32}$$

This reaction is achieved in practice by having a platinum–platinum black catalytic electrode dipping in a solution of hydrogen ions. Gaseous hydrogen is bubbled over the electrode. For standardization the hydrogen gas is at 1 atm (its standard state) and the hydrogen ion concentration is one molar or, more precisely, adjusted to unit activity. Equation (15-29) may be rewritten

$$\Delta G_0 + RT \ln \frac{a_{H^+} a_{Re_2}}{a_{Ox_2} a_{H_2}^{1/2}} + \mathscr{F}\mathscr{E} = 0 \tag{15-33}$$

Since a_{H^+} and a_{H_2} are each unity for the standard cell, we can rewrite eq. (15-33) as

$$\mathscr{E} = \frac{-\Delta G_0}{\mathscr{F}} + \frac{RT}{\mathscr{F}} \ln \frac{a_{Ox_2}}{a_{Re_2}} \tag{15-34}$$

The standard oxidation–reduction potential of a couple is measured when both oxidant and reductant are at the same activity and is therefore

$$\mathscr{E}_{\text{standard}} = \frac{-\Delta G_0}{\mathscr{F}} \tag{15-35}$$

The quantity ΔG_0 is the standard free energy of the reaction

$$\tfrac{1}{2}H_2 + Ox_2 \rightleftarrows H^+ + Re_2 \tag{15-36}$$

A large positive value of $\mathscr{E}$ means a large negative value of ΔG_0 which indicates a high affinity for electrons. On the other hand, a negative value of $\mathscr{E}$ (positive ΔG_0) means that the spontaneous direction of reaction (15-36) is from right to left. More negative standard potential corresponds to more reducing couples; more positive potentials correspond to more oxidizing couples.

A number of conventions are employed in describing electrolytic cells and we shall introduce them with the description of a standard hydrogen half-cell connected to a saturated calomel half-cell:

$$- \text{Pt}, H_2(1 \text{ atm}) \,|\, \text{HCl}(1.0\ M) \,||\, \text{KCl(sat)}\ Hg_2Cl_2(s) \,|\, \text{Hg} + \tag{15-37}$$

The overall reaction is

$$H_2 + Hg_2^{++} \rightleftarrows 2H^+ + 2Hg \tag{15-38}$$

and mercurous ion (Hg_2^{++}, a complex ion) is reduced to mercury. Expression (15-37) indicates the following: A platinum electrode in contact with hydrogen gas at 1 atm is dipping in a solution of hydrochloric acid at unit activity. The vertical line signifies a boundary between phases and a double vertical line signifies a liquid junction. Thus the hydrochloric acid solution is connected by a liquid junction to a saturated solution of KCl and Hg_2Cl_2 in contact with solid Hg_2Cl_2 (to insure saturation) connected to a mercury electrode. The minus and plus signs indicate the electrode voltages. Generally the half-cells are arranged so the right-hand electrode is positive and the reaction as written moves spontaneously from left to right. A modification of the cell as shown allows us to discuss the measurement of pH. Consider the following electrolytic system:

$$- \text{Pt}, \text{H}_2(1\ \text{atm})\,|\,\text{H}^+ \text{ unknown activity}\,||\,\text{KCl(sat)}\,\text{Hg}_2\text{Cl}_2(s)\,|\,\text{Hg} + \qquad (15\text{-}39)$$

Equation (15-29) may be written as

$$\Delta G_0 + RT \ln \frac{(a_{\text{H}^+})^2 (a_{\text{Hg}})^2}{(a_{\text{Hg}_2^{++}})(a_{\text{H}_2})} + 2\mathscr{F}\mathscr{E} = 0 \qquad (15\text{-}40)$$

Hydrogen gas is at unit activity as is solid mercury. We adopt saturated mercurous ion in saturated potassium chloride as the standard state of $a_{\text{Hg}_2^{++}}$. Equation (15-40) then reduces to

$$\mathscr{E} = \frac{-\Delta G_0}{2\mathscr{F}} - \frac{RT}{\mathscr{F}} \ln a_{\text{H}^+} \qquad (15\text{-}41)$$

But pH is defined as $-\log a_{\text{H}}{}^+$, or $-\dfrac{\ln a_{\text{H}^+}}{2.303}$

$$p\text{H} = \frac{(\mathscr{E} - \mathscr{E}_0)\mathscr{F}}{2.303RT} \qquad (15\text{-}42)$$

where $\mathscr{E}_0$ is simply $-\Delta G_0/2\mathscr{F}$.

Thus the redox system provides a practical method of pH measurement made possible because the behavior of the H_2, H^+ couple depends strongly on the H^+ activity. In modern pH meters the hydrogen electrode is replaced by a glass membrane electrode which has similar properties with respect to its pH potential behavior. The glass and calomel electrodes are connected to a potentiometer circuit which is calibrated by dipping the electrodes into solutions of known pH.

The glass electrode consists of a thin glass membrane separating two liquid phases of different hydrogen activity. It is an empirical generalization that *the potential across such a system is proportional to the logarithm of the ratio of the hydrogen ion activities.* The mechanism of such a device is not completely understood and there are two competing theories. However, the results are very reproducible and enable us to use glass electrodes in the practical measure of pH in spite of our uncertainties as to the mode of operation.

Returning to oxidation–reduction measurements, the reported potentials are directly proportional to the negative of the standard free energies of reaction (15-36). Redox potentials are another way of reporting Gibbs free energies, which is of particular convenience for electron transfer reactions.

Consider next two oxidation–reduction half-cells which are short circuited and allowed to react until equilibrium is reached. At equilibrium the cell will have completely reacted and the potential goes to zero. Thus Eq. (15-29) becomes

$$\ln \frac{a_{\mathrm{Ox}_1} a_{\mathrm{Re}_2}}{a_{\mathrm{Ox}_2} a_{\mathrm{Re}_1}} = \frac{-\Delta G_0}{RT} \tag{15-43}$$

If we replace activities by concentrations we note the following:

$$C_{\mathrm{Ox}_1} + C_{\mathrm{Re}_1} = C_1 \tag{15-44}$$

$$C_{\mathrm{Ox}_2} + C_{\mathrm{Re}_2} = C_2 \tag{15-45}$$

If 1 and 2 are the only oxidants and reductants in the system then

$$C_{\mathrm{Ox}_1} + C_{\mathrm{Ox}_2} = \text{constant} \qquad C_{\mathrm{Re}_1} + C_{\mathrm{Re}_2} = \text{constant} \tag{15-46}$$

Equations (15-43)–(15-46) can be solved for the equilibrium concentrations of the components of two reacting couples. The same result would obtain if the two couples were mixed in a single cell and allowed to react to completion. In the first case the energy of the reaction is available as electrical energy; in the second case it is only available as heat or, in more complex cases, for coupling to other reactions.

C. SURFACE CHEMISTRY

We now proceed to a third case where the addition of a term to the Gibbs free energy allows us to explore another range of phenomena, in this case surface chemistry. We consider a system where the amount of energy involved in the formation of surfaces is of sufficient magnitude to be included in the total energy. The Gibbs free energy then becomes

$$dG = -S\,dT + V\,dP + \sum_i \mu_i\,dn_i + \sigma\,d\mathscr{A} \tag{15-47}$$

where σ is the surface energy (the amount of energy required to change the surface of the system by one unit of area, all other quantities being held constant).

A simple and highly idealized case will illustrate the preceding, although an additional term must be considered. Suppose our system is a large spherical drop of pure liquid sitting on the bottom of a beaker as in fig. 15-4. If the droplet breaks up into smaller spherical droplets, it will decrease its gravitational energy and increase its surface energy. The problem is to determine the equilibrium droplet size. For this case we need to introduce the Gibbs free energy including gravitational and surface terms. At constant pressure and temperature this becomes

$$dG = d\mathscr{G} + \sigma\,d\mathscr{A} \tag{15-48}$$

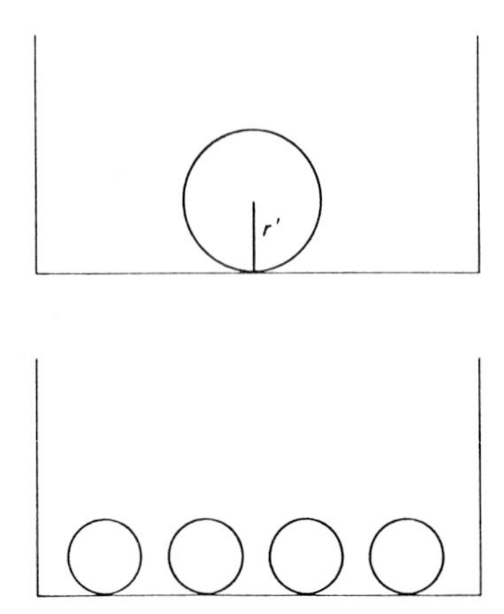

FIGURE 15-4. Droplet of radius r on the bottom of a beaker. In the lower representation the droplet breaks up into a number of smaller droplets.

If the initial droplet has a volume V and density ρ, the mass is $V\rho$ and the gravitational potential when the droplets have a radius r is

$$\mathscr{G} = V\rho r g \tag{15-49}$$

Each droplet acts with respect to gravitational potential as if its mass were concentrated at the center of the droplet. If the system breaks up into N droplets, the area will be

$$\mathscr{A} = 4\pi r^2 N \tag{15-50}$$

Since the total volume must be conserved regardless of the droplet size, we can write

$$\tfrac{4}{3}\pi r^3 N = V \tag{15-51}$$

Combining (15-50) and (15-51) we can express the area in more convenient form as

$$\mathscr{A} = \frac{3V}{r} \tag{15-52}$$

If we now substitute eqs. (15-49) and (15-52) into (15-48) and set $dG = 0$, which is the equilibrium condition, we get

$$dG = V\rho g\, dr - \frac{3V\sigma\, dr}{r^2} = 0 \tag{15-53}$$

Solving eq. (15-53) for the equilibrium radius, we get

$$r = \sqrt{\frac{3\sigma}{\rho g}} \tag{15-54}$$

Equation (15-54) is then a solution to our original problem. The equilibrium droplet size leads to a minimum Gibbs free energy for the system. This general type of analysis is applicable to colloidal suspensions and other problems where the surface-to-volume ratio is large enough to include terms involving surface energy. Note that we have not included an entropy of mixing term which could be significant if there were a large enough number of droplets.

Next move to a somewhat more complex surface problem where we need to consider both chemical potentials and surface terms. We wish to study the accumulation of dissolved molecules at surfaces. For a system at constant temperature and pressure the Gibbs free-energy equation becomes

$$dG = \sum \mu_i \, dn_i + \sigma \, d\mathscr{A} \tag{15-55}$$

The μ_i designates the chemical potential of the bulk phase and n_i refers to the number of moles of the ith component in both the bulk phase of volume V and the surface phase of area $\mathscr{A}$. From the properties of partial differentials we note

$$\mu_i = \frac{\partial G}{\partial n_i}, \qquad \sigma = \frac{\partial G}{\partial \mathscr{A}} \tag{15-56}$$

and

$$\frac{\partial \mu_i}{\partial \mathscr{A}} = \frac{\partial \sigma}{\partial n_i} \tag{15-57}$$

If the ith component is adsorbed (enriched or depleted) at the surface, we may designate the amount adsorbed per unit area as Γ_i. The true bulk phase concentration is

$$C_i = \frac{n_i - \Gamma_i \mathscr{A}}{V} \tag{15-58}$$

while the apparent concentration is

$$C_i^* = \frac{n_i}{V} \tag{15-59}$$

We may then note (if concentration approximates activity)

$$\mu_i - \mu_{i0} = RT \ln \frac{n_i - \Gamma_i \mathscr{A}}{V} \tag{15-60}$$

$$\frac{\partial \mu_i}{\partial \mathscr{A}} = \frac{-RT\Gamma_i}{n_i - \Gamma \mathscr{A}} \cong \frac{-RT\Gamma_i}{n_i} \tag{15-61}$$

The right-hand term of this equation results from the fact that in most cases of interest the total amount of material in solution is large compared with that absorbed at interfaces. Combining (15-57), (15-59), and (15-61) we get

$$\Gamma_i = \frac{-n_i}{RT}\frac{\partial\sigma}{\partial n_i} = \frac{-C_i^*}{RT}\frac{\partial\sigma}{\partial C_i^*} \tag{15-62}$$

Note that $\partial\sigma/\partial C_i^*$ is the change in surface energy as the ith component is added to the system. It may be either positive or negative so that Γ, the "surface excess," will be of either sign. Thus a substance which reduces the surface tension will accumulate at the interface, whereas one which raises the surface tension will be depleted at the boundary. Equation (15-62) is known as the Gibbs adsorption isotherm and is an example of the kind of problem that can be elucidated by the approach of the Gibbs free energy. Experimental studies on the absorption isotherm are reviewed in Chapter 3 of the book by N. K. Adam (see Bibliography).

D. THE PHASE RULE

Next we come to the phase rule which allows one to make certain predictions about multiphase multicomponent systems. If an equilibrium system exists in two phases (liquid–gas, liquid–liquid, solid–liquid, etc.) at constant temperature and pressure, then we can represent the chemical potential of the ith substance in the two phases as μ_{i_1} and μ_{i_2}. At equilibrium the transfer of dn moles from phase 1 to phase 2 will have a free-energy change of

$$dG = dn(\mu_{i_2} - \mu_{i_1}) \tag{15-63}$$

Since the condition of equilibrium is a free-energy minimum, dG must be zero and $\mu_1 = \mu_2$. The chemical potential must be the same in all phases. This can be seen to be generally true for any two phases in contact; if one of the components has a smaller free energy in one phase than another, it will move into the lower phase and so reduce the free energy of the entire system. Equilibrium can only exist when all components have the same chemical potential in all phases.

Assume that a system has A phases and B components. To specify each phase requires $B - 1$ composition variables (mole fractions—only $B - 1$ are independent since the sum of the mole fractions must be unity). Therefore to completely specify the state requires $A(B - 1) + 2$ variables since we must also specify the temperature and pressure of the whole. Since all phases are in contact, one value of the temperature and one value of the pressure will characterize the entire system. However, there are a number of restrictions which come about since the chemical potential of a component is the same in all phases and is itself a function of the compositional variables, pressure and temperature. Thus for each component i there are $A - 1$ relations of the type $\mu_{ij} = \mu_{ik}$. For all B components there are $B(A - 1)$ such equations of constraint. Each of these may be used to eliminate one variable, so that we are left with the following expression for the number of independent variables:

$$A(B - 1) + 2 - B(A - 1) = B - A + 2 \qquad (15\text{-}64)$$

The relationship between the number of independent variables or degrees of freedom and the number of phases and components is called the phase rule.

Consider some examples; for a pure substance which can exist in different phases, B is 1 and the number of degrees of freedom is $3 - A$. For a one-phase system, there are $3 - 1$ or 2 degrees of freedom and any set of temperatures and pressures consistent with that phase can be maintained. For two-phases such as liquid and gas there are $3 - 2 = 1$ degree of freedom. If the temperature is specified, the vapor pressure of a solid or liquid is completely determined. For a three-phase, one-component system there are $3 - 3$ or zero degrees of freedom; that is, there is a unique and completely determined triple point. In retrospect this is why the triple point of water proved to be such a good choice for a temperature standard.

Next consider a two-component system, which will then have $2 + 2 - A = 4 - A$ degrees of freedom. Where the two exist in one phase there are $4 - 1 = 3$ degrees of freedom—temperature, pressure, and mole fraction can be varied independently. Next consider the system existing in two phases; think of a water–phenol solution and its equilibrium vapor mixture. There are then $4 - 2 = 2$ degrees of freedom, so that if temperature and pressure are specified all compositional variables are fixed for both phases. Next consider three phases, such as gas and two immiscible liquid phases (two immiscible water–phenol phases in equilibrium with the vapor). There is then one degree of freedom, and once temperature is fixed all other variables are set. Numerous other applications have been made of the phase rule.

E. OSMOTIC PRESSURE

Finally consider a thermodynamic treatment of osmotic pressure. A semi-permeable membrane separates two compartments which are in contact with an isothermal reservoir. On the left-hand side of the membrane there is a pure solvent and on the right side is a solution using the same solvent (fig. 15-5). The membrane is freely permeable to solvent and impermeable to solute. The left-hand compartment is at atmospheric pressure, while a

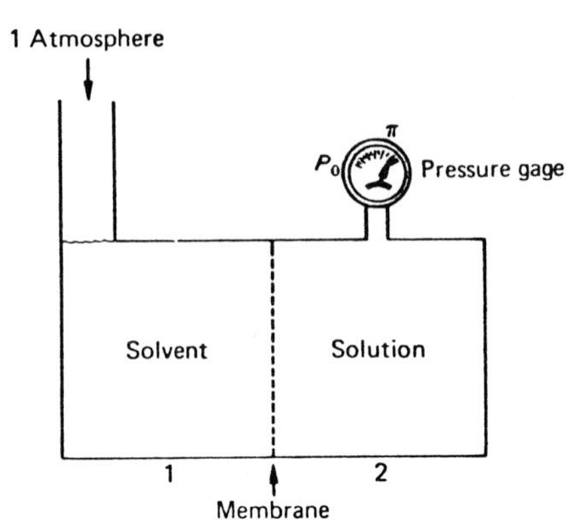

FIGURE 15-5. Two compartments separated by a semipermeable membrane. The solvent is maintained at atmospheric pressure while the osmotic pressure of the solution rises to the value Π.

pressure must be maintained on the right-hand compartment in order to maintain equilibrium; that is, to prevent the flow of solvent across the membrane. Since the system is maintained at constant volume (the liquids are essentially incompressible) the condition of equilibrium is that the Helmholtz free energy be a minimum. At constant volume and temperature this condition reduces to

$$dA = \sum \mu_i \, dn_i = 0 \tag{15-65}$$

The only change dn_i that is possible is the transport of dn moles of solvent across the membrane:

$$dA = (\mu_2 - \mu_1)\, dn = 0 \tag{15-66}$$

The requirement for equilibrium is then that $\mu_2 = \mu_1$. The chemical potential of the solvent on the right side is given by

$$\mu_2 = \mu_0 + RT \ln a + \int_0^{\Pi} \left(\frac{\partial \mu}{\partial P}\right) dP \tag{15-67}$$

That is, we first formulate the chemical potential at standard pressure, and then formulate the change in chemical potential in going from the standard state to the final state at osmotic equilibrium. However, from the expression for the Gibbs free energy, we can show that

$$\frac{\partial \mu_i}{\partial P} = \frac{\partial V}{\partial n_i} = \bar{V}_i \tag{15-68}$$

$\bar{V}_i$ is partial molar volume which, to a first approximation, is a constant for most solvents. For ideal solutions we have already shown that the activity is equal to the mole fraction

$$a = \frac{n_{\text{solvent}}}{n_{\text{solvent}} + n_{\text{solute}}} \tag{15-69}$$

where the n's are the numbers of moles. We can expand the activity term for dilute solutions as follows:

$$RT \ln a = -RT \ln \frac{1}{a} = -RT \ln\left(1 + \frac{n_{\text{solute}}}{n_{\text{solvent}}}\right)$$

$$\cong -RT \frac{n_{\text{solute}}}{n_{\text{solvent}}} \tag{15-70}$$

In arriving at eq. (15-70) we have used the approximation that $\ln(1 + x)$ can be represented by x for small values of the variable. The chemical potential μ_1 is μ_0 since the solvent in the left-hand compartment is in its standard state (pure liquid at 1 atm pressure). We thus get the following relation, based on the equality of the chemical potential on both sides of the membrane:

$$\mu_0 = \mu_0 + RT \ln a + \int_0^{\Pi} \bar{V}\, dP \tag{15-71}$$

Using the results of eq. (15-70) and integrating the last term of eq. (15-71), we arrive at the following formulation:

$$RT \frac{n_{\text{solute}}}{n_{\text{solvent}}} = \Pi \bar{V} \tag{15-72}$$

For dilute solutions the volume of the entire system is very closely approximated by n_{solvent} times $\bar{V}$. Therefore

$$\Pi V = n_{\text{solute}} RT \tag{15-73}$$

Equation (15-73) is the usual form for the osmotic pressure of dilute perfect solutions. It has been formulated entirely from thermodynamic arguments. The treatment once again exemplifies the utility of using free-energy functions for the solution of actual problems.

While osmotic phenomena are well known, a somewhat less familiar case illustrates the power of the method. Referring again to fig. 15-5, assume now the solution is on the left side and the solvent is on the right side of the membrane. Equation (15-71) becomes

$$\mu_0 + RT \ln a = \mu_0 + \int_0^{\Pi} \bar{V}\, dP \tag{15-74}$$

and (15-73) would consequently become

$$\Pi \bar{V} = -n_{\text{solute}} RT \tag{15-75}$$

Chamber 2 would be under a negative pressure or a tension state. Such states have been experimentally observed.

Negative pressure states in liquids appear to be of frequent occurrence in nature, and account for the rise of sap in trees—a phenomenon of utmost ecological significance. Thus negative values of pressure which occur in thermodynamic formalism correspond to an actual state of the liquid. Such phenomena have not been extensively exploited and may turn out to have exciting engineering applications as is noted in the essay by Hayward in the Bibliography. The generality of the Gibbsian approach to problems of this nature may be seen in the work on osmotic pressure of liquid-helium solutions. Experimental verification of negative pressures is discussed in the paper of Wilson and Tough which is cited in the Bibliography.

TRANSITION

The equations of electrochemistry are derived from the fundamental thermodynamic relations by the addition of an energy term for the transport of charge around an external circuit. This leads to relations between the electrical potential and the activities of various electron donors and receptors in the system. Thus the particular class of reactions involving electron transfers (oxidation–reduction) is especially open to study by these methods.

The equations of surface chemistry follow from adding a surface energy term to the differential of the Gibbs free energy. The most widely used of these is the Gibbs adsorption isotherm which relates the surface excess of a solute to its effect in changing the surface tension of a solution.

Osmotic pressure is analyzed by including in the free energy a term indicating the change of chemical potential with hydrostatic pressure. This leads to the usual relations between osmotic pressure and concentration.

This chapter completes our formal treatment of equilibrium thermal physics and we now move on to a discussion of fluctuations and the processes by which a system decays to equilibrium. This material will begin to point the way from equilibrium to nonequilibrium theory.

BILIOGRAPHY

Adam, N. K., "The Physics and Chemistry of Surfaces." Oxford Univ. Press, London and New York, 1946.

A review of both theory and experiment.

Bates, R. G., "Determination of pH; Theory and Practice." Wiley, New York, 1973.

This is the standard work on pH and its measurement.

Clark, W. M., "Oxidation–Reduction Potentials of Organic Systems." Krieger, Huntington, New York, 1972.

A very thorough step-by-step development of a complex formalism. This book also contains a good deal of experimental detail on measurements in this field.

Dole, M., "Principles of Experimental and Theoretical Electrochemistry." McGraw-Hill, New York, 1935.

A clear statement of the basic theory of electrochemical phenomena.

Gibbs, J. W., "The Scientific Papers of J. Willard Gibbs." Longmans, Green, New York, 1906.

Volume I contains the extensive paper, "On the Equilibrium of Heterogeneous Substances;" first published in the Transactions of the Connecticut Academy III, pp. 108–248, 343–524. These were published in 1875, 1876, and 1878. Gibbs laid out the fundamentals of chemical thermodynamics, which have had such wide application. For comments on the role of Gibbsian thermodynamics in biology, see Morowitz, H. J., A biologist's view of Gibbs' contributions, J. Wash. Acad. Sci. **64**, *209, 1974.*

Harned, H. S., and Owen, B. B., "The Physical Chemistry of Electrolytic Solutions." Van Nostrand-Reinhold, Princeton, New Jersey, 1958.

An encyclopedic work on the thermodynamic and kinetic properties of ions in solution.

Hayward, A. T. J., Negative pressure in liquids: Can it be harnessed to serve man? *Amer. Sci.* **59,** 434 (1971).

Reviews experimental evidence for negative pressure states.

Lewis, G. N., and Randall, M., "Thermodynamics" (revised by Pitzer, K. S., and Brewer, L.). McGraw-Hill, New York, 1961.

The reduction of Gibbsian methods to practical applications was summarized and codified by Lewis and Randall in 1923. Pitzer and Brewer completely revised the original work and brought it up to date in 1961. This classic is necessary for a contemporary understanding of Gibbsian thermodynamics.

Mauro, A., Osmotic flow in a rigid porous membrane, *Science* **149**, 867 (1965).

A report on the measurement of negative osmotic pressures.

Morris, J. G., "A Biologist's Physical Chemistry." Addison-Wesley, Reading, Massachusetts, 1974.

True to its title, this book develops physical chemistry as is needed in biology. Clear presentation is achieved with minimum mathematical sophistication.

Wilson, M. F., and Tough, J. T., Osmotic pressure of dilute solutions of He^3 in He^4, *Phys. Rev. A* **1**, 914 (1970).

Negative osmotic pressures are reported in the temperature range below 2K.

16 Thermal Energy

A. ORDER AND DISORDER

Thermal energy, diffusion, and Brownian motion are three related topics of thermal physics which prove to be of great significance in understanding much of biology in terms of its physical foundations. Noise is a general term covering all random processes and manifests itself not only as Brownian motion but in light scattering and spontaneous fluctuating electrical potentials across a resistance. Thermal noise and fluctuations will be discussed further in Chapter 20.

From what has been said in previous chapters it is clear that all molecules and parts of molecules are in ceaseless thermal motion at all temperatures other than absolute zero. Temperature at the molecular level means the moving, twisting, turning, writhing, oscillating, and general jumping around of molecular structures. Indeed, one deceptive thing about the elegant atomic models that we build today is that they present a false static view of molecules. In drawing structures and building models, we tend to lose sight of the constant random motion which is a necessary part of the description of matter. This thermal energy is responsible for:

1. The Decay of Ordered Systems to a State of Maximum Disorder or Maximum Entropy. As a result of the apparent randomness, hot spots develop and the existence of these hot spots leads to the breakdown of ordered structures. From the Maxwell–Boltzmann distribution we known that the probability of a molecular degree of freedom having an energy

E_c or greater is proportional to

$$\int_{E_c}^{\infty} e^{-E/kT}\, dE$$

If T is large enough, such a molecule can acquire energy sufficient to dissociate covalent bonds and destroy ordered structures. At thermal equilibrium molecules are constantly acquiring large amounts of kinetic energy. Consequently, the decay is ceaseless and can only be countered by doing the requisite work to rebuild the altered structures. The biosphere represents a kind of steady state between the buildup of ordered macromolecular structures using photosynthetically acquired energy and the breakdown of these structures owing to random thermal energy.

2. *Chemical Reactions.* In general, for a reaction to occur the reacting species must have an energy ΔE in order to get over a potential-energy barrier. This ΔE comes from the kinetic energy of the reacting species. The higher the temperature the more molecules will have the requisite energy on collision and the faster the process will proceed. This is reflected in the temperature dependence of the absolute reaction rate which we studied in Chapter 13. At absolute zero, the transitions do not take place.

3. *Diffusion.* If one examines the trajectory of a single molecule, there appears a type of random motion due to its thermal energy and its collisions with other molecules. Such a trajectory is shown in fig. 16-1. If we have a collection of such molecules, each will have the same probability of moving from left to right as from right to left. Consequently, if there is a concentration difference between two neighboring points, there will be a net flow from the higher to the lower concentration. This random movement of molecules under a concentration gradient is called diffusion.

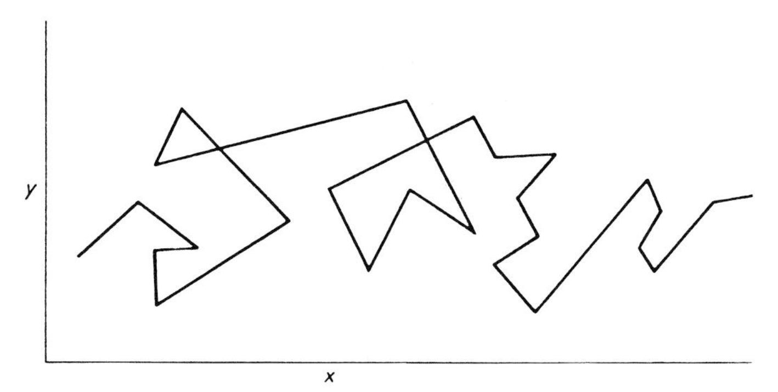

FIGURE 16-1. The trajectory of a particle undergoing random thermal motion. Represented is a projection on the xy plane.

4. *Brownian Motion.* A microscopically observable particle suspended in a liquid is seen to undergo an irregular trajectory similar to that shown in fig. 16-1. This motion results from the collision of large numbers of liquid molecules with the particle. Since these collisions are random in nature and due to the thermal energy, at any instant there may be an imbalance of the force on one side of the particle compared to the opposite side. The motion under this force is Brownian motion. A collection of particles undergoing Brownian motion will also exhibit diffusion with a net flow from the higher to the lower concentration.

What the previous discussion indicates is that at the atomic level things are very active and tend to get disordered. Erwin Schrödinger in his very perceptive book "*What is Life*?" stressed that most physical laws that we deal with are macroscopic and average over a sufficient number of molecules so that the type of disorder we have discussed does not invalidate the precision of the law. Only when we observe events on a very micro scale, such as Brownian motion, do the random fluctuations come into play. Schrödinger then posed the following question: How do biological systems, which, in the limit, consist of events at a very micro level (a single molecule in the case of genetic events), manage to exhibit such orderly behavior under the ceaseless bombardment of random thermal energy? This question has not been completely answered, although our increasing knowledge of molecular biology and cellular control mechanisms suggests numerous ways out of the dilemma posed by Schrödinger. The rest of this chapter will examine in more detail a few aspects of thermal energy which prove to be of importance in biology.

B. DIFFUSION

If a given sample consists of molecules in random motion, then one would expect a net flux from a region of higher concentration to one of lower concentration. This follows from a straightforward probabilistic argument. The question has been approached from the empirical side leading to a relation known as Fick's first law, which states that the flow per unit area across a surface in the yz plane is given by

$$\text{flow} = -D\frac{dC}{dx} \tag{16-1}$$

The quantity dC/dx is the concentration gradient in the x direction and the flow is in fact a vector quantity. Thus Fick's law when extended to three dimensions becomes

$$\text{flow} = -D\left(\mathbf{i}\frac{\partial C}{\partial x} + \mathbf{j}\frac{\partial C}{\partial y} + \mathbf{k}\frac{\partial C}{\partial z}\right) \tag{16-2}$$

where **i**, **j**, and **k** are unit vectors along the x, y, and z axes. The constant of proportionality D is known as the diffusion coefficient or, under the appropriate conditions, the diffusion constant to reflect its independence of C. The minus sign expresses the fact that flow is from a higher to lower concentration. Expression (16-2) can be regarded as an empirical generalization or, for dilute solutions or low-pressure gases, we could have derived the relations by summing over the random motion.

For gases kinetic theory may be used to derive an expression for the diffusion coefficient in terms of molecular size, temperature, and various interaction parameters between molecules. For macromolecules in solution or particles suspended in liquids, the diffusion coefficient can be related to the hydrodynamic properties of the particle, and this will be carried out in a later section.

In our discussion we will confine ourselves to diffusion in one dimension. The results can always be generalized to three dimensions; this complicates the mathematics but does not introduce conceptual innovations.

We will start with eq. (16-1) and derive a different form of the diffusion equation more useful in the solution of certain problems. Consider a slab of thickness dx perpendicular to the x axis (see fig. 16-2). Consider an area $\mathscr{A}$ of this slab. The diffusion flow into the slab is given from eq. (16-1) as $-D\mathscr{A}\,\partial C/\partial x$. The diffusion flow across the corresponding face on the other side of the slab is given by the following equation:

$$\text{flow} = -D\mathscr{A}\,\frac{\partial C(x+dx)}{\partial x} = -D\mathscr{A}\left[\frac{\partial C}{\partial x} + \frac{\partial^2 C}{\partial x^2}\,dx\right] \tag{16-3}$$

The net rate of change of the amount of material in the slab by diffusion is the flow in through the left-hand face minus the flow out through the

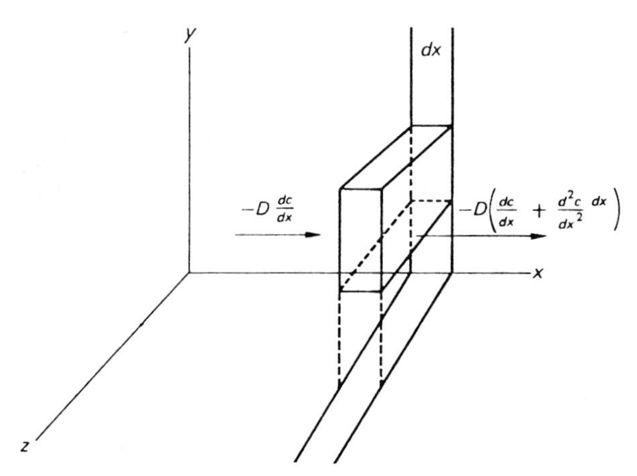

FIGURE 16-2. Diffusion through a thin slab of thickness dx.

right-hand face:

$$\text{net rate of change} = -D\mathscr{A}\frac{\partial C}{\partial x} + \left[D\mathscr{A}\left(\frac{\partial C}{\partial x} + \frac{\partial^2 C}{\partial x^2}\,dx\right)\right]$$

$$= D\mathscr{A}\frac{\partial^2 C}{\partial x^2}\,dx \qquad (16\text{-}4)$$

We can also calculate the rate of change of material in the slab by a different procedure. The quantity we wish to determine is the rate of change of the number of moles of the diffusing component. This is $(\partial C/\partial t)\mathscr{A}\,dx$ which is the rate of change of concentration times the volume. We can equate this to eq. (16-4), getting

$$\frac{\partial C}{\partial t}\mathscr{A}\,dx = D\mathscr{A}\frac{\partial^2 C}{\partial x^2}\,dx \qquad (16\text{-}5)$$

Dividing through by $\mathscr{A}\,dx$ yields

$$\frac{\partial C}{\partial t} = D\frac{\partial^2 C}{\partial x^2} \qquad (16\text{-}6)$$

The above is known as the diffusion equation and is sometimes referred to as Fick's second law of diffusion. Equation (16-6) may now be used to estimate the distance that a particle or molecule will diffuse in a given time. Start with a planar source of particles. (This is an idealization; what we actually start with is a very thin slab, located at the origin, with a very high concentration of particles. It is in the center of a large container that has a zero concentration of particles everywhere else at zero time.) What we require is a solution of eq. (16-6) that satisfies the following boundary conditions:

$$\begin{aligned} C &= \infty \quad \text{at} \quad x = 0, \quad t = 0 \\ C &= 0 \quad \text{at} \quad x \neq 0, \quad t = 0 \end{aligned} \qquad (16\text{-}7)$$

The time-dependent solution of eq. (16-6) then describes the outward movement of molecules from the initial very thin slab at zero time. We shall propose a solution and show that it has the required properties. The solution is

$$C = \frac{\mathfrak{K}}{t^{1/2}}e^{-x^2/4Dt} \qquad (16\text{-}8)$$

The coefficient $\mathfrak{K}$ is a constant of integration. To show that eq. (16-8) is indeed a solution, we must substitute it into eq. (16-6) and show that the resultant expression is an identity. This task is left to the reader. Anticipating the results, we note that

$$C(0,0) = \frac{\Re e^{-0}}{(0)^{1/2}} = \infty \tag{16-9}$$

This accords with our boundary conditions. For finite values of x we must show that $C(x, 0)$ goes to zero. Rather than doing this analytically, which is somewhat difficult, let us assign $\Re$ and $-x^2/4D$ each values of 1 and examine the behavior of the function as $t \to 0$ in table 16-1. This demonstrates the point that the function also fits the second boundary condition, since for finite x it very rapidly goes to zero as $t \to 0$.

TABLE 16-1
Limit of C

t	$C(x, t)$
1	0.367
0.25	0.0276
0.1	0.000135
0.01	10^{-42}

The solution set of eq. (16-6) as a function of time is shown in fig. 16-3. The particles diffuse out uniformly, getting on the average farther and farther from the point of origin. We can use eq. (16-8) to calculate the average distance that a particle has diffused in time t. For reasons of mathematical convenience we will calculate the mean-square distance that a particle diffuses out from the origin. The fraction of particles lying in a plane slab

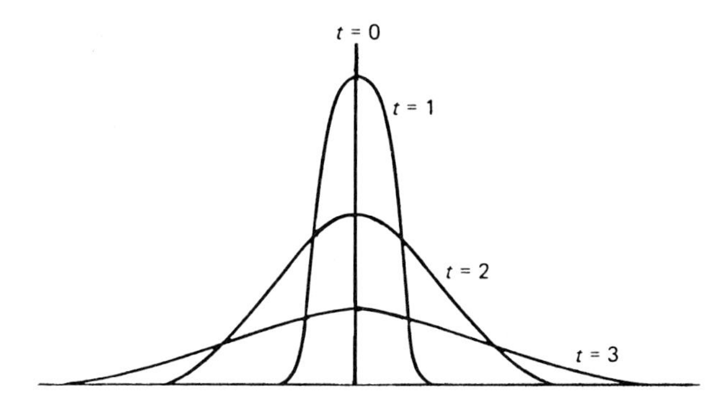

FIGURE 16-3. The concentration of diffusing solute as a function of time. All the material is originally concentrated in a plane at the origin and spreads out in time.

between x and $x + dx$ is

$$df(x, t) = \frac{C(x, t)\,dx}{\int_{-\infty}^{\infty} C(x, t)\,dx} \tag{16-10}$$

The mean-square distance is given by

$$\langle x^2 \rangle = \int_{-\infty}^{\infty} x^2\,df(x, t) = \frac{\int_{-\infty}^{\infty} x^2 C(x, t)\,dx}{\int_{-\infty}^{\infty} C(x, t)\,dx} \tag{16-11}$$

If we substitute eq. (16-8) into eq. (16-11) and integrate, we can evaluate $\langle x^2 \rangle$ in terms of D and t:

$$\langle x^2 \rangle = 2Dt \tag{16-12}$$

Equation (16-12) is perfectly general, and applies to liquids and gases under any conditions where the diffusion equation applies.

Some additional insights might be obtained into eq. (16-12) if we consider the problem of a one-dimensional random walk, sometimes referred to as "the drunkard's walk." An individual starts on a line at $x = 0$ and takes N steps, each of $\mathfrak{L}$ units, but randomly steps either to the right ($\mathfrak{L}_i = +|\mathfrak{L}|$) or to the left ($\mathfrak{L}_i = -|\mathfrak{L}|$). The net displacement is given by

$$x = \sum_{i=1}^{N} \mathfrak{L}_i \tag{16-13}$$

If we had an ensemble of drunks the average displacement would be

$$\langle x \rangle = \langle \mathfrak{L}_1 + \mathfrak{L}_2 \cdots \mathfrak{L}_N \rangle = \langle \mathfrak{L}_1 \rangle + \langle \mathfrak{L}_2 \rangle \cdots \langle \mathfrak{L}_N \rangle = 0 \tag{16-14}$$

The zero value of each $\langle \mathfrak{L}_i \rangle$ comes from the fact that it will have positive and negative values with equal frequency. A more interesting result might be obtained from the mean-square displacement. The square of the displacement is

$$x^2 = \sum_{i=1}^{N} \mathfrak{L}_i \sum_{j=1}^{N} \mathfrak{L}_j = \sum_{i=1}^{N} \mathfrak{L}_i^2 + \sum_{i \neq j} \mathfrak{L}_i \mathfrak{L}_j \tag{16-15}$$

Taking the mean, we get

$$\langle x^2 \rangle = \sum_{i=1}^{N} \langle \mathfrak{L}_i^2 \rangle + \sum_{i \neq j} \langle \mathfrak{L}_i \mathfrak{L}_j \rangle \tag{16-16}$$

The second term vanishes because $\mathfrak{L}_i$ and $\mathfrak{L}_j$ are uncorrelated and are positive and negative with equal frequency. Thus we get

$$\langle x^2 \rangle = N\mathfrak{L}^2 \tag{16-17}$$

This formula gives immediate insight into $\sqrt{N}$ dependencies in random processes. If the walker takes $\mathscr{U}$ steps per second, then $N = \mathscr{U}t$ and the equation becomes

$$\langle x^2 \rangle = \mathfrak{L}^2 \mathscr{U} t \tag{16-18}$$

Comparison of (16-18) and (16-12) gives a clue as to the relation of the random walk and the Fick formalism for diffusion.

C. BROWNIAN MOTION

We will now consider a more restrictive case, the random motion of particles which are large in comparison to the size of solvent molecules. This will apply to particles undergoing Brownian motion as well as to the diffusion of macromolecules such as proteins. For such particles we can use dynamical methods and write down the equation of motion for the one-dimensional case

$$X(t) - \mathfrak{f}\frac{dx}{dt} = m\frac{d^2x}{dt^2} \tag{16-19}$$

$X(t)$ is a time-varying fluctuating force due to an asymmetry in the collisions with solvent molecules. Equation (16-19) is just a statement of Newton's law that force is equal to mass times acceleration. Two forces act on each particle. The first is the random fluctuating force due to thermal energy from unbalanced collisions with solvent molecules. The second is the frictional force which is experienced by any particle moving through a viscous fluid. The force is directed opposite to the velocity vector and for small velocities is given by $-\mathfrak{f}v$ where v is the velocity, dx/dt, in one dimension, and $\mathfrak{f}$ is the frictional coefficient. The quantity $\mathfrak{f}$ can in general be computed from hydrodynamics and for a sphere of radius r it is $6\pi\eta r$, where η is the coefficient of viscosity.

The approach we shall follow is to use eq. (16-19) to evaluate $\langle x^2 \rangle$ by dynamical methods. We then compare the result to eq. (16-12) to see the meaning of the diffusion constant in terms of hydrodynamic variables. We begin by undertaking a set of transformations to recast eq. (16-19) in terms of x^2. We introduce a new variable $\mathscr{X} = x^2$. It then follows that

$$\begin{aligned} \frac{d\mathscr{X}}{dt} &= 2x\frac{dx}{dt} \\ \frac{d^2\mathscr{X}}{dt^2} &= 2\left(\frac{dx}{dt}\right)^2 + 2x\frac{d^2x}{dt^2} \end{aligned} \tag{16-20}$$

We then substitute (16-20) into (16-19) and rearrange:

$$\tfrac{1}{2}m\frac{d^2\mathscr{L}}{dt^2} + \tfrac{1}{2}\mathfrak{f}\frac{d\mathscr{L}}{dt} - m\left(\frac{dx}{dt}\right)^2 = xX(t) \tag{16-21}$$

The next step is to integrate the equation of motion with respect to time. (Note that dx/dt is the velocity of the particle, which we will represent as v.)

$$\tfrac{1}{2}m\frac{d\mathscr{L}}{dt} + \tfrac{1}{2}\mathfrak{f}\mathscr{L} - \int_0^t mv^2\,dt = \int_0^t xX\,dt + \text{constant} \tag{16-22}$$

The constant in eq. (16-22) is zero since we assume that $x = 0$ at $t = 0$. Since X is a random fluctuating force, over a long time it will be uncorrelated with the displacement x, so that the right-hand side of eq. (16-21) is zero. We now have a differential equation in x^2, while we would like to evaluate $\langle x^2\rangle$, which is x^2 averaged over a large number of particles. The averaging leads to the following correspondence:

$$\begin{gathered}\tfrac{1}{2}m\frac{d\mathscr{L}}{dt} \to \tfrac{1}{2}m\left\langle\frac{d\mathscr{L}}{dt}\right\rangle \\ \tfrac{1}{2}\mathfrak{f}\mathscr{L} \to \tfrac{1}{2}\mathfrak{f}\langle\mathscr{L}\rangle \\ \int_0^t mv^2\,dt = m\overline{v^2}t = kTt\end{gathered} \tag{16-23}$$

The first two transformations are obvious. The third comes about since the thermal energy of translation of a particle for any degree of freedom is $\tfrac{1}{2}kT$. These transformations lead to

$$\left\langle\frac{d\mathscr{L}}{dt}\right\rangle + \frac{\mathfrak{f}}{m}\langle\mathscr{L}\rangle = \frac{2kTt}{m} \tag{16-24}$$

Equation (16-24) is a differential equation in $\langle\mathscr{L}\rangle$ which can be directly integrated to give

$$\langle\mathscr{L}\rangle = \langle x^2\rangle = \chi e^{-(\mathfrak{f}/m)t} + \frac{2kT}{\mathfrak{f}}\left(t - \frac{m}{\mathfrak{f}}\right) \tag{16-25}$$

In usual cases encountered in biological application $m/\mathfrak{f}$ is small compared to the times of observation, so that the first and third terms vanish. We can examine the ratio $\mathfrak{f}/m$ for a sphere, for which $\mathfrak{f}$ is $6\pi\eta r$ and m is $\tfrac{4}{3}\pi r^3\rho$, where ρ is the density. The ratio is then $9\eta/2r^2\rho$; ρ is the order of 10^3 kg m^{-3}; η for water at 293 K is 0.001 in SI units. The ratio is then $4.5 \times 10^{-6}r^{-2}$. For a 1-μm particle this is 4.5×10^6 s^{-1}, a very large number as pointed out above. Equation (16-25) then reduces to

$$\langle x^2\rangle = \frac{2kTt}{\mathfrak{f}} \tag{16-26}$$

We can now compare eqs. (16-12) and (16-26). This leads to the result

$$D = \frac{kT}{\mathfrak{f}} \tag{16-27}$$

This expression relates Brownian motion, thermal energy, and the hydrodynamic properties of particles. Equation (16-26) along with a determination of $\mathfrak{f}$ allows us to measure k from a series of observations of particles undergoing Brownian motion. Such experiments were carried out by Perrin in one of the classical determinations of physical constants.

D. EQUIPARTITION OF ENERGY

One more powerful result which we have already used emerges from a consideration of thermal energy: the principle of equipartition of energy, which states that *the mean kinetic energy associated with each degree of freedom of a molecule is* $\frac{1}{2}kT$ (conclusion). This principle follows directly from the Maxwell–Boltzmann distribution in the classical form with the introduction of a postulate from mechanics and an approximation.

We introduce without proof (see L. Page in Bibliography) a result from advanced mechanics which states that, for a molecule of N degrees of freedom, the total energy can be expressed as

$$E = \mathscr{V}(q_1 \cdots q_N) + \sum_{i=1}^{N} \zeta_i p_i^2 \tag{16-28}$$

where the q are generalized coordinates and the p are generalized momenta. $\mathscr{V}$ is the potential energy and the ζ_i are proportionality parameters. Using classical statistical mechanics the probability that the values of each p and q lie between p_i and $p_i + dp_i$ and $q_i + dq_i$ is

$$\mathscr{P} = \frac{e^{-E/kT}\, dq_1 \cdots dq_N\, dp_i \cdots dp_N}{\int_{\text{all } p,q} e^{-E/kT}\, dq_1 \cdots dq_N\, dp_i \cdots dp_N} \tag{16-29}$$

Consider the jth degree of freedom with its associated momentum p_j and its associated kinetic energy $\zeta_j p_j^2$. We wish to obtain the average value of this kinetic energy which is

$$\langle \zeta_j p_j^2 \rangle = \frac{\int_{\text{all } p,q} e^{-E/kT} \zeta_j p_j^2\, dq_1 \cdots dq_N\, dp_1 \cdots dp_N}{\int_{\text{all } p,q} e^{-E/kT}\, dq_1 \cdots dq_N\, dp_1 \cdots dp_N} \tag{16-30}$$

Introducing (16-28) into (16-30) and rearranging, we get

$$\langle \zeta_j p_j^2 \rangle = \frac{\int_{\text{all } p,q \text{ except } p_j} \exp -\left(\frac{\mathscr{V}}{kT} + \frac{1}{kT} \sum_{i=1}^{N(\text{not } j)} \zeta_j p_j^2 \right) dq_1 \cdots dq_N \frac{dp_1 \cdots dp_N}{dp_j}}{\int_{\text{all } p,q \text{ except } p_j} \exp -\left(\frac{\mathscr{V}}{kT} + \frac{1}{kT} \sum_{i=1}^{N(\text{not } j)} \zeta_j p_j^2 \right) dq_1 \cdots dq_N \frac{dp_1 \cdots dp_N}{dp_j}}$$

$$\times \frac{\int_{-\infty}^{\infty} \exp -\left(\frac{\zeta_j p_j^2}{kT} \right) \zeta_j p_j^2 \, dp_j}{\int_{-\infty}^{\infty} \exp -\left(\frac{\zeta_j p_j^2}{kT} \right) dp_j} \tag{16-31}$$

The integrals over all values except p_j are identical in the numerator and denominator and drop out. We substitute $x = \sqrt{\zeta_j} p_j$ and get

$$\langle \zeta_j p_j^2 \rangle = \frac{\int_{-\infty}^{\infty} e^{-x^2/kT} x^2 \, dx}{\int_{-\infty}^{\infty} e^{-x^2/kT} \, dx} \tag{16-32}$$

The quantity x as used above is not a coordinate but is introduced simply as a variable of integration. Referring to tables of definite integrals, eq. (16-32) leads to

$$\langle \zeta_j p_j^2 \rangle = \frac{\frac{1}{2}\sqrt{\pi (kT)^3}}{\sqrt{\pi kT}} = \tfrac{1}{2} kT \tag{16-33}$$

Since the final average is independent of ζ_j, all degrees of freedom have the same average kinetic energy or thermal energy. Thus kinetic energy is equipartitioned through all degrees of freedom of a molecular system.

TRANSITION

Random thermal motion is a universal property of matter which is manifested in such phenomena as thermal denaturation, chemical reactions, diffusion, and Brownian motion. Diffusion theory allows us to study the time course of solute concentration in a nonuniform system. Brownian motion can be subjected to statistical analysis and, along with diffusion theory, allows us to relate the macroscopic diffusion constant to the microscopic frictional coefficient. As another aspect of thermal energy, the principle of equipartition of energy is derived from the Maxwell–Boltzmann distribution.

To make some of the abstractions that we have been dealing with more concrete, we now turn to an examination of the experimental determination of thermodynamic variables. These methods provide the ultimate data for our analyses in the various applications of thermal physics.

BIBLIOGRAPHY

Einstein, A., "Brownian Movement." Dover, New York, 1956.

An English version of the original papers by the master himself.

Jost, W., "Diffusion in Solids, Liquids, Gases." Academic Press, New York, 1952.

An advanced monograph on the theory and application of diffusion phenomena.

MacDonald, D. K. C., "Noise and Fluctuations." Wiley, New York, 1962.

An excellent compact introductory volume dealing with the physics of noise.

Page, L., "Introduction to Theoretical Physics." Van Nostrand-Reinhold, Princeton, New Jersey, 1952.

The mechanics and classical statistical mechanics used in this chapter are developed in this book.

Perrin, J. B., "Atoms." Constable Press, London, 1916.

A detailed report on the experiments on Brownian motion that provided support for Einstein's theory thus lending credence to the atomic theory of matter. This is an English translation of "Les Atomes" published in 1913 by F. Alcan, Paris.

Rashevsky, N., "Mathematical Biophysics." Univ. of Chicago Press, Chicago, Illinois, 1938.

Part I contains many interesting examples of the application of diffusion theory to biological systems.

17 Thermal Measurements

Since we have employed varying levels of abstraction throughout our discussion of thermodynamics, it is easy to lose sight of the fact that our formal structure ultimately rests on empirical generalizations. At each stage we must be able to relate our constructs to laboratory operations, experimental procedures which each of us can in principle carry out. In order to make our subject more concrete, it is worth pausing at this point and considering how measurements are actually made in thermal physics.

A review of the previous chapters will show that the empirical parameters of thermodynamics fall into two categories: those that are derivable from mechanics and analytical chemistry, and those that are exclusively thermodynamic. Among the former are quantities like volume, pressure, surface tension, mass, density, mole number, gravitational potential, and many others. We will not discuss the measurement of these mechanical quantities, as we can take their determination as something given apart from thermodynamics.

The purely thermodynamic quantities all relate to the measurement of temperature and ultimately derive from the temperature scale. Energy is a mixed case. It is clear from our discussion of the measurability of internal energy that all energy values can be reduced to mechanical (electrical) terms; nevertheless, data are frequently expressed in calorimetric form, which is dependent on temperature measurement. This situation has been described by Rossini† as follows:

† F. D. Rossini, "Chemical Thermodynamics." Wiley, New York, 1950. Copyright © by John Wiley and Sons.

> Notwithstanding the fact that practically all accurate calorimetric measurements made after about 1910 were actually based on electrical energy, most investigators continued until about 1930 to express their final results in such a way as to make it appear that the unit of energy was in some way still connected with the heat capacity of water. Actually, what they did was to convert their values, determined in international joules, into one or more of the several calories based on the heat capacity of water, usually for comparison with older values reported in calories in the literature. This procedure should have been reversed; that is, the older data should have been converted to the modern unit of energy.

Our task in this chapter is then threefold: (a) to discuss the experimental methods of measuring temperature, (b) to discuss the actual measurement of energy in thermodynamics and to relate the measures to calorimetric concepts, and (c) to discuss briefly the calorimeters and other devices which are used to determine thermochemical quantities such as heat of reaction, specific heats, thermal coefficients, and other parameters of interest.

By international agreement the triple point of water has a temperature of 273.16 K exactly. This triple point is produced for thermometric measurement using a triple-point cell as shown in fig. 17-1. The cells are glass cylinders 5 cm in diameter with reentrant coaxial walls for thermometers about 39 cm long and 1.3 cm inside diameter. Extending up from the cell is a tube which is sealed off above the cell after filling. After very exhaustive cleaning the cell is filled with very high purity water to within about 2 cm of the top and is then sealed off. To prepare the cell for measurement, it is first immersed in an ice bath. One should note that, since the triple point is only 0.0100 K above the ice point, once the triple point is achieved there will be very little heat exchange between the cell and the bath. Crushed dry ice is then placed in the well until a mantle of ice 3–10 mm thick forms on the outside of the well. The dry ice is then removed and the cell immersed below the water level of an ice bath so that ice-cold water fills the well. After equilibration the inner well constitutes a bath for thermometry at the triple point. These cells provide the experimental basis for the fixed point of the absolute thermodynamic temperature scale.

Given a fixed point, we next proceed to the fabrication of a thermometer. As we have proved, an ideal-gas thermometer should correspond to the Kelvin scale. Such thermometers are in fact used for this purpose. A device of this type is used at the United States National Bureau of Standards and consists of a 500-cm^3 spherical shell made of platinum–20% rhodium. This bulb is immersed in a very elaborate furnace made of four concentric copper shells. The furnace is designed to maintain very constant temperature and to keep the pressure outside and inside the bulb equal. The gas container

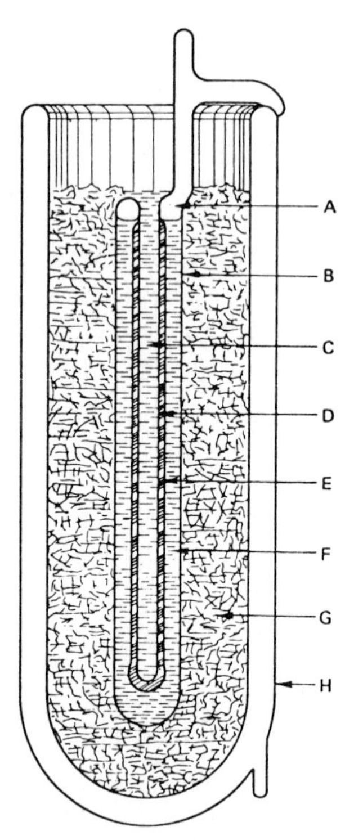

FIGURE 17-1. Triple-point cell; *A*, water vapor; *B*, Pyrex cell; *C*, water from ice bath; *D*, thermometer well; *E*, ice mantle; *F*, air-free water; *G*, flaked ice and water; *H*, insulated container.

connects to a precision mercury manometer which has a sensitivity of 2–3 nm of mercury. Helium is generally the gas of choice in this type of thermometry, although extensive comparisons between gases have been carried out.

The two devices we have just outlined are in principle sufficient for complete thermometry. They are not particularly easy to use and are in general employed only in standards laboratories. In practice each temperature range has its measurement of choice and a number of secondary standards are introduced. Thus, below 1 K, the magnetic susceptibility of a paramagnetic substance such as chromic potassium alum is used for accurate temperature measurement. Between 1 and 5.2 K, the vapor pressure of liquid helium has been used for thermometry. Platinum resistance thermometers are used over a wide temperature range.

Secondary standards are measured from the primary thermometric definitions and then used in practical thermometry. Thus the boiling point of oxygen is 90.168 K, the boiling point of water is 373.15 K, the boiling point of sulfur is 717.750 K, and the boiling point of silver is 1233.950 K.

For purposes of calorimetry the platinum resistance thermometer is probably the most frequently used measuring device. The instrument is in principle very simple and consists of very fine platinum wire usually helically wound on an electrically nonconducting rod. If the electrical resistance $\mathcal{R}$ is known at some calibration temperature T_0 to be $\mathcal{R}_0$, then we can always write an equation of the form

$$\mathcal{R} = \mathcal{R}_0 + \mathcal{R}_1(T - T_0) + \mathcal{R}_2(T - T_0)^2 + \mathcal{R}_3(T - T_0)^3 \cdots \quad (17\text{-}1)$$

Equation (17-1) follows from the fact that any reasonably well behaved function can be approximated arbitrarily closely by a power-series expansion. Each available standard temperature allows the determination of one more coefficient in the expansion of eq. (17-1).

The usefulness of the platinum resistance thermometer is then determined by

(a) the ease with which electrical resistance can be measured either by bridge circuits or by potentiometric methods;
(b) the very high degree of accuracy with which electrical quantities can be measured;
(c) the extremely good reproducibility of measurement on platinum coils.

The actual type of electrical measurement carried out varies considerably. To make the situation more concrete, consider the circuit in fig. 17-2. This will be familiar as a form of the Wheatstone bridge circuit. The thermometer

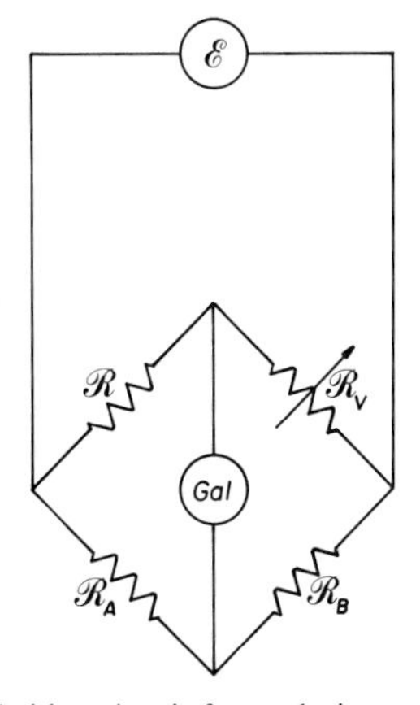

FIGURE 17-2. Wheatstone bridge circuit for a platinum resistance thermometer. $\mathcal{R}$ is the thermometer element and $\mathcal{E}$ is the applied voltage.

is designated $\mathscr{R}$ and is inserted in the bridge circuit as shown. $\mathscr{E}$ is a very steady source of electrical potential such as a battery and the galvanometer is designated *Gal.* $\mathscr{R}_A$ and $\mathscr{R}_B$ are two resistors of very precisely determined resistance. $\mathscr{R}_V$ is a variable resistance which can be controlled with great accuracy. To take a measurement, the platinum coil is placed in contact with the object whose temperature is being measured. $\mathscr{R}_V$ is then varied until the galvanometer reads no current. The condition for zero current is

$$\frac{\mathscr{R}}{\mathscr{R}_V} = \frac{\mathscr{R}_A}{\mathscr{R}_B} \tag{17-2}$$

Since $\mathscr{R}_A$, $\mathscr{R}_B$, and $\mathscr{R}_V$ are known with high precision, we have an accurate determination of $\mathscr{R}$ and, by use of eq. (17-1), a precise measurement of temperature.

As already mentioned, in the actual present-day applications of thermodynamics the measure of energy depends upon the application of electrical measurements. National Standards Laboratories maintain standard cells and resistances so that the appropriate quantities may be measured in absolute volts and absolute ohms. When a voltage of one absolute volt is placed across a resistance of one absolute ohm for one second, the energy dissipated is one absolute joule. Measurements based on the absolute joule now form the basis of calorimetry, and data are converted into the older thermal nomenclature by the conversion factor: one calorie equals 4.1840 absolute joules. All thermodynamic energy measurements can now be stated in these standards.

Once having settled on the measurement of temperature and energy, we can now discuss the measurement of specific heats. This exemplifies a general class of thermodynamic measurements and at the same time is of special importance since the determination of absolute entropies is based on heat-capacity measurements.

A sample is placed in a cell with a heating coil and a platinum resistance thermometer. The cell is adiabatically isolated. In practice this can be achieved by placing the cell in the middle of an evacuated chamber to minimize conductive and convective heat transfer. Radiative transfer is minimized by keeping the chamber walls at the same temperature as the cell. This is done with a feedback circuit operating heating coils in the wall. A measured current is made to flow at a preset voltage through the heating coil for a measured time and the temperature is determined. This gives $\Delta Q/\Delta T$ or the heat capacity. The calorimeter must first be run without a sample so that the heat capacity of the sample chamber can be computed and subtracted out. If the mass of the sample is then known, its specific heat can be computed. Variations of the apparatus are used over various temperature ranges but the basic principle of using the electrical measurements directly is usually followed.

Most of thermochemistry depends on the experimental determination of heats of reaction, of which the heat of combustion in oxygen is probably the most frequently used quantity. These reaction heats are determined in a bomb calorimeter. The reaction vessel or bomb is surrounded by a jacket of stirred calorimetric liquid, usually water. A weighed sample of substance is placed in the bomb along with high-pressure oxygen and the reaction is triggered with a hot electrical coil in the sample. The temperature rise of the calorimetric fluid is measured and this ΔT is a function of the heat of reaction. The calorimeter is calibrated either by reacting a standard substance of known ΔH or by heating the calorimetric liquid directly with a heating coil and measuring the temperature rise versus energy input as precisely determined from electrical measurements on the heater. For heats of combustion the most frequently used standard is the heat of combustion of high purity benzoic acid. This material is supplied by the U.S. National Bureau of Standards and certified to have a combustion heat of 26.433_8 absolute kilojoules per gram mass.

The actual carrying out of calorimetric experiments requires great care and the very brief outline we have presented does not indicate the elaborate experimental setup, the precise instrumentation, and the sophisticated computing programs which now go into the obtaining of thermodynamic data. The purpose of our survey has been to convey a sense of experimental reality to the concepts we have been discussing. Thermal physics, like all other sciences, stands or falls on experimental data, a fact which should always underlie the abstractions that are created to provide insightful analysis of the subject.

TRANSITION

Temperature measurements are ultimately referred to the triple point of water which is experimentally achieved in a triple-point cell. Ideal gas thermometers and defined fixed points provide a high-temperature scale. A number of secondary methods cover the available range of temperatures. Energy inputs for specific heat determinations are based on resistive heating and the measurements of voltages, currents, and resistance. Combustion calorimetry is referred to basic energy measurements by the use of such standards as benzoic acid.

In the next chapter we move from laboratory measurement to a much more philosophical aspect of bioenergetics, the biological implications of the information entropy formalism.

BIBLIOGRAPHY

Plumb, H. H. (ed.), "Temperature—Its Measurement and Control in Science and Industry" Volume 4. Instrument Society of America, Pittsburgh, Pennsylvania, 1972.

An encyclopedic work covering all aspects of temperature. Volumes 1, 2, and 3 are earlier versions. Not all the earlier material is found in volume 4.

Weber, R. L., "Heat and Temperature Measurement." Prentice-Hall, Englewood Cliffs, New Jersey, 1950.

A one-volume review of classical methods of measurement.

18 Information–Entropy Formalism in Biology

Earlier we developed the relation between entropy and information. We will now explore some biological implications of this relation. We proceed by a conceptual experiment which provides a focal point for our approach. We start with a living cell: as an example we will choose a bacterium. It can be grown in nutrient medium to give rise to an arbitrarily large number of essentially identical cells. By essentially identical, we mean that they all have, to a rough approximation, the same genome, volume, energy, and the same atomic composition. If we proceed to place each cell in a rigid adiabatic box of volume V, we will have an ensemble of isoenergetic systems. Allowing this ensemble to age for a very long time, all or most all of the cells will die, since a living cell is a very unlikely quantum state of the equilibrium system. In fact, in any real experiment that we could do, it would be impossible to get enough cells to actually have any living cells in the final ensemble. What we are carrying out is a purely thought experiment since no ensemble on the surface of the earth could ever be large enough for our purposes. What happens is that, as members of the ensemble age, the biochemical structures break down and the various adiabatic boxes will be filled with CO_2, CH_4, graphite, N_2, and other breakdown products of the cell. The temperature of most boxes will rise since, as equilibrium is approached, much of the covalent bond energy will be converted to thermal energy. In any case we are led to the concept of an ensemble having the atomic composition, volume, the energy of a known living cell.

As already indicated, all quantum states of the same energy are assumed to have the same *a priori* probability of occurring in the ensemble. Suppose that for the ensemble under consideration there are $\mathscr{W}$ possible quantum states and $\mathscr{X}$ of these correspond to the system being a living cell. From the experimental failure of life to ever spontaneously generate in an equilibrium ensemble we can conclude that $\mathscr{W}$ is very much greater than $\mathscr{X}$:

$$\mathscr{W} \gg \mathscr{X} \tag{18-1}$$

The probability of an ensemble member being in a living state is $\mathscr{X}/\mathscr{W}$, so that the amount of information we have in knowing that a cell is alive is

$$I = -\log_2 \frac{\mathscr{X}}{\mathscr{W}} \tag{18-2}$$

A somewhat more natural way of forming our ensemble would have been to isolate each cell in a rigid impermeable container and place it in contact with an infinite isothermal reservoir. The probability of such an ensemble member being alive would then be given by

$$\mathscr{P}_a = \frac{\sum_i \delta_{ia} e^{-\varepsilon_i/kT}}{\sum e^{-\varepsilon_i/kT}} \tag{18-3}$$

The symbol δ_{ia} is similar to the Kronecker δ and has a value one if the ith state is alive and a value zero if the ith state is not alive. Equation (18-3) follows directly from quantum statistical mechanics where the ε_i represent eigenenergies of all possible quantum states of the system. Now, all living states will have very close to the same energy, which we will designate as ε_a. If there are $\mathscr{X}$ such states and if we represent the partition function as Z, we can rewrite eq. (18-3) as

$$\mathscr{P}_a = \frac{\mathscr{X} e^{-\varepsilon_a/kT}}{Z} \tag{18-4}$$

The information content in knowing that a cell is alive is then

$$I = -\log_2 \mathscr{P}_a = -\frac{\ln \mathscr{P}_a}{0.693} \tag{18-5}$$

We can immediately substitute eq. (18-4) into eq. (18-5) and get

$$I = \frac{1}{0.693}\left[\frac{\varepsilon_a}{kT} - \ln \mathscr{X} + \ln Z\right] \tag{18-6}$$

If we now utilize eq. (8-22) which gives the relationship between Helmholtz free energy, partition function, and the other thermodynamic functions, we see that

$$I = \frac{1}{0.693}\left[\frac{\varepsilon_a - kT \ln \mathscr{X} - U + TS}{kT}\right] \tag{18-7}$$

Equation (18-7) may be rewritten as

$$I = \frac{1}{0.693}\left[\left(\frac{\varepsilon_a - U}{kT}\right) - \left(\frac{k \ln \mathscr{X} - S}{k}\right)\right] \tag{18-8}$$

The information richness of the biological system (large I) comes from two factors in eq. (18-8), one of which we can comprehend relatively easily and the second of which remains something of a mystery from the point of view of physics. The first term involves $(\varepsilon_a - U)$, the difference in energy between the living system and an equilibrium system at the same temperature. Living systems are energy rich. The process of photosynthesis leads to a storage of covalent bond energy. The difference $\varepsilon_a - U$ cannot be in thermal modes because it would very rapidly transfer to the reservoir if it existed in a kinetic form. The only way to maintain systems of the type we are discussing in an energy-rich state is to store the energy in potential form in molecules.

The chemistry of the overall photosynthetic process is usually represented as

$$6CO_2 + 6H_2O \rightarrow C_6H_{12}O_6 + 6O_2 \tag{18-9}$$

Equation (18-9) properly represents the reactants and products but neglects a very complex scheme of intermediate steps. The transformation produces the following changes in the various state functions (assuming the reactants and products are in their standard states):

$$\Delta G = 2872.3 \quad \text{kJ}, \qquad \Delta H = 2803.9 \quad \text{kJ}$$

$$\Delta S = -238.5 \quad \text{J/K}, \qquad \Delta U = 2816.3 \quad \text{kJ}$$

The optical pumping of photosynthesis thus provides the $\varepsilon_a - U$; that is, this process brings the biosphere to a high potential energy, which accounts for the first term in eq. (18-8).

The second term, which may be rewritten $(S/k - \ln \mathscr{X})$, is always positive, as S/k is large compared to $\ln \mathscr{X}$. The reasoning is the same as that involved in eq. (18-1); the total number of states is large compared with the number of living states. It is this term that is at the moment quite beyond the realm of thermodynamics since we have no real explanation of why, of all the possible quantum states, the biosphere is restricted to such a small subset. In any case, the first term in eq. (18-7) measures the energetic improbability and the second term represents the configurational improbability of the living state.

The very ordered state of a biological system would, if left to itself, decay to the most disordered possible state. For this reason work must constantly be performed to order the system. The continuous performance of this work requires a hot source and cold sink, which are ordinarily provided on the earth's surface by the heat of the sun and the cold of outer space.

Consider next the entropy changes which occur in the growth of a cell. The process we envision starts with a constant-volume flask containing sterile nutrient medium. The flask is in contact with an infinite isothermal reservoir. Into this flask we place a living cell and follow the processes which take place in cell growth and division so that the final product is two cells. The process may be represented as follows:

$$\text{cell} + \text{nutrients} \rightarrow 2 \text{ cells} + \text{waste products}$$

The initial cell emerges in the same state that it began in, so that it plays a catalytic role and we can rewrite the previous scheme as

$$\text{nutrients} \rightarrow \text{cell} + \text{waste products}$$

This process will in general be accompanied by a transfer of heat to or from the reservoir. The entropy change can then be divided into two parts, the entropy change of the system and the entropy change of the reservoir:

$$\Delta S_{\text{total}} = \Delta S_{\text{system}} + \Delta S_{\text{reservoir}} > 0 \tag{18-10}$$

Since the reaction is irreversible, we require that the total entropy change is positive. We may also write

$$\begin{aligned} \Delta S_{\text{system}} &= S_{\text{cell}} + S_{\text{waste products}} - S_{\text{nutrient}} \\ \Delta S_{\text{reservoir}} &= \frac{\Delta Q_{\text{reservoir}}}{T} = -\frac{\Delta Q_{\text{system}}}{T} \end{aligned} \tag{18-11}$$

In actual cases where the energy comes from nutrient molecules, cell growth appears to be exothermic (empirical generalization). The growth of photosynthetic cells must be treated by a different formalism. Equations (18-10) and (18-11) can be combined to give

$$\Delta S_{\text{total}} = \Delta S_{\text{system}} - \frac{\Delta Q_{\text{system}}}{T} > 0 \tag{18-12}$$

As long as the total entropy change is positive, the growth of cells is an example of the operation of the second law of thermodynamics. The source is this case is the stored energy of the nutrients and the sink is the reservoir. Work is done in ordering the precursors into a cell.

At all levels life is subject to the second law of thermodynamics. The sun as a high-temperature source and outer space as a low-temperature sink provide two effectively infinite isothermal reservoirs for the continuous performance of work. On the surface of the earth a portion of that work continually goes into building up ordered biological structures out of simple molecules such as CO_2, H_2O, N_2, NH_3, etc. Dissipative processes inherent in the random distribution of thermal energy act to constantly degrade biological structures and return the material to the small molecule pool. This tension between photosynthetic buildup and thermal degradation drives the global operation of the biosphere and leads to the great ecological cycles. The entire process generates entropy owing to the flow of energy from the sun to outer space, but the local changes may lead to great order.

TRANSITION

The information formalism allows us to express the information richness of living systems in terms of (a) an improbable storage of energy in high-energy electronic states, (b) a configurational unlikeliness of the living state when compared to states of similar energy. This takes us to the limit of a purely thermodynamic description of the system. To proceed further we must enter the study of irreversible processes and in the next chapter we shall commence that study.

BIBLIOGRAPHY

Morowitz, H. J., Some order–disorder considerations in living systems, *Bull. Math. Biophys.* **17**, 81 (1955).

A detailed view of the information content of bacterial cells.

Morowitz, H. J., "Energy Flow in Biology." Academic Press, New York, 1968.

A monograph on biological organization as a problem in thermal physics.

19 Irreversible Thermodynamics

A. NONEQUILIBRIUM SYSTEMS

Up until this point in our study, the notion of a thermodynamic system has been constrained largely to a time-invariant, spatially homogeneous entity. These are the characteristic features of equilibrium, the guiding construct of classical thermodynamics. No biological system is at equilibrium, so that traditional theory can only be used to obtain limiting statements about such objects rather than detailed predictions. Of course, parts of a living organism may be very close to equilibrium with respect to some processes while being far from equilibrium with respect to others. For example, the hydrostatic pressure across a membrane may have come to osmotic equilibrium, while both sides may be far from equilibrium with respect to chemical reactions which are taking place slowly compared to the rate of transmembrane water movement. As another example, a small poikilothermic (cold-blooded) animal may come into thermal equilibrium with its environment while remaining a far-from-equilibrium open system with respect to the uptake of food and discharge of waste.

In an effort to expand the useful domain of thermodynamics, the basic approach has been extended by the development of nonequilibrium theory and the thermodynamics of steady states, which allows further insights into near-to-equilibrium systems by such concepts as local variables, entropy flow, and entropy generation. The development of phenomenological relations between flows and forces and the Onsager reciprocity relations among cross coefficients then provide an approach to many problems of interest. The next few chapters will outline the structure of a near-to-equilibrium approach to nonequilibrium systems and discuss some uses of this type of theory in problems of biological interest.

At the outset we should point out certain limitations and difficulties. Real biological systems, because of their inherent complexity and inhomogeneity, do not conform to the idealized restrictions which are always demanded by theoretical analysis. A general theory does not yet exist which encompasses arbitrary systems regardless of their heterogeneity or condition with respect to equilibrium. In addition, thermodynamics, because of its focus on macroscopic variables, does not contribute directly to problems of molecular specifics; the main thrust is to explore what can be said of systems independent of microscopic detail. Fluctuation theory and the study of thermal noise is somewhat intermediate between the macroscopic and microscopic views. Once the macroscopic properties have been determined, one is free to raise the question as to whether a given molecular model is consistent with the thermodynamic constraints. This caveat is included because a number of people have tried to use irreversible thermodynamics, without molecular detail, to generate biological structure (structure in the broadest sense) and this has led to a sense of disappointment with the formalism. Complex systems require detailed kinetic and molecular models. Thermodynamics in any form cannot supply these minutiae, but can only deal with overall features, constraints, and consequences of particular models. It cannot, for example, tell us that proteins are enzymes or nucleic acids are the storers of genetic information. Such features are too detailed for this method of approach and must be either introduced *ad hoc* or derived from molecular chemistry and its quantum-mechanical base.

It is also important to realize that the areas of thermal physics we will discuss in the next few chapters are still in the formative stage and no universally accepted theory exists for all aspects of the subject. Different researchers approach the problems with quite varied viewpoints. This is part of the excitement, and constitutes a challenge to students not only to master what has been done but to develop new and improved outlooks. At this point in the book our task changes from mastering a generally used method to exploring and trying to comprehend ideas nearer to the forefront of scientific development.

In classical theory we deal with certain extensive variables such as mass, energy, and entropy which are dependent on the total size of the system. We also introduce a number of intensive variables such as pressure $(-\partial U/\partial V)$, temperature $(\partial U/\partial S)$, and chemical potential $(\partial U/\partial n_i)$, which, because they are derivatives of one extensive variable with respect to another, are independent of the system's size. Both the extensive and intensive variables of a classical equilibrium do not vary with time or with position in the homogeneous system so they may be regarded as global variables. This can be more explicitly seen if we recall that equilibrium requires long-time aging either in adiabatic isolation or in contact with an isothermal isobaric reservoir. Under these conditions all gradients eventually disappear and the system approaches a state of spatial homogeneity and time independence. Equilibrium thermodynamics is therefore sometimes referred to as thermostatics, reserving the more general term for time-dependent processes.

The simplest type of sustained nonequilibrium systems is one which is in contact with two reservoirs: one of which acts as a source, and the other as a sink, for energy in some form. For such cases we will develop two assymmetry theorems which indicate the most general type of organizational behavior for steady states where the flow of energy to the sink is by heat transfer.

Theorem I. *Sustained nonequilibrium systems are spatially inhomogeneous.*

In order to maintain a nonequilibrium system we require, as an irreducible minimum condition, that the system be in contact with a source and a sink. Both sources and sinks must be macroscopic, by definition; that is, the irreversible character of the reservoir-system-transfers demand macroscopic reservoirs. The fact that source and sink are macroscopic means that they cannot be homogeneously distributed in space. Therefore the source and sink interfaces must be separated by macroscopic distances. Since there are fluxes from the source to the sink, gradients must exist in the system and spatial homogeneity cannot obtain. Spatial assymetry is thus present in every sustained nonequilibrium system where the transfer to the sink is by heat flow.

Theorem II. *Sustained nonequilibrium systems are asymmetric in momentum space.*

Starting with the arguments of Theorem I, consider a surface across which a flux is occurring. The flux is given by

$$\mathbf{J}_{\mathfrak{w}} = \int_{-\infty}^{\infty} \frac{\mathbf{p}_n}{m} \mathfrak{W}(\mathbf{p})\mathscr{P}(\mathbf{p})\, dp_x\, dp_y\, dp_z \tag{19-1}$$

The momentum is $\mathbf{p}$, m is the mass, $\mathbf{p}_n$ is the component of momentum normal to the surface, and $\mathfrak{W}(\mathbf{p})$, the property being transported, is a scalar which may be a function of the momentum. For the kinds of fluxes involved in sustained nonequilibrium in material systems, $\mathfrak{W}(\mathbf{p})$ is always a scalar (i.e., mass, energy, charge). The quantity $\mathscr{P}(\mathbf{p})\, d\mathbf{p}$ is the number of molecules with momenta between $\mathbf{p}$ and $\mathbf{p} + d\mathbf{p}$. If $\mathscr{P}(\mathbf{p})$ is symmetric, that is if $\mathscr{P}(\mathbf{p}) = \mathscr{P}(-\mathbf{p})$, then the integral vanishes. Since this is contradictory to the assumption of flux we must assume $\mathscr{P}(\mathbf{p}) \neq \mathscr{P}(-\mathbf{p})$.

B. LOCAL VARIABLES

The nonequilibrium condition is, as noted, characterized by contact with more than one reservoir or contact with reservoirs that are changing as a function of time. Either of these conditions leads to gradients and spatial inhomogeneity so that if we wish to formulate a thermodynamics for such states we must be able to establish local values for all the parameters of interest. Since extensive variables cannot have local values we need first to develop a scheme for converting these variables to corresponding local ones. Intensive variables can be regarded as simply taking on local values.

Consider a system of volume V' which is small, but still large enough for macroscopic thermodynamic variables to have meaning. Associated with this volume will be the extensive quantities S, U, and m as well as other thermodynamic quantities. We now define

$$s_V = \frac{S}{V'}, \qquad u_V = \frac{U}{V'}$$
$$m_{iV} = \frac{m_i}{V'} = \rho_i, \qquad \rho = \frac{m}{V'} \tag{19-2}$$

m_i is the mass of the ith component in the volume element. Note that the local variable associated with mass is the density, ρ. The variable ρ_i designates the density of the ith component. As an alternative to the above formulation, we could convert global to local variables by dividing by the mass instead of the volume. Thus

$$s = \frac{S}{m} = \frac{s_V}{\rho}, \qquad u = \frac{U}{m} = \frac{u_V}{\rho}$$
$$\mathfrak{v} = \frac{V'}{m} = \frac{1}{\rho}, \qquad \mathfrak{N}_i = \frac{m_i}{m} \tag{19-3}$$

To evaluate local variables we focus on some arbitrary nonequilibrium system and conceptually divide it up into small-volume elements as above. Surround one of these volume elements by a rigid adiabatic wall and allow it to age to equilibrium. The quantities U and m are defined for the equilibrium system and, since they are conserved, the values are applicable for any nonequilibrium case as well. The method of isolation will fix V'. The quantities m_i can also be determined from the concentration of the chemical species. Concentration is a concept of universal validity not limited to equilibrium cases. We now make the assumption of local equilibrium which consists of two parts. First we start with the result of classical thermodynamics that a relation exists such that entropy is a homogeneous first-order function of the extensive parameters:

$$S = S(U, V, m_i) \tag{19-4}$$

Because of the formal properties of the function (the Euler relation), we can write (see Chapter 10)

$$s = \frac{S(U, V, m_i)}{m} = S\left(\frac{U}{m}, \frac{V}{m}, \frac{m_i}{m}\right) = S(u, \mathfrak{v}, \mathfrak{N}_i) \tag{19-5}$$

Next we assume that each small subvolume is close enough to equilibrium that the intensive values that can be computed from equation (19-5) have their usual meaning (postulate):

$$\left(\frac{\partial s}{\partial u}\right)_{\mathfrak{v},\mathfrak{N}_i} = \frac{1}{T}, \quad \left(\frac{\partial s}{\partial v}\right)_{u,\mathfrak{N}_i} = \frac{P}{T}, \quad \left(\frac{\partial s}{\partial \mathfrak{N}_i}\right)_{u,\mathfrak{v},\mathfrak{N}_j} = \frac{-\mu_i}{T} \tag{19-6}$$

The T, P and μ_i defined above are local variables. The μ_i in eq. (19-6) differ slightly from the chemical potentials used earlier. The N_i are mass fractions where previously the derivatives were taken with respect to mole numbers. The μ_i of eq. (19-6) is thus the usual chemical potential divided by the molecular weight.

These are the crucial assumptions of near-to-equilibrium thermodynamics. They embody two ideas: first, there exists a near-to-equilibrium entropy which may be defined by eq. (19-4); second, eq. (19-6) computes intensive variables which are experimentally meaningful. We establish a hypothesis that, although T, P, and μ_i may vary with position, the variation is sufficiently slow that at any point an approximate equilibrium condition exists. Our first departure from classical thermodynamics is thus from a global to a local equilibrium theory. Given (19-6), we can write

$$\begin{aligned} ds &= \frac{\partial s}{\partial u}\,du + \frac{\partial s}{\partial \mathfrak{v}}\,d\mathfrak{v} + \sum_i \left(\frac{\partial s}{\partial \mathfrak{N}_i}\right) d\mathfrak{N}_i \\ &= \frac{du}{T} + \frac{P}{T}\,d\mathfrak{v} - \sum \frac{\mu_i}{T}\,d\mathfrak{N}_i \end{aligned} \tag{19-7}$$

We thus have an analytical relation between the local near-to-equilibrium variables. Placing this in slightly different form we have

$$T\,ds = du + P\,d\mathfrak{v} - \sum \mu_i\,d\mathfrak{N}_i \tag{19-8}$$

This is a local formulation of the Gibbs equation and is the starting point of much of near-to-equilibrium theory. The validity of the approach must ultimately rest in experimentally verifying that the local entropy defined in eq. (19-5) will indeed yield eq. (19-6) when T, P, and μ_i are locally defined as indicated above and can be measured in some suitable way. If the system described above is allowed to decay to equilibrium, the preceding relations will yield classical thermostatics.

C. CONTINUUM FORMULATION

The primary purpose of nonequilibrium theory is to deal with processes, changes occurring within a system. In general a simple process can be described as a flow of matter, energy, electric charge, or some other quantity of interest. Mechanics deals with flows as trajectories of particles, while hydrodynamics extends this treatment to continuous media which are the objects of interest in many of the applications in biology. Most uses of irreversible theory have been in continuum type problems. To provide the background for this type of analysis we will develop two fundamental expressions of hydrodynamics: the equation of motion, and the equation of continuity. We will also introduce some notation from vector analysis that will be useful in subsequent studies.

Consider a small cube of fluid of dimensions dx, dy, and dz, whose center is located at a point in space x, y, z (fig. 19-1). The equation of motion is Newton's second law:

$$\mathbf{F} = m \frac{d^2\mathbf{r}}{dt^2} \tag{19-9}$$

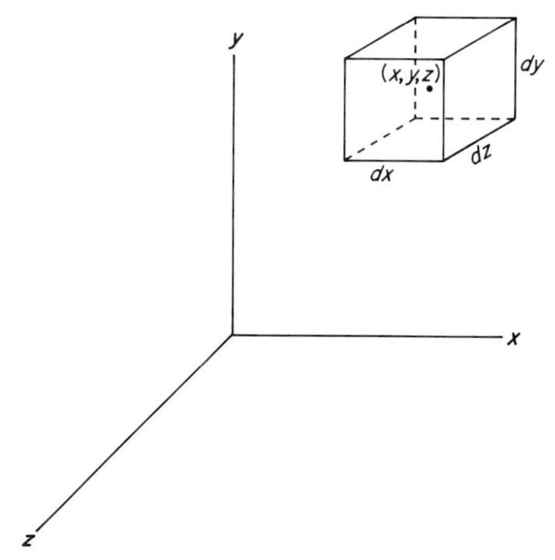

FIGURE 19-1. Element of volume $dV = dx\ dy\ dz$ with the center located at the point x, y, z.

The mass of the cube is $\rho\, dV$ where ρ is the density and dV is the volume, $dx\, dy\, dz$. The acceleration may be written $d\mathbf{v}/dt$, where $\mathbf{v}$ is the velocity. The total force may be split into two components: $\mathbf{F}_e$ due to external fields such as gravity, and $\mathbf{F}_P$ due to pressure in the fluid. $\mathbf{F}_e$ will be defined per unit mass so the net external force on the cube is $\mathbf{F}_e \rho\, dV$. Equation (19-9) may thus be written

$$\mathbf{F}_e \rho\, dV + \mathbf{F}_P\, dV = \rho\, dV \frac{d\mathbf{v}}{dt} \tag{19-10}$$

To determine $\mathbf{F}_P$ for an isotropic substance consider first the force in the x direction acting on the faces of the cube in the yz plane. The force on the left-hand face is

$$\mathbf{i} P\left(x - \frac{dx}{2}\right) dy\, dz \tag{19-11a}$$

and on the right-hand face

$$-\mathbf{i} P\left(x + \frac{dx}{2}\right) dy\, dz \tag{19-11b}$$

The symbol $\mathbf{i}$ stands for a unit vector along the x axis. Force is a vector quantity and the pressure force is perpendicular to the surface $dy\, dz$. Expanding each expression by the Taylor series,

$$\begin{aligned} P\left(x - \frac{dx}{2}\right) &= P(x) - \frac{\partial P}{\partial x}\frac{dx}{2} \\ -P\left(x + \frac{dx}{2}\right) &= -\left[P(x) + \frac{\partial P}{\partial x}\frac{dx}{2}\right] \end{aligned} \tag{19-12}$$

Adding the two terms to calculate the x component of the total force on the cube we get

$$\mathbf{F}_{Px}\, dV = -\mathbf{i}\frac{\partial P}{\partial x}\, dx\, dy\, dz = -\mathbf{i}\frac{\partial P}{\partial x}\, dV \tag{19-13}$$

The total pressure force on the cube can be obtained by adding similar terms in the y and z directions:

$$\mathbf{F}_P\, dV = -\left[\mathbf{i}\frac{\partial P}{\partial x} + \mathbf{j}\frac{\partial P}{\partial y} + \mathbf{k}\frac{\partial P}{\partial z}\right] dV = -\nabla P\, dV \tag{19-14}$$

where $\mathbf{j}$ and $\mathbf{k}$ are unit vectors along the y and z axes, and ∇ is the vector operator

$$\nabla = \mathbf{i}\frac{\partial}{\partial x} + \mathbf{j}\frac{\partial}{\partial y} + \mathbf{k}\frac{\partial}{\partial z} \tag{19-15}$$

Substituting (19-14) into (19-10) and rearranging we get

$$\mathbf{F}_e\rho - \nabla P = \rho \frac{d\mathbf{v}}{dt} \tag{19-16}$$

which is the equation of motion in the hydrodynamics of isotropic fluids.

Consider next the flow of mass in the x direction across the faces of the cube in the yz plane (fig. 19-1). At the left-hand face, it is

$$\rho v_x\left(x - \frac{dx}{2}\right)dy\,dz \tag{19-17a}$$

and, at the right-hand face, it is

$$\rho v_x\left(x + \frac{dx}{2}\right)dy\,dz \tag{19-17b}$$

The quantity v_x is the component of fluid velocity perpendicular to the yz plane. Since both v and ρ are functions of the coordinates, $\rho v_x(x)$ is a shorthand notation for $\rho(x)v_x(x)$. The flow in on the left minus the flow out on the right is the rate of change in the amount of material in the cube due to flow in this direction. Subtracting the two terms of (19-17) and setting the sum equal to the rate of change of mass in the volume leads to

$$-\frac{\partial(\rho v_x)}{\partial x}\,dx\,dy\,dz = \left(\frac{\partial \rho}{\partial t}\right)_x dV \tag{19-18}$$

Adding y and z contributions, we get

$$-\frac{\partial(\rho v_x)}{\partial x} - \frac{\partial(\rho v_y)}{\partial y} - \frac{\partial(\rho v_z)}{\partial z} = -\nabla \cdot (\rho \mathbf{v}) = \frac{\partial \rho}{\partial t} \tag{19-19}$$

Equation (19-19) has been designated the equation of continuity and along with eq. (19-16) provides the basis of solving many flow problems. Equation (19-19) is somewhat more general than was indicated above. Since ρ may be charge density or density of any other conserved quantity, an equation of this form may be used in many situations.

Another relation of a more formal character will prove useful in developing a thermodynamic theory of flow. Consider some intensive property of the fluid which usually will be a function of time, $\mathfrak{O}(x, y, z, t)$.

In general if we consider the time derivatives $d\mathfrak{O}/dt$, we note that

$$\frac{d\mathfrak{O}}{dt} = \frac{\partial \mathfrak{O}}{\partial t}\frac{dt}{dt} + \frac{\partial \mathfrak{O}}{\partial x}\frac{dx}{dt} + \frac{\partial \mathfrak{O}}{\partial y}\frac{dy}{dt} + \frac{\partial \mathfrak{O}}{\partial z}\frac{dz}{dt} \tag{19-20}$$

Since we are in a hydrodynamic flow field the derivatives dx/dt, dy/dt, and dz/dt are simply the components of the velocity v_x, v_y, and v_z. The expression may then be written

$$\frac{d\mathfrak{D}}{dt} = \frac{\partial\mathfrak{D}}{\partial t} + v_x \frac{\partial\mathfrak{D}}{\partial x} + v_y \frac{\partial\mathfrak{D}}{\partial y} + v_z \frac{\partial\mathfrak{D}}{\partial z} \tag{19-21}$$

This is usually expressed in vector operator notation as

$$\frac{d\mathfrak{D}}{dt} = \left(\frac{\partial\mathfrak{D}}{\partial t}\right)_{x,y,z} + \mathbf{v} \cdot \nabla\mathfrak{D} \tag{19-22}$$

This equation distinguishes between $\partial\mathfrak{D}/\partial t$, the rate of change of $\mathfrak{D}$ at some fixed point in space x, y, z, and $d\mathfrak{D}/dt$, the change in the value of $\mathfrak{D}$ at a point moving along in the hydrodynamic flow stream. For a vector quantity a similar relation may be constructed. Consider the vector $\boldsymbol{\Upsilon}$:

$$\frac{d\boldsymbol{\Upsilon}}{dt} = \frac{\partial\boldsymbol{\Upsilon}}{\partial t}\frac{dt}{dt} + \frac{\partial\boldsymbol{\Upsilon}}{\partial x}\frac{dx}{dt} + \frac{\partial\boldsymbol{\Upsilon}}{\partial y}\frac{dy}{dt} + \frac{\partial\boldsymbol{\Upsilon}}{\partial z}\frac{dz}{dt} \tag{19-23}$$

$$\frac{d\boldsymbol{\Upsilon}}{dt} = \frac{\partial\boldsymbol{\Upsilon}}{\partial t} + (\mathbf{v} \cdot \nabla)\boldsymbol{\Upsilon} \tag{19-24}$$

Use of eqs. (19-22) and (19-24) will allow us to recast some of our relations in a more useful form. For example using (19-24), eq. (19-16) may be rewritten

$$\frac{\partial\mathbf{v}}{\partial t} + (\mathbf{v} \cdot \nabla)\mathbf{v} = \mathbf{F}_e - \frac{1}{\rho}\nabla P \tag{19-25}$$

Another useful expression comes from multiplying eq. (19-22) by the density:

$$\rho\frac{d\mathfrak{D}}{dt} = \rho\frac{\partial\mathfrak{D}}{\partial t} + \rho\mathbf{v} \cdot \nabla\mathfrak{D} \tag{19-26}$$

The form of the derivative of a product then establishes

$$\frac{\partial(\mathfrak{D}\rho)}{\partial t} = \rho\frac{\partial\mathfrak{D}}{\partial t} + \frac{\mathfrak{D}\partial\rho}{\partial t} \tag{19-27}$$

Substituting (19-27) into (19-26) yields

$$\rho\frac{d\mathfrak{D}}{dt} = \frac{\partial(\mathfrak{D}\rho)}{\partial t} - \frac{\mathfrak{D}\partial\rho}{\partial t} + \rho\mathbf{v} \cdot \nabla\mathfrak{D} \tag{19-28}$$

The equation of continuity, (19-19), allows us to substitute for $\partial\rho/\partial t$:

$$\rho\frac{d\mathfrak{D}}{dt} = \frac{\partial(\mathfrak{D}\rho)}{\partial t} + \mathfrak{D}\nabla \cdot \rho\mathbf{v} + \rho\mathbf{v} \cdot \nabla\mathfrak{D} \tag{19-29}$$

We now introduce a general relation from vector analysis:

$$\boldsymbol{\nabla}\cdot \boldsymbol{\Upsilon}\mathfrak{D} = \mathfrak{D}\,\boldsymbol{\nabla}\cdot \boldsymbol{\Upsilon} + \boldsymbol{\Upsilon}\cdot\boldsymbol{\nabla}\mathfrak{D} \tag{19-30}$$

This allows us to rewrite eq. (19-29) as

$$\rho\,\frac{d\mathfrak{D}}{dt} = \frac{\partial(\mathfrak{D}\rho)}{\partial t} + \boldsymbol{\nabla}\cdot\mathfrak{D}\rho\mathbf{v} \tag{19-31}$$

The scalar product $\boldsymbol{\nabla}\cdot\boldsymbol{\Upsilon}$ is well known as the divergence of $\boldsymbol{\Upsilon}$. We will subsequently utilize eq. (19-31) in developing expressions for entropy flow.

In deriving the equation of continuity we made two assumptions: (a) the quantity of interest was conserved; (b) the quantity was transported only by hydrodynamic flow. The equation can be generalized by relaxing these conditions. Consider next a scalar quantity $\mathfrak{D}$ which is not conserved. If we designate the rate at which $\mathfrak{D}$ is produced or consumed per unit volume as $\sigma(x, y, z)$, then eq. (19-19) becomes

$$\frac{\partial\mathfrak{D}}{\partial t} = -\boldsymbol{\nabla}\cdot(\mathfrak{D}\mathbf{v}) + \sigma \tag{19-32}$$

Examples of conserved and nonconserved quantities are given in table 19-1.

TABLE 19-1

Conserved	Nonconserved
Mass	Moles of the ith substance
Charge	Heat
Total energy	Entropy

Relaxing condition (b) means the consideration of vector flows of $\mathfrak{D}$ due to processes other than being carried along in the hydrodynamic flow stream. Examples are given in table 19-2. If $\mathbf{J}$ represents the flow vector, then by an

TABLE 19-2

Quantity being transported	Process
Ions	Electrophoresis
Heat	Conduction
Molecules	Diffusion
Molecules	Centrifugation

argument similar to that used in the development of (19-19) we can show that the increase of $\mathfrak{D}$ due to such processes is $-\boldsymbol{\nabla}\cdot\mathbf{J}$. The full generalized equation is then

$$\frac{\partial \mathfrak{D}}{\partial t} = -\nabla(\mathfrak{D}\mathbf{v}) + \sigma - \nabla \cdot \mathbf{J} = \sigma - \nabla \cdot (\mathfrak{D}\mathbf{v} + \mathbf{J}) \tag{19-33}$$

If $\mathfrak{D}$ represented heat, σ would be generation of heat by chemical reactions, $-\nabla(\mathfrak{D}\mathbf{v})$ would be accumulation by hydrodynamic flow (convection), and $-\nabla \cdot \mathbf{J}$ would be accumulation by Fourier heat conduction. For conserved quantities σ is always zero.

Corresponding to intensive variable $\mathfrak{D}$ there is some extensive variable $\mathcal{O}$ so that $\mathfrak{D}$ is $\mathcal{O}/V'$ as indicated above. Equation (19-33) then contains (a) a source term $\sigma(\mathcal{O})$ which is the rate of production of $\mathcal{O}$ per unit volume per unit time; (b) a hydrodynamic term indicating the change due to fluid flow; (c) a flow term due to other processes.

The vector $\mathbf{J}$ is the flux vector and its form will depend on the particular parameter under consideration. For example, for diffusion $\mathbf{J} = D\,\nabla C$ where D is the diffusion constant and C the concentration.

D. ENTROPY FLOW AND ENTROPY PRODUCTION

The final form of writing eq. (19-33) sums the flow terms into a single term, the divergence of the total flow vector. The development of irreversible theory now proceeds by considering the local entropy as a physical quantity subject to the type of hydrodynamic treatment we have just been discussing. The formal way of making this assumption is to write eq. (19-33) for the local entropy and to formulate the local time derivative of the Gibbs equation. Thus since the entropy per unit volume is ρs we first write

$$\frac{\partial}{\partial t}(\rho s) = \sigma_s - \nabla \cdot (\rho s \mathbf{v}) - \nabla \cdot \mathbf{J}_s' = \sigma_s - \nabla \cdot \mathbf{J}_s \tag{19-34}$$

where $\mathbf{J}_s$ is the total entropy flow vector and $\mathbf{J}_s'$ is the vector due to other than hydrodynamic transport terms. The local time derivative of the Gibbs equation may now be written as

$$T\frac{\partial s}{\partial t} = \frac{\partial u}{\partial t} + P\frac{\partial \mathfrak{v}}{\partial t} - \sum \mu_i \frac{\partial \mathfrak{N}_i}{\partial t}$$

$$= \frac{\partial u}{\partial t} - \frac{P}{\rho^2}\frac{\partial \rho}{\partial t} - \sum \mu_i \frac{\partial \mathfrak{N}_i}{\partial t} \tag{19-35}$$

The second form of eq. (19-35) comes from substituting the reciprocal of the density, $1/\rho$, for the local volume per unit mass, $\mathfrak{v}$. To connect eqs. (19-34) and (19-35) we recall the features of eq. (19-33), the generalized equation of continuity, and write

$$\frac{\partial(\rho u)}{\partial t} + \nabla \cdot \mathbf{J}_u = 0 \tag{19-36}$$

$$\frac{\partial(\rho \mathfrak{N}_i)}{\partial t} + \nabla \cdot \mathbf{J}_i = \sigma_i \tag{19-37}$$

$$\frac{\partial \rho}{\partial t} + \nabla \cdot \mathbf{J}_m = 0 \tag{19-38}$$

Equation (19-36) is the conservation of energy relation in flow terms. The σ term is zero because total energy can be neither created nor destroyed. Equation (19-37) is the continuity equation for the ith chemical species, and σ_i is the local rate of production or consumption of that species in chemical reactions. Equation (19-38) is an extension of eq. (19-19) and is the conservation of mass relation. We now proceed to substitute eqs. (19-36), (19-37), and (19-38) into eq. (19-35) and multiply both sides by ρ/T. This leads to

$$\rho \frac{\partial s}{\partial t} = \frac{u}{T} \nabla \cdot \mathbf{J}_m - \frac{\nabla \cdot \mathbf{J}_u}{T} + \frac{P}{T\rho} \nabla \cdot \mathbf{J}_m - \sum \frac{\mu_i}{T} \mathfrak{N}_i \nabla \cdot \mathbf{J}_m$$
$$+ \sum \frac{\mu_i \nabla \cdot \mathbf{J}_i}{T} - \sum \frac{\mu_i \sigma_i}{T} \tag{19-39}$$

We then add $(s\, \partial\rho/\partial t)$ to the left side and the equal term $(-s\nabla \cdot \mathbf{J}_m)$ to the right side and collect terms:

$$\frac{\partial(\rho s)}{\partial t} = \rho \frac{\partial s}{\partial t} + s \frac{\partial \rho}{\partial t}$$
$$= \frac{(u + Pv - Ts - \sum \mu_i \mathfrak{N}_i)\nabla \cdot \mathbf{J}_m}{T} - \frac{\nabla \cdot \mathbf{J}_u}{T} + \sum \frac{\mu_i}{T} \nabla \cdot \mathbf{J}_i - \sum \frac{\mu_i \sigma_i}{T} \tag{19-40}$$

The term in parentheses multiplying $\nabla \cdot \mathbf{J}_m$ is identically zero as may be seen from eq. (10-36). We now use the vector relation (19-30) to recast (19-40) into somewhat different form. Thus

$$\frac{\partial(\rho s)}{\partial t} = -\nabla \cdot \frac{\mathbf{J}_u}{T} + \mathbf{J}_u \cdot \nabla \frac{1}{T} + \sum \nabla \cdot \frac{\mathbf{J}_i \mu_i}{T} - \sum \mathbf{J}_i \cdot \nabla \frac{\mu_i}{T} - \sum \frac{\mu_i \sigma_i}{T} \tag{19-41}$$

Collecting terms and comparing eqs. (19-41) and (19-34) we get

$$\mathbf{J}_s = \frac{1}{T}(\mathbf{J}_u - \sum \mu_i \mathbf{J}_i)$$
$$\sigma_s = \mathbf{J}_u \cdot \nabla \frac{1}{T} - \sum \mathbf{J}_i \cdot \nabla \left(\frac{\mu_i}{T}\right) - \sum \frac{\mu_i \sigma_i}{T} \tag{19-42}$$

Using the Gibbs equation and the fluid view of entropy allows us to get explicit expressions for the entropy flux $\mathbf{J}_s$ and the entropy production σ_s.

The last term in the entropy production can be cast in a somewhat different form. Consider that the reacting species are linked by stoichiometric constraints such as

$$\nu_A A + \nu_B B \leftrightharpoons \nu_C C + \nu_D D \tag{19-43}$$

The capital letters designate the chemical species and the ν's the stoichiometric coefficients. If the reactions proceed so that masses dm_A and dm_B are consumed and dm_C and dm_D are produced, then the stoichiometric constraints demand that

$$\frac{|dm_A|}{\nu_A M_A} = \frac{|dm_B|}{\nu_B M_B} = \frac{|dm_C|}{\nu_C M_C} = \frac{|dm_D|}{\nu_D M_D} \tag{19-44}$$

where the M's are the molecular weights of the various constituents. The ratio in eqs. (19-44) is designated $d\xi$ where ξ is defined as the degree of advancement of the reaction. The time derivative of ξ is simply the reaction velocity. If we define the stoichiometric coefficients to be negative on the left-hand side of (19-43) and positive on the right, we can then write

$$\sigma_i = \nu_i \frac{d\xi}{dt} \tag{19-45}$$

The last term in eq. (19-42) then becomes

$$-\sum \frac{\mu_i \sigma_i}{T} = -\frac{1}{T}\sum_i \mu_i \nu_i \frac{d\xi}{dt} = \frac{\mathfrak{A}}{T}\frac{d\xi}{dt} \tag{19-46}$$

where $\mathfrak{A}$, the affinity of the reaction, has been defined as $-\sum \mu_i \nu_i$. This quantity can be seen to be the difference in Gibbs free energy between the products and reactants. At equilibrium it goes to zero and in general may be regarded as a measure of how far a given chemical reaction is from an equilibrium state. The entropy production is

$$\sigma_s = \mathbf{J}_u \nabla \frac{1}{T} - \sum \mathbf{J}_i \nabla \frac{\mu_i}{T} + \frac{\mathfrak{A}}{T}\frac{d\xi}{dt} \tag{19-47}$$

We can thus regard entropy as a nonconserved fluid, since σ_s is a nonvanishing quantity.

The approach of eqs. (19-34) to (19-47) seems somewhat contrived and artificial and indeed it is. This stems from two factors; one is the postulational nature of eqs. (19-34) and (19-35), and the second is the arbitrariness in the choice of variables that we end up using in eqs. (19-42). The postulational nature might have been stressed by choosing new symbols for s, μ_i, T, P, etc., in the equations above since these are now the variables of nonequilibrium thermodynamics, and are not identical with the equilibrium variables. We could then have had a correspondence principle indicating that as the system approached equilibrium the quantities from eq. (19-4) and (19-8) would go over into the corresponding equilibrium values.

The arbitrariness in the choice stems from the fact that there is not a unique set of flow variables, but various linear combinations serve equally well. When we come to the Onsager formulation we will see a more natural choice.

Reference to eq. (19-47) might serve to stress the flexibility in the choice of variables. In the absence of bulk hydrodynamic flow, $\mathbf{J}_u$ may be split into two terms, heat flow and diffusion flow. Thus

$$\mathbf{J}_u = \mathbf{J}_q + \sum \mu_i \mathbf{J}_i \tag{19-48}$$

Equation (19-47) can now be rewritten as

$$\begin{aligned}\sigma_s &= \mathbf{J}_q \cdot \nabla \frac{1}{T} + \sum \mu_i \mathbf{J}_i \cdot \nabla \frac{1}{T} - \sum \mu_i \mathbf{J}_i \cdot \nabla \frac{1}{T} - \sum \frac{\mathbf{J}_i}{T} \cdot \nabla \mu_i + \frac{\mathfrak{A}}{T} \frac{d\xi}{dt} \\ &= \mathbf{J}_q \cdot \nabla \frac{1}{T} - \sum \frac{\mathbf{J}_i}{T} \cdot \nabla \mu_i + \frac{\mathfrak{A}}{T} \frac{d\xi}{dt}\end{aligned} \tag{19-49}$$

The entropy flow then becomes $\mathbf{J}_s = \mathbf{J}_q / T$. The bilinear form between the **J**'s and the gradients is still preserved in (19-49) although we have changed to a new set of variables.

It is well to stress, however, that irreversible thermodynamics does not follow as a deduction from classical thermodynamics. We therefore introduce new postulates and new methods of analysis. The ultimate validity, as always, is the correspondence between theory and experiment. Since the material we are now discussing is relatively new, the full range of experimental validity is untested and we are justified in maintaining reservations. However, nothing ventured, nothing gained, so we move ahead.

With the preceding in mind we examine the entropy generation in eq. (19-49). We note that σ_s equals a series of bilinear terms which are of the form of a flow or rate of change multiplied by a term which we can recognize as conjugate force. This is easily seen in the first two terms which are the product of the heat flow times the gradient of $1/T$ and the flow of each chemical species $\mathbf{J}_i$ multiplied by the gradient of the chemical potential.

The final term in eq. (19-42) is of different character than the others. Rather than being the scalar product of a vector force and vector flow, it is the product of two scalars, affinity and reaction rate. The bilinear form is, however, preserved with affinity playing the role of a force and reaction rate playing the role of a flow. Processes can be either scalar or vectorial in nature (or even of high tensor rank), but all enter into the production of entropy.

The entropy production, σ_s, results from processes within the system. These processes such as heat flow along a temperature gradient, matter flow along an activity gradient, and chemical reactions, give a clue to the detailed events involved in the entropy increase in nonequilibrium systems. The local statement of the second law of thermodynamics is that $\sigma_s > 0$, the local entropy generation is always positive. This does not mean that $\partial(\rho s)/\partial t$ need be positive since there is no restriction on the sign of $\nabla \cdot \mathbf{J}_s$.

We frequently find it convenient to use $T\sigma_s$, the product of the entropy production and the temperature. This quantity is defined as the dissipation. Equation (19-42) can be generalized to include all processes occurring in the system. Thus if we have a charge flow in an electrical potential $\mathscr{E}$ there is an additional term $-\mathbf{J}_e\mathscr{E}/T$ and the entropy production is increased by $-\mathbf{J}_e \cdot \nabla(\mathscr{E}/T)$

E. PHENOMENOLOGICAL EQUATIONS

The forms of the various vectors must be independently available from some considerations outside of thermodynamics. Thus if there are no hydrodynamic flows, the transport of chemical species will be entirely governed by Fick's first law of diffusion:

$$\mathbf{J}_i = D_i \nabla C_i \tag{19-50}$$

Equalities of the form (19-50) are called phenomenological equations. They relate flows to conjugate thermodynamic forces. Other examples of phenomenological relations are Fourier's law of heat conduction and Ohm's law of electrical current flow. These are usually written

$$\mathbf{J}_q = -\mathfrak{q}\nabla T \tag{19-51}$$

$$\mathbf{J}_e = -\mathfrak{g}\nabla\mathscr{E} \tag{19-52}$$

where $\mathfrak{q}$ and $\mathfrak{g}$ are the thermal and electrical conductivities, respectively.

The general phenomenological equations for the cases given above can be formulated in one dimension as

$$\mathbf{J}_i = L_i\mathbf{X}_i \tag{19-53}$$

where the subscript i now covers all possible flows including energy, mass, charge, and numbers of moles of each species. Each flow is proportional to an $\mathbf{X}_i$ term in the gradient of some intensive quantity. These $\mathbf{X}_i$ may be regarded as generalized thermodynamic forces which give rise to flows in nonequilibrium situations. However, it is found that a given flux may be due to more than one cause. Thus the flow of matter may result from a pressure gradient, a concentration gradient, or a thermal gradient. Equation (19-53) can be extended to read

$$\mathbf{J}_i = \mathbf{J}_i(\mathbf{X}_j) \tag{19-54}$$

For small values of $\mathbf{X}_j$ the Taylor expansion yields

$$\mathbf{J}_i = \mathbf{J}_i(0) + \sum_j \left(\frac{\partial \mathbf{J}_i}{\partial \mathbf{X}_j}\right)_0 \mathbf{X}_j + \text{higher-order terms} \tag{19-55}$$

We are considering isotropic cases where the component of $\mathbf{J}_i$ due to $\mathbf{X}_j$ is parallel to $\mathbf{X}_j$ so that $\partial \mathbf{J}_i/\partial \mathbf{X}_j$ is a scalar quantity. For the anisotropic case the derivative would have to be expressed in tensor form. $\mathbf{J}(0)$ is zero, since fluxes vanish in the absence of forces. We define a set of coefficients L_{ij} as

$$L_{ij} = \left(\frac{\partial \mathbf{J}_i}{\partial \mathbf{X}_j}\right)_0 \tag{19-56}$$

Equation (19-55) then becomes

$$\mathbf{J}_i = \sum L_{ij} \mathbf{X}_j \tag{19-57}$$

Equation (19-53) is thus seen as a special case of eq. (19-57) where all $\mathbf{X}_j$ but one are equal to zero and L_{ii} is written as L_i. The linear forms of eq. (19-57) stems from taking the expansion in the domain of very small $\mathbf{X}_j$. It has been experimentally ascertained, however, that for many phenomena the linearity holds over a rather wide range and eq. (19-57) is of general validity. This range is designated the linear domain and it is of special interest since it is simpler to analyze and subject to particular relations such as Onsager's reciprocity formula (see next chapter).

The choice of $\mathbf{X}_j$ in the first instance can be made by considering the equations such as (19-49) which can be cast in a very simple bilinear form if we define

$$\mathbf{X}_q = \nabla \frac{1}{T}, \qquad \mathbf{X}_i = -\frac{\nabla \mu_i}{T}, \qquad X_c = \frac{\mathfrak{A}}{T} \tag{19-58}$$

We can then write

$$\sigma_s = \mathbf{J}_q \cdot \mathbf{X}_q + \sum \mathbf{J}_i \cdot \mathbf{X}_i + X_c \frac{d\xi}{dt} \tag{19-59}$$

and the entropy production can always be represented as a sum of the products of thermodynamic forces and thermodynamic fluxes. Combining (19-51), (19-53), and (19-58) we can see that

$$\mathbf{J}_q = L_{qq} \mathbf{X}_q = L_{qq} \nabla \frac{1}{T} = -\mathfrak{q} \nabla T \tag{19-60}$$

from which it follows that

$$L_{qq} = \text{q}T^2 \tag{19-61}$$

We then have an algorithm for choosing the functional form of the forces and relating the phenomenological equations to the more abstract representation in eq. (19-57). As previously indicated, the Onsager formalism will provide us with a less arbitrary way of choosing forces and fluxes.

We now turn attention to the meaning of the L_{ij}'s for i not equal to j. A concrete case is a single chemical substance in a thermal gradient. Thus

$$\begin{aligned} \mathbf{J}_q &= L_{qq}\nabla\left(\frac{1}{T}\right) - L_{q1}\frac{\nabla\mu_1}{T} \\ \mathbf{J}_1 &= -L_{11}\frac{\nabla\mu_1}{T} + L_{1q}\nabla\left(\frac{1}{T}\right) \end{aligned} \tag{19-62}$$

The first equation asserts that the flow of heat depends on contributions due to the temperature gradient and the gradient of chemical potential. Even if the temperature gradient went to zero, a flow of matter due to a gradient of μ would carry along its thermal energy and a heat flow would result. Similarly, a temperature gradient, even at constant concentration, would lead to a flow of molecules from the hotter to the colder region. The processes are linked because mass flow and heat flow are different properties of the same molecule. This necessary coupling is an important aspect of irreversible thermodynamics. It led to Onsager's formulation that L_{ij} equals L_{ji}. In the case that two processes are not coupled both L_{ij} and L_{ji} are zero. Going back to eq. (19-59) we note that $\mathbf{J}_q$ and $\mathbf{J}_i$, which are coupled, are both vector quantities. On the other hand, the velocity of chemical reaction, $d\xi/dt$, is a scalar quantity that, from symmetry conditions, cannot be coupled to a vector quantity. Thus L_{cq} and L_{qc} will both be zero. Of course, in dealing with chemical reactions that are not in homogeneous phase due to interfaces, oriented catalysts, or other directing forces, there can emerge quantities analogous to ξ that are vectorial. Such oriented chemical reactions may be of importance in problems such as oxidative phosphorylation at the inner mitochondrial membrane.

The entropy production which was expressed for a special case in (19-59) can be generalized and we are able to write

$$\sigma_s = \sum_j \mathbf{J}_j \cdot \mathbf{X}_j \tag{19-63}$$

where the summation is over all the forces and fluxes operative in the system. Equations (19-57) and (19-63) are then the starting points for analysis of near-to-equilibrium systems.

TRANSITION

The move from classical equilibrium thermodynamics to a formalism for near-to-equilibrium systems starts with a method for assigning local values to extensive quantities. We relate these local variables through the Gibbs equation. To deal with generalized flow problems we develop the hydrodynamic equation of motion and equation of continuity. The latter allows us to express mass and energy conservation in terms of local variables. When thermodynamics is recast into this hydrodynamic form we can define an entropy flow vector $\mathbf{J}_S$ and an entropy production σ. Entropy production involves the internal flows under forces within the system. The second law of thermodynamics can be locally formulated as $\sigma \geqslant 0$.

To derive certain relations between forces and fluxes within the system, we need to consider fluxuations at equilibrium. These time-varying changes provide a bridge between equilibrium and kinetic aspects of thermal physics. In Chapter 20 we start with experimental aspects of fluctuations and move to a consideration of the linkages of various fluxes under nonequilibrium conditions.

BIBLIOGRAPHY

Glansdorff, P., and Prigogine, I., "Structure, Stability and Fluctuations." Wiley (Interscience), New York, 1974.

An advanced treatment stressing the relation of thermodynamics and the point of view of hydrodynamics.

Prigogine, I., "Introduction to Thermodynamics of Irreversible Processes." Wiley, New York, 1955.

A short introduction to near-to-equilibrium thermodynamics.

Yourgrau, W., Van der Merwe, A., and Raw, G., "Treatise on Irreversible and Statistical Thermophysics." Macmillan, New York, 1966.

New concepts in nonclassical thermodynamics are discussed from the physicist's point of view.

20 Fluctuations

A. FLUCTUATING VARIABLES

Before we can undertake a discussion of Onsager's formulation of near-to-equilibrium processes, we will examine the fluctuations which occur in a macroscopic system. This involves an extension of some ideas first introduced in our discussion of thermal noise in Chapter 16.

Consider a volume V of a dilute aqueous solution of molecules. If there are n_i moles of solute, the corresponding intensive variable, the concentration or C_i, is given by n_i/V. Next assume that we are able to isolate a number of subvolumes $\mathfrak{v}$. We ask the question: will they all have the same value of C_i? The answer is no, since all of the solute molecules are in constant thermal motion so there is an ongoing random transport of molecules in and out of each subvolume. This leads to a fluctuating value of C_i so that if we could measure C_i in a single subvolume as a function of time we would get a result like that shown in fig. 20-1. Curves of this type contain a great deal of information and we shall consider some of the features.

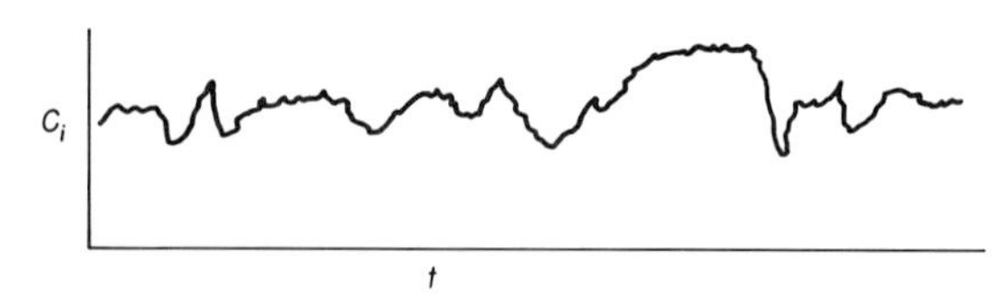

FIGURE 20-1. C_i, a random variable, fluctuates around some mean value, $\bar{C}_i$.

(1) The mean value of C_i which we will designate as $\bar{C}_i$. This is the quantity that would emerge from a thermodynamic measurement and would also be the ensemble average from statistical mechanics. In fluctuation analysis $\bar{C}_i$ is the time average. In designating averages we sometimes use a bar over the symbol as in $\bar{C}_i$ and sometimes use brackets as in $\langle C_i \rangle$. It is convenient to retain these two methods for typography and in the case of the Onsager relations to conform to the original papers.

(2) The mean-square deviation $\overline{(C_i - \bar{C}_i)^2}$. This time average is a measure of the magnitude of fluctuations about the mean. The relative size of the deviation gives information about the number of events involved or the "graininess" of the phenomenon.

(3) The frequency spectrum. A graph such as fig. 20-1 is made up of a continuum of different frequencies deriving from many components of the fluctuation process. It therefore contains information about the kinetics of the system: the response time, the viscous damping, and other such features.

(4) Correlation information. A study of quantities such as the time average of $C_i(t + \tau)C_i(t)$, which is known as the autocorrelation function, contains additional information about the kinetics of the system and the processes taking place within the system. The quantity τ is a time parameter separating the two observations of C_i.

Thinking about the preceding quantities gives some insights as to why fluctuation analysis can take us far beyond the equilibrium view. Equilibrium thermodynamics deals only with the average $\bar{C}_i$. The mean-square deviation is available from the ensemble distribution of values and thus relates to statistical mechanics. In general, to extend the treatment to include the deviation, we must specify something about the molecular nature of the system. Frequency spectroscopy and autocorrelation analysis take us completely beyond equilibrium theory and make contact with the kinetics and molecular details of the processes taking place in the system. The analyses of these phenomena therefore provide the bridge between equilibrium and nonequilibrium theories.

B. LIGHT SCATTERING

Before proceeding to a more detailed view of the theories, we will consider a case where we can view the macroscopic consequences of fluctuations; that is, where we carry out a measurement that is proportional to $(C_i - \bar{C}_i)^2$. Light scattering from a binary solution provides just such an example. The experimental setup is shown in fig. 20-2. Polarized light impinges upon the sample. Most of the light passes through and is measured by photocell 1.

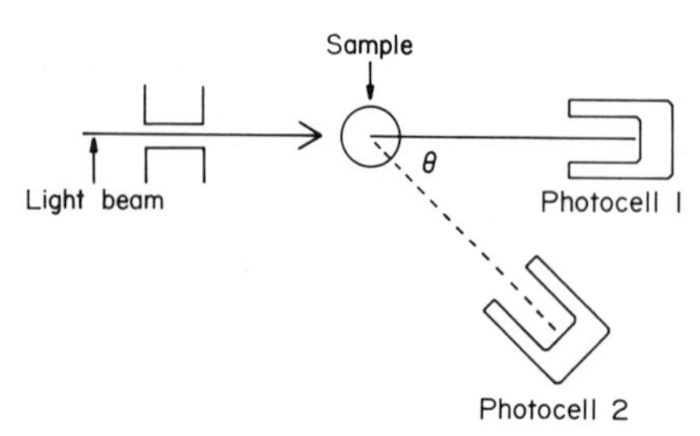

FIGURE 20-2. A narrow beam of polarized light impinges on sample. Most of the light is directly transmitted and is measured by a photocell. A small fraction is scattered at each angle, θ, and may be measured by an appropriately placed second photocell.

This is designated I_0. The amount of light scattered at angle θ is measured by photocell 2 and is designated I_θ.

Each small subvolume of the sample scatters light according to the formula (from electromagnetic theory)

$$\frac{I_\theta}{I_0} = \frac{16\pi^4 \alpha^2 \sin^2\theta_1}{\lambda^4 r^2} \tag{20-1}$$

where α is the polarizability, θ_1 the angle between the incident and scattered beam, λ the wavelength, and r the distance from the scattering center.

If the sample is a perfect crystal, α is everywhere the same and the situation is shown in fig. 20-3. Subvolume 1 scatters as much light to subvolume 2 as subvolume 2 scatters back since $\sin^2\theta_1 = \sin^2(\theta_1 + \pi)$. The two beams are exactly out of phase and cancel so there is no net scattering from the crystal.

If the sample is a binary solution, then α for each subvolume is $\alpha + \delta\alpha$ where the fluctuations in the polarizability α are due to the fluctuations in

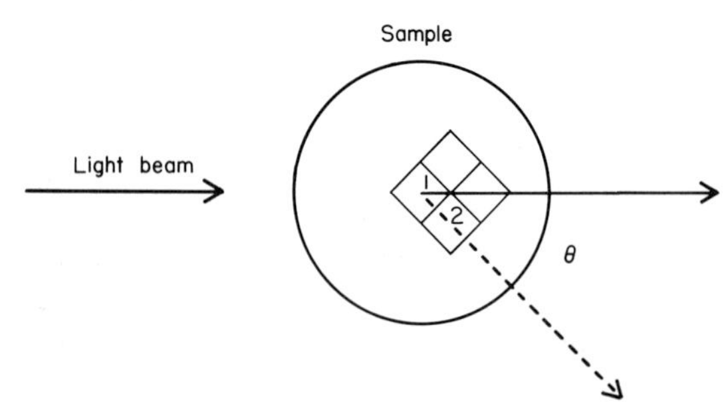

FIGURE 20-3. Detail of the light scattering from a sample. For a perfect crystal volume 1 will scatter the same amount of light to volume 2 as the reverse process.

concentration of the solute which will be designated by the subscript 2. Thus we have

$$\delta\alpha = \frac{\partial \alpha}{\partial C_2}\,\delta C_2 = \frac{\partial \alpha}{\partial C_2}(C_2 - \bar{C}_2) \tag{20-2}$$

Thus each subvolume contributes to the scattering

$$\frac{I_\theta}{I_0} = \frac{16\pi^4 \sin^2\theta_1[\alpha^2 + 2\alpha\,\delta\alpha + (\delta\alpha)^2]}{\lambda^4 r^2} \tag{20-3}$$

If we add up all the subvolumes, they form an ensemble. The α^2 term is the symmetric term identical to that in the perfect crystal case and leads to no net scattering. Since fluctuations are random and symmetrical about the mean, the average of $\delta\alpha$ vanishes. This leaves

$$\frac{I_\theta}{I_0} = \frac{16\pi^4 \sin^2\theta_1 N\left(\dfrac{\partial \alpha}{\partial C_2}\right)^2 \overline{(C_2 - \bar{C}_2)^2}}{\lambda^4 r^2} \tag{20-4}$$

where N is the number of subvolumes in the sample. Without going into further detail, we note that the scattered light depends on the mean-square deviation of the concentration. Since all the terms in (20-4) except $\overline{(C_2 - \bar{C}_2)^2}$ are measurable, we can get the fluctuation values directly from experiment.

In some favorable cases, it is possible to measure $C_i(t)$ directly. Consider for example that C_i is a fluorescent species. If we irradiate a subvolume with a saturating amount of exciting light, the fluorescent emission will be directly proportional to $C_i(t)$. The constant of proportionality can be measured from macroscopic experiments. The experiments Koppel *et al.* referred to in the Bibliography provide an example of this technique. A fluorescent dye was adsorbed into a lipid membrane where it is free to diffuse in two dimensions. A tiny spot of laser light the order of $(1\ \mu\text{m})^2$ illuminated the membrane, and the fluorescence was measured with a photomultiplier and associated circuitry. The measured intensity then provides the input data for fluctuation spectroscopy analysis.

Similarly, Brownian motion (see Chapter 16) depends on a fluctuation in the collisions of a visible particle with the surrounding molecules. In Brownian motion we are observing some statistical consequences of fluctuations occurring at the molecular level. Indeed Perrin (see Bibliography) was able to use Brownian motion to measure Avogadro's number. This is a striking example of how the structure of matter shows up in noise analysis.

The fluctuations of electrons in a resistor leads to a time-varying voltage across that circuit element. Sensitive enough instrumentation to measure these fluctuations was developed in the 1920s, and in 1928 J. B. Johnson reported on a series of such measurements. These fluctuations are often referred to as *Johnson noise*. In parallel with this experimental work, H. Nyquist developed a general theory of random thermal noise in electrical components. The theoretical expressions involved the Boltzmann constant, and the value of 1.27×10^{-16} ergs/deg obtained by Johnson was close enough to other values to support the Nyquist theory. The Bibliography refers to these two important early papers on electrical fluctuations.

C. *FLUCTUATION ANALYSIS OF AN ISOMERIZATION REACTION*

We will examine the case of a simple chemical reaction to demonstrate the kinds of information that can be obtained from the study of fluctuations. Consider the isomerization

$$A \underset{k_2}{\overset{k_1}{\rightleftharpoons}} B \tag{20-5}$$

where k_1 and k_2 are the rates of the forward and reverse reactions. Assume that we allow the reaction to run to equilibrium and then measure $n_A(t)$. We assume that n_A can be continuously monitored spectroscopically or by some other technique and we anticipate a curve similar to fig. 20-1. At equilibrium the result of classical thermodynamics is

$$\frac{\langle n_B \rangle}{\langle n_A \rangle} = K = e^{-\Delta G/RT} \tag{20-6}$$

where ΔG is the Gibbs free-energy difference between products and reactants and K is the equilibrium constant. $\langle n_B \rangle$ and $\langle n_A \rangle$ are the long-time averages and are therefore the equilibrium thermodynamic variables. A further postulate usually made is that detailed balance obtains at equilibrium and, assuming mass action kinetics,

$$k_1 \langle n_A \rangle = k_2 \langle n_B \rangle \tag{20-7}$$

This is an "extra-thermodynamic" assumption and in more complex cases is important to our analysis of nonequilibrium systems. The postulate will be discussed in greater detail later in the chapter. If we further note that $n_A + n_B = n_T$, the total number of molecules, then the measurement of $\langle n_A \rangle$ gives the ratio of k_2 to k_1:

$$\frac{\langle n_A \rangle}{n_T - \langle n_A \rangle} = \frac{k_2}{k_1} \tag{20-8}$$

Suppose we perturb the equilibrium system a substantial amount by Δn_A. We note, from the usual kinetic analysis,

$$\frac{dn_A}{dt} = -k_1 n_A + k_2 n_B$$
$$\frac{d(\langle n_A \rangle + \Delta n_A)}{dt} = -k_1(\langle n_A \rangle + \Delta n_A) + k_2(\langle n_B \rangle - \Delta n_B) \tag{20-9}$$

which from eq. (20-7) and the fact that $\Delta n_A + \Delta n_B = 0$ leads to

$$\Delta n_A = \Delta n_A(0) e^{-(k_1 + k_2)t} \tag{20-10}$$

Thus a perturbed system decays back to equilibrium with a time constant of $1/(k_1 + k_2)$.

We now consider two quantities that can be computed from $n_A(t)$: the mean-square deviation $\langle (n_A - \langle n_A \rangle)^2 \rangle$ and the autocorrelation function for the deviation $\langle (n_A(t) - \langle n_A \rangle)(n_A(t + \tau) - \langle n_A \rangle) \rangle$. We define δn_A as $(n_A - \langle n_A \rangle)$, the fluctuation of n_A, and proceed to compute the preceding two quantities:

$$\langle (\delta n_A)^2 \rangle = \langle n_A{}^2 \rangle - 2\langle n_A \rangle \langle n_A \rangle + \langle n_A \rangle^2 = \langle n_A{}^2 \rangle - \langle n_A \rangle^2 \tag{20-11}$$

and

$$\mathscr{U}(\tau) = \langle \delta n_A(t)\, \delta n_A(t + \tau) \rangle \tag{20-12}$$

The symbol $\mathscr{U}(\tau)$ is introduced to designate the autocorrelation function. To evaluate the autocorrelation function we need one further assumption regarding the regression of fluctuations. First consider $\mathscr{U}(\tau)$ for very large τ. The two values of δn_A will be uncorrelated and the product will be either positive or negative with equal probability and, because of symmetry considerations, it will have a zero average. As τ approaches zero $\mathscr{U}(\tau)$ can be seen to approach $\langle (\delta n_A)^2 \rangle$. *For intermediate value of τ we assume that a fluctuation will, on the average, regress with the same time constant as would be observed for a macroscopic perturbation.* This is the postulate that relates fluctuation theory to macroscopic kinetics. Its validity ultimately depends on agreement of predictions with experiment. We argue that the processes driving a perturbed system to equilibrium are the same as those causing the regression of fluctuations. Although fluctuations are unpredictable we can meaningfully talk about their average behavior. We have

$$\mathscr{U}(\tau) = \langle (\delta n_A)^2 \rangle e^{-(k_1 + k_2)\tau} \tag{20-13}$$

The autocorrelation function will thus give us the kinetic constant $k_1 + k_2$. Along with eq. (20-8), the fluctuation behavior allows us to determine all the equilibrium as well as kinetic information. A study of fluctuations at equilibrium conveys a great deal of information about the dynamics of the system. This result is general and we can expect fluctuation spectroscopy to become an increasingly valuable tool in biological research.

The use of fluctuation spectroscopy to study chemical reaction kinetics has been explored both experimentally and theoretically by Feher and Weissman. A report of their work is listed in the Bibliography. Their work forms a foundation for this field and should be consulted by readers wishing a more detailed treatment than has been presented above.

The preceding discussion indicates the possibility of representing a non-equilibrium system as a fluctuation of some corresponding equilibrium system. Thus any nonequilibrium state is an ensemble member of the micro-canonical ensemble having the same total energy. The rate process will then correspond to a regression of a fluctuation. We can define the system in terms of a number of macroscopic parameters $\mathcal{O}$ which have equilibrium reference values of $\langle \mathcal{O} \rangle$. A new set of parameters $\alpha_i = \mathcal{O} - \langle \mathcal{O} \rangle$ will now be chosen to characterize the system. The α_i are zero at equilibrium which presents a convenience in computations.

D. ENTROPY IN NONEQUILIBRIUM SYSTEMS

The next problem is to find a suitable function to represent the entropy when the system is not at equilibrium. This is a new problem, not contained in classical thermodynamics, and therefore involves an extension of that theory. In order to provide a heuristic device for this development we start with a microcanonical ensemble. The entropy according to the Boltzmann formulation is

$$S = k \ln \mathscr{W} \tag{20-14}$$

where $\mathscr{W}$ is the total number of possible microstates corresponding to the macroscopic constraints. Next consider a nonequilibrium state of the same systems characterized by a set of fluctuation variables $\alpha_1, \alpha_2, \alpha_3, \ldots, \alpha_n$. Individual variables will be designated α_i or α_j. The state either could occur as a fluctuation in a microcanonical ensemble or could be established by boundary conditions. Out of the $\mathscr{W}$ possible states, a small subset $\mathscr{W}_\alpha$ corresponds to the conditions listed above. We then define the entropy of the state $\alpha_1, \ldots, \alpha_n$ as $S_1 = k \ln \mathscr{W}_\alpha$. The change of entropy in going to the fluctuation state is

$$\Delta S = S_\alpha - S = k \ln \mathscr{W}_\alpha / \mathscr{W} \tag{20-15}$$

Since $\mathscr{W}_\alpha$ is a subset of the $\mathscr{W}$ possible states, the ratio $\mathscr{W}_\alpha / \mathscr{W}$ is less than one and ΔS is always negative. Thus, the entropy of the equilibrium state is a maximum and any states in the neighborhood must be of lower entropy.

The above treatment does not recognize that the α's, being macroscopic variables, are classical and should be able to vary continuously and, in addition, constrains us to consideration of microcanonical ensembles. We can rephrase our inquiry by asking what is the probability that the state α_1 to $\alpha_1 + d\alpha_1$, α_2 to $\alpha_2 + d\alpha_2$, etc., will occur in the equilibrium ensemble. We call this probability $\mathscr{P}(\alpha_1, \ldots, \alpha_n)\, d\alpha_1, \ldots, d\alpha_n$. Corresponding to the ratio $\mathscr{W}_\alpha/\mathscr{W}$ we now have

$$\frac{\mathscr{W}_\alpha}{\mathscr{W}} \rightarrow \frac{\int_{\alpha}^{\alpha+\Delta\alpha} \mathscr{P}(\boldsymbol{\alpha})\, d\boldsymbol{\alpha}}{\int_{\text{all values of } \alpha} \mathscr{P}(\boldsymbol{\alpha})\, d\boldsymbol{\alpha}} \tag{20-16}$$

In eq. (20-16) we represent the fluctuations $\alpha_1, \alpha_2, \alpha_3, \ldots, \alpha_n$ by the vector $\boldsymbol{\alpha}$. In this notation a differential element of volume $d\boldsymbol{\alpha}$ will correspond to $d\alpha_1\, d\alpha_2\, d\alpha_3 \cdots d\alpha_n$.

$\Delta\boldsymbol{\alpha}$ is a measure of the size of the subset of states. The fine-grained character of the distribution will depend on the precision with which $\Delta\boldsymbol{\alpha}$ is chosen. If $\Delta\boldsymbol{\alpha}$ is small enough, $\mathscr{P}(\boldsymbol{\alpha})$ is constant over the integration and we can rewrite (20-16) as

$$\frac{\int_{\alpha}^{\alpha+\Delta\alpha} \mathscr{P}(\boldsymbol{\alpha})\, d\boldsymbol{\alpha}}{\int_{\text{all values of } \alpha} \mathscr{P}(\boldsymbol{\alpha})\, d(\boldsymbol{\alpha})} = \frac{\mathscr{P}(\boldsymbol{\alpha})\, \Delta\boldsymbol{\alpha}}{\sum_k \mathscr{P}(\boldsymbol{\alpha}_k)\, \Delta\boldsymbol{\alpha}} = \frac{\mathscr{P}(\boldsymbol{\alpha})}{\sum_k \mathscr{P}(\boldsymbol{\alpha}_k)} = \frac{\mathscr{P}(\boldsymbol{\alpha})}{\mathscr{P}_0} \tag{20-17}$$

$\boldsymbol{\alpha}_k$ represents a possible value of the vector $\boldsymbol{\alpha}$ and we sum over all such values. Equations (20-15)–(20-17) can now be combined to yield

$$e^{\Delta S/k} = \frac{\mathscr{P}(\boldsymbol{\alpha})}{\mathscr{P}_0} \tag{20-18}$$

Since $\mathscr{P}(\boldsymbol{\alpha})$ is not constrained to a microcanonical ensemble, relation (20-18) is quite general and may be used to define a nonequilibrium entropy function:

$$S = S_{eq} + \Delta S \tag{20-19}$$

As the system goes to equilibrium, the numerator of eq. (20-16) will go over all possible values of $\boldsymbol{\alpha}$, that is, $\mathscr{P}(\boldsymbol{\alpha}) \rightarrow \mathscr{P}_0$, and ΔS will go to zero. Thus the nonequilibrium formalism is continuous with classical thermodynamics.

Next consider how $\mathscr{P}(\boldsymbol{\alpha})$ might be evaluated. As a solution to the equilibrium state of classical statistical mechanics we represent the system in a $6N$-dimensional space, $3N$ values of the coordinates q and $3N$ values of the momenta p, and solve for the probability density at any point in the space. The fluctuation state $\alpha_1, \ldots, \alpha_n$ in $\boldsymbol{\alpha}$ space corresponds to some subvolume in pq space. We therefore require that

$$\frac{\int_{\mathbf{p},\mathbf{q}}^{\mathbf{p}+\Delta\mathbf{p},\mathbf{q}+\Delta\mathbf{q}} \mathscr{P}(\mathbf{p},\mathbf{q})\,d\mathbf{p}\,d\mathbf{q}}{\int_{\substack{\text{all possible}\\ \text{values of } \mathbf{p},\mathbf{q}}} \mathscr{P}(\mathbf{p},\mathbf{q})\,d\mathbf{p}\,d\mathbf{q}} = \frac{\int_{\boldsymbol{\alpha}}^{\boldsymbol{\alpha}+\Delta\boldsymbol{\alpha}} \mathscr{P}(\boldsymbol{\alpha})\,d\boldsymbol{\alpha}}{\int_{\substack{\text{all possible}\\ \text{values of } \boldsymbol{\alpha}}} \mathscr{P}(\boldsymbol{\alpha})\,d\boldsymbol{\alpha}} \tag{20-20}$$

$\mathscr{P}(\mathbf{p},\mathbf{q})$ is the probability distribution function of classical statistical mechanics. A mapping of the $\boldsymbol{\alpha}$ space on the $\mathbf{pq}$ space allows us in principle to evaluate $\mathscr{P}(\boldsymbol{\alpha})$. We do not actually carry out this operation, but establish that $\mathscr{P}(\boldsymbol{\alpha})$ exists and is in principle measurable. We may then conclude that ΔS exists and is a well-behaved function of the $\boldsymbol{\alpha}$'s. We now consider the physical nature of the state of the system described by the state vector $\boldsymbol{\alpha}$. In establishing the values of thermodynamic parameters, we assumed the measurement time sufficiently long for a mean value to be obtained. This is equivalent to $\langle\boldsymbol{\alpha}\rangle = \mathbf{0}$. To consider a fluctuation state $\boldsymbol{\alpha}$ we must speed up the measurement process so that nonzero $\boldsymbol{\alpha}$'s may be considered. The average is over the measurement time. Thus, a consideration of fluctuations immediately introduces kinetic factors in the system because the time of measurement is explicitly introduced. Indeed, the distribution $\mathscr{P}(\boldsymbol{\alpha})$ depends on the instrument's response time and the distribution, used in eq. (20-16), assumes a very fast measuring process.

For nonequilibrium systems we can maintain $\boldsymbol{\alpha} = \boldsymbol{\alpha}_0$ by external constraints. Such a system still fluctuates, but this time around $\boldsymbol{\alpha}_0$ rather than $\mathbf{0}$. When we release the constraints, the system will move from $\boldsymbol{\alpha} = \boldsymbol{\alpha}_0$ to $\boldsymbol{\alpha} = \mathbf{0}$. This release of a macroscopic constraint will later be compared with the regression of a fluctuation.

Having established that ΔS is a well-behaved function of $\boldsymbol{\alpha}$, we can use the Taylor expansion to write

$$\Delta S \cong \Delta S_0 + \sum_i \left(\frac{\partial\,\Delta S}{\partial \alpha_i}\right)_0 \alpha_i + \frac{1}{2}\sum_i\sum_j \left(\frac{\partial^2\,\Delta S}{\partial\alpha_i\,\partial\alpha_j}\right)_0 \alpha_i\alpha_j \tag{20-21}$$

To evaluate ΔS_0 refer to eq. (20-18) and (20-17):

$$e^{\Delta S_0/k} = \frac{\mathscr{P}(\mathbf{0})}{\mathscr{P}(\mathbf{0}) + \sum_k \mathscr{P}(\boldsymbol{\alpha}_k)} \tag{20-22}$$

Because the distribution is very sharply peaked around $\boldsymbol{\alpha} = \mathbf{0}$, $\mathscr{P}(\mathbf{0}) \gg \mathscr{P}(\boldsymbol{\alpha}_k)$ for all values of $\boldsymbol{\alpha}_k$ except those in the immediate neighborhood of $\mathbf{0}$. As a consequence, ΔS_0 is a constant whose value approaches zero. The term $\partial\,\Delta S/\partial\alpha_i$ vanishes since at equilibrium the entropy is a maximum which implies that all first derivatives are zero. We next define the coefficients $\mathfrak{b}_{ij}$:

$$\mathfrak{b}_{ij} = -\left(\frac{\partial^2\,\Delta S}{\partial\alpha_i\,\partial\alpha_j}\right)_0 \tag{20-23}$$

Equation (20-21) can now be written as

$$\Delta S = -\frac{1}{2}\sum_i \sum_j \mathfrak{b}_{ij}\alpha_i\alpha_j \tag{20-24}$$

The absence of third- and higher-order terms in the expansion (20-21) is justified provided we are dealing in the near-to-equilibrium domain of small $\boldsymbol{\alpha}$'s.

The next step is to introduce generalized forces (sometimes referred to as Onsager forces) which will be defined in the following way:

$$F_i = \frac{\partial\,\Delta S}{\partial\alpha_i} = -\sum_j \mathfrak{b}_{ij}\alpha_j \tag{20-25}$$

The entropy production is the time derivative of ΔS and may be written

$$\Delta\dot{S} = -\sum_i \sum_j \mathfrak{b}_{ij}\dot{\alpha}_i\alpha_j = \sum_i \dot{\alpha}_i F_i \tag{20-26}$$

All quantities have now been defined so that the entropy production in a very general way can be expressed as a bilinear form in the thermodynamic forces F_i and the thermodynamic fluxes $\dot{\alpha}_i$.

If we identify $\Delta\dot{S}$ with σ, the entropy production, $\dot{\alpha}_i$ with J_i, the fluxes, and F_i with X_i, the forces, then we make contact between eq. (20-26) and eq. (19-53). Equation (19-53) was arrived at from an assumption of local equilibrium while eq. (20-26) assumes a near-to-equilibrium system. The two assumptions lead to very similar formalisms, although the latter one is of considerably greater generality.

In subsequent analysis we will require the long-time average of $\alpha_i F_j$. Since we are considering fluctuations at equilibrium we may use the ergodic hypothesis (the assumption that the time average is equal to the ensemble average) and compute the ensemble average of $\alpha_i F_j$, which will later be most useful:

$$\begin{aligned}
\langle \alpha_i F_j \rangle &= \int_{\boldsymbol{\alpha}} \alpha_i F_j \mathscr{P}(\boldsymbol{\alpha})\,d\boldsymbol{\alpha} \\
&= \int_{\boldsymbol{\alpha}} \alpha_i \frac{\partial\,\Delta S}{\partial\alpha_j}\mathscr{P}_0 e^{\Delta S/k}\,d\boldsymbol{\alpha} \\
&= \int_{\boldsymbol{\alpha}} k\mathscr{P}_0\alpha_i \frac{\partial e^{\Delta S/k}}{\partial\alpha_j}\,d\alpha_1\,d\alpha_2 \cdots d\alpha_n \\
&= \int_{\boldsymbol{\alpha}} \frac{d\boldsymbol{\alpha}}{d\alpha_j}\int_{-\infty}^{\infty} k\mathscr{P}_0\alpha_i \frac{\partial e^{\Delta S/k}}{\partial\alpha_j}\,d\alpha_j
\end{aligned} \tag{20-27}$$

We are now in position to integrate by parts with respect to the variable α_j. Noting that

$$\int_a^b u\,dv = [uv]_a^b - \int_a^b v\,du \tag{20-28}$$

we rewrite eq. (20-27) as

$$\langle \alpha_i F_j \rangle = \int \frac{d\boldsymbol{\alpha}}{d\alpha_j} \left\{ [k\mathscr{P}_0 \alpha_i e^{\Delta S/k}]_{-\infty}^{\infty} - \int k\mathscr{P}_0 \frac{\partial \alpha_i}{\partial \alpha_j} e^{\Delta S/k}\, d\alpha_j \right\} \tag{20-29}$$

$\partial\alpha_i/\partial\alpha_j$ can be replaced by δ_{ij}, the Kronecker delta, an operator which has a value one when $i = j$ and zero when $i \neq j$. For very large values of α_i, the probability $\mathscr{P}_0 e^{\Delta S/k}$ tends very rapidly toward zero so that the first term in the brackets of (20-29) vanishes at both limits. This leaves

$$\langle \alpha_i F_j \rangle = -k\delta_{ij} \int_{\boldsymbol{\alpha}} \mathscr{P}_0 e^{\Delta S/k}\, d\boldsymbol{\alpha} \tag{20-30}$$

If $\mathscr{P}(\boldsymbol{\alpha})$ is a normalized probability distribution then we note

$$\int_{\text{all values of } \boldsymbol{\alpha}} \mathscr{P}(\boldsymbol{\alpha})\, d\boldsymbol{\alpha} = \int \mathscr{P}_0 e^{\Delta S/k}\, d\boldsymbol{\alpha} = 1 \tag{20-31}$$

Equation (20-30) can then be cast in the final very simplified form

$$\langle \alpha_i F_j \rangle = -k\delta_{ij} \tag{20-32}$$

E. THE ONSAGER RECIPROCAL RELATIONS

We next proceed to derive a set of reciprocal relations based on two postulates:

(1) The principle of microscopic reversibility.
(2) The relation between the kinetics of macroscopic processes and the regression of fluctuations.

We first examine a macroscopic and chemically oriented postulate which suggested to Onsager the application of the principle of microscopic reversibility in irreversible thermodynamics. The related postulate, called the principle of detailed balance, has been introduced in Chapter 13. We now return to the three isomer reactions shown in fig. 13-3. If we have a vat in which this isomerization reaction is taking place and allow the system to come to equilibrium in contact with an isothermal reservoir, then the final activities (concentrations) of A, B, and C are time independent and governed by the thermodynamic relations

$$\frac{a_B}{a_A} = e^{-(G_B - G_A/RT)}, \qquad \frac{a_C}{a_B} = e^{-(G_C - G_B/RT)} \tag{20-33}$$

where a's are activities and G's are Gibbs free energies. The principle of detailed balance goes one step further and asserts that at equilibrium

$$k_1 a_A = k_2 a_B, \qquad k_3 a_B = k_4 a_C, \qquad k_5 a_C = k_6 a_A \tag{20-34}$$

Thus at each step the forward and reverse processes for each reaction are taking place at equal rates at equilibrium. This is of course a time average since slight imbalances in the rates will lead to the fluctuations we have been discussing. An important feature of detailed balance is that it is an extra thermodynamic principle not derivable from previous postulates. There is no thermodynamic reason why equilibrium could not be maintained by a constant flow around the loop ABC. The principle was first introduced as an independent postulate and was later shown to be related to certain features of mechanics and the principle of microscopic reversibility.

If we have a system of particles moving under Newton's laws of motion, then each particle will trace out a trajectory $\mathbf{r}_i(t)$ and will have a velocity $d\mathbf{r}_i/dt$. If we were to suddenly change the velocity of each particle to $-d\mathbf{r}_i/dt$, the same trajectories would be traced out in reverse with the velocity having the reverse sign at each point. This result is a consequence of the symmetry properties of Newtonian mechanics, and can be illustrated by a simple example. Consider a particle moving in one dimension under a constant force

$$m\frac{d^2x}{dt^2} = F \tag{20-35}$$

at $t = 0$, $x = x_0$, $v = 0$. Solving the equation leads to

$$v = \frac{Ft}{m}, \qquad x = x_0 + \frac{Ft^2}{2m}, \qquad x = \frac{v^2 m}{2F} + x_0 \tag{20-36}$$

The graph of x versus v, a state space representation of the system, is shown in fig. 20-4. The trajectory goes from point I to point II during time from $t = 0$ to $t = t_1$. At $t = t_1$ change the velocity from $v = (F/m)t$ to $v = -(F/m)t$.

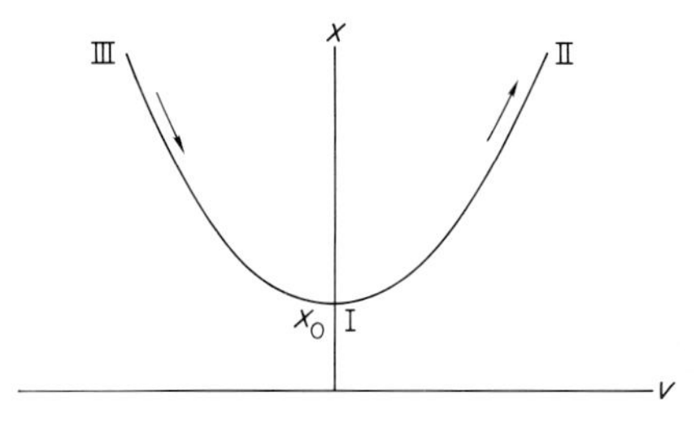

FIGURE 20-4. Particle trajectories in xv space.

The state space representation moves from point II to point III. The governing law of motion is still eq. (20-35) which can be solved subject to the new boundary conditions:

$$\frac{dx}{dt} = v = \frac{Ft}{m} - \frac{Ft_1}{m}, \qquad x = x_0 + \frac{mv^2}{2F} \tag{20-37}$$

The system follows the trajectory from point III to point I in fig. 20-4.

The symmetry of the curve about the X axis indicates that the curve III → I retraces the curve I → II with $v = -v$ at each point.

In equilibrium statistical mechanics, all distributions are symmetrical with respect to velocity. Therefore, because of the symmetry of mechanics, any forward trajectory has the same likelihood of occurring at equilibrium as the reverse trajectory. This extension of mechanics to statistical mechanics has been designated the principle of microscopic reversibility. The principle leads directly to detailed balance. Consider the transition (forward reaction) A → B. Such a transformation may occur along a very large number of trajectories in some phase space representing this molecule. The reverse trajectories B → A (reverse reaction) involve a simple replacement of all the v's by their negatives. Since microscopic reversibility shows that forward and reverse trajectories are equally likely, the macroscopic consequence is that forward and reverse reactions are also equally likely.

The principle of microscopic reversibility may be expressed in a more formal way. Consider a very large ensemble of systems at equilibrium characterized by macroscopic fluctuation variables $\alpha_1, \ldots, \alpha_n$. At time t select a subensemble with definite values of the α's given by $\alpha_1', \alpha_2', \alpha_3', \alpha_4', \ldots, \alpha_n'$. Assuming we had been following the trajectories of all of these ensemble members in α space both before and after we choose the subensemble, the principle of microscopic reversibility states

$$\langle \alpha_j(t + \tau) \rangle_{\alpha_1', \alpha_2', \ldots, \alpha_n'} = \langle \alpha_j(t - \tau) \rangle_{\alpha_1', \alpha_2', \ldots, \alpha_n'} \tag{20-38}$$

Equation (20-38) is averaged over the subensemble and asserts that since all trajectories occur with the same frequency as their velocity reversal trajectories, then if we start with a given point in α space, the average configuration τ seconds later would have been the same as is the average configuration τ seconds earlier. The significance of eq. (20-38) is that it gives us a mathematical form of a general principle that we can use in deriving further results.

The next postulate deals with the regression of fluctuations and states that *there is a linear domain over which the average rate of decay of the fluctuations is proportional to their magnitude.* Formally this may be expressed as

$$\dot{\alpha}_i = \sum_j \kappa_{ij} \alpha_j \tag{20-39}$$

First note that $\dot{\alpha}_i$ is an ensemble average over the subensemble chosen above for which all members have the same value of $\boldsymbol{\alpha} = \boldsymbol{\alpha}'$ at some fixed time. Second, the finite value of κ_{ij} when $i \neq j$ indicates a coupling between various fluctuations which is a characteristic feature of systems. Equation (20-39) is a postulate of irreversible thermodynamics. While it is a very reasonable hypothesis over some range of $\boldsymbol{\alpha}$, there is a residual question as to how large a range can be encompassed by the simple linear relation. The quantity $\dot{\alpha}_i$ is the ensemble average decay rate of fluctuations and can be written as

$$\dot{\alpha}_i = \frac{1}{\tau} \langle \alpha_i(t + \tau) - \alpha_i(t) \rangle_{\alpha_1', \ldots, \alpha_n'} \tag{20-40}$$

Again we select a subensemble with $\boldsymbol{\alpha}$ values of $\alpha_1', \alpha_2', \ldots, \alpha_n'$ and examine the α_i values τ seconds later. The $\dot{\boldsymbol{\alpha}}$ are thus rates of decay of fluctuations starting from some point in $\boldsymbol{\alpha}$ space.

Since we have already established eq. (20-25) as defining the thermodynamic forces, we can use a matrix inversion on eq. (20-25) to solve for the α_i and substitute into (20-38) and (20-39), getting

$$\frac{1}{\tau} \langle \alpha_i(t + \tau) - \alpha_i(t) \rangle = \sum_k L_{ik} F_k \tag{20-41}$$

The L_{ik} are a new set of constants involving the κ_{ij} of eq. (20-39) and the $\mathfrak{b}_{ij}$ of eq. (20-25). We now wish to draw some conclusions from eq. (20-41) subject to the constraints of microscopic reversibility [eq. (20-38)]. To do this we recast (20-38) in a different form by multiplying both sides by $\alpha_i(t)$:

$$\alpha_i(t) \langle \alpha_j(t + \tau) \rangle = \alpha_i(t) \langle \alpha_j(t - \tau) \rangle \tag{20-42}$$

We now average (20-42) over all initial configurations $\boldsymbol{\alpha}$ that is over the entire ensemble rather than the subensemble. This leads to

$$\langle \alpha_i(t) \alpha_j(t + \tau) \rangle = \langle \alpha_i(t) \alpha_j(t - \tau) \rangle \tag{20-43}$$

Note that the left side is of the form of a correlation function with ensemble averaging rather than time averaging. On the right side the average should be independent of the exact value of t, since we are at equilibrium and averaging over all possible values of α_j. We can thus substitute $t + \tau$ for t wherever t occurs on the right side and get

$$\langle \alpha_i(t) \alpha_j(t + \tau) \rangle = \langle \alpha_j(t) \alpha_i(t + \tau) \rangle \tag{20-44}$$

This symmetry relation in the correlation function is the consequence of microscopic reversibility. Note that we are now going beyond the ergodic hypothesis by first representing a system by an ensemble and then concerning ourselves with the time course of fluctuating macroscopic variables in members of the ensemble. We are using both time and ensemble averages to derive our results. We can get away with the approach because we are in $\boldsymbol{\alpha}$ space, and $\boldsymbol{\alpha}$, being macroscopic, permits (in principle) the measurement of the time course of the system.

We next multiply eq. (20-41) by $\alpha_j(t)$ and extend our average to the entire ensemble. On multiplying we get

$$\alpha_j(t)\langle \alpha_i(t+\tau) - \alpha_i(t)\rangle = \tau \sum_l L_{il} F_l \alpha_j(t) \tag{20-45}$$

On averaging we get

$$\langle \alpha_j(t)\alpha_i(t+\tau)\rangle - \langle \alpha_j(t)\alpha_i(t)\rangle = -\tau k L_{ij} \tag{20-46}$$

The right-hand average has been computed using the result obtained in eq. (20-32).

Equation (20-41) could have been formulated for the fluctuating variable α_j:

$$\frac{1}{\tau}\langle \alpha_j(t+\tau) - \alpha_j(t)\rangle = \sum_l L_{jl} F_l \tag{20-47}$$

If we multiply (20-47) by α_i and perform the averaging as in (20-45) and (20-46), we then get

$$\langle \alpha_i(t)\alpha_j(t+\tau)\rangle - \langle \alpha_i(t)\alpha_j(t)\rangle = -\tau k L_{ji} \tag{20-48}$$

Next we compare eqs. (20-48) and (20-46). The first terms of the two equations are equal by the symmetry condition of eq. (20-44). The two second terms are identically equal. Therefore the two left-hand sides are equal, hence, the right-hand sides must be equal. Thus

$$L_{ji} = L_{ij} \tag{20-49}$$

Equations (20-49) are the celebrated Onsager reciprocity relations. Remembering the following identifications,

$$J_i = \dot{\alpha}_i \quad \text{fluxes correspond to the time rate of regression of fluctuations} \tag{20-50}$$

$$X_i = F_i \quad \text{macroscopic forces correspond to Onsager forces} \tag{20-51}$$

we also note that

$$\Delta S = \Delta S(\boldsymbol{\alpha}) \tag{20-52}$$

and

$$\frac{dS}{dt} = \frac{d\,\Delta S}{dt} \quad \text{since } S = S_{eq} + \Delta S \text{ and } S_{eq} \text{ is a constant} \tag{20-53}$$

We then note that

$$\frac{dS}{dt} = \sum_i \frac{\partial \Delta S}{\partial \alpha_i} \frac{d\alpha_i}{dt}$$

$$= \sum_i F_i \dot{\alpha}_i = \sum X_i J_i \tag{20-54}$$

subject to the reciprocity conditions. Thus the fluctuation regression point of view is consistent with the entropy flow formalism and additionally provides a new set of constraints, the Onsager reciprocity relations.

TRANSITION

We started this chapter with a review of the potential uses of fluctuation theory in developing insights into small systems. This theory is a powerful tool and where applicable deepens our understanding of the molecular events which are taking place in available experimental material. Its increasing use is most promising.

The next step (which starts with Einstein's work in 1905) is the definition of an entropy function which is useful for near-to-equilibrium cases. Again this cannot come from classical theory which averages over all possible states of the ensemble but must come from high-speed, small-volume measurements which respond to fluctuations in the variables. The magnitude and frequency of these fluctuations reflects the molecular processes going on in the sample. We utilize fluctuation variables to define an instantaneous entropy for any fluctuation state and we are then able to relate that entropy measure to the probability of the fluctuation in an equilibrium ensemble.

These considerations establish the formal apparatus to consider the relationship between regression of fluctuations and macroscopic rate constants for the decay of induced perturbations in the system. Microscopic reversibility is introduced as a postulate strongly suggested by the symmetry properties of the laws of mechanics and the symmetry of the velocity distribution in statistical mechanics. Using the preceding ideas we are able to generalize the relations between forces and fluxes and emerge with three very general formulas:

$$J_i = \sum_j L_{ij} X_j \tag{20-55}$$

$$L_{ij} = L_{ji} \tag{20-56}$$

$$\sigma_s = \sum_i X_i J_i \tag{20-57}$$

These equations then provide the basic tools of near-to-equilibrium thermodynamics. We will now move on to some applications of this formalism.

BIBLIOGRAPHY

Debye, P. J. W., "The Collected Papers." Wiley (Interscience) New York, 1954.

The papers on light scattering develop aspects of the fluctuation point of view.

Feher, G., and Weissman, M., *Proc. Nat. Acad. Sci. U.S.* **70**, 870–875 (1973).

This paper reviews the theory and describes experimental approaches to the study of chemical reaction kinetics by fluctuation spectroscopy. This is an essential paper in the understanding of this area of research.

Johnson, J. B., *Phys. Rev.* **32**, 97–109 (1928).

The first accurate experimental demonstration of voltage fluctuations in a resistor.

Katchalsky, A., and Curran, P. F., "Nonequilibrium Thermodynamics in Biophysics." Harvard Univ. Press, Cambridge, Massachusetts, 1965.

The monograph develops both the theory and applications of irreversible thermodynamics.

Koppel, D., Axelrod, D., Schlessinger, J., Elson, E. L., and Webb, W. W., *Biophys. J.* **16**, 1315–1329 (1976).

This paper on dynamics of fluorescent marker concentration as a probe of mobility describes the instrumentation for use in this type of fluctuation analysis and then describes the investigation of a number of biological materials.

Nyquist, H., *Phys. Rev.* **32**, 110–113, 1928.

The theoretical treatment of thermal fluctuations of voltages and currents in electrical impedance elements.

Perrin, J. B., "Atoms." Constable Press, London, 1916.

A detailed report on the experiments on Brownian motion that provided support for Einstein's theory, thus lending credence to the atomic theory of matter. This is an English translation of "Les Atomes" published in 1913 by F. Alcan, Paris.

21 Applications of Irreversible Theory

A. THERMOELECTRIC EFFECTS

Some of the more detailed applications of irreversible thermodynamics have been in the field of thermoelectricity. The major areas of usefulness have been in those problems where the reciprocal relations provide an additional constraint to add to the analysis. The existence of nonzero L_{ij} and L_{ji} comes about from real physical connections between the ith and jth fluxes. In the case of thermoelectricity the flow of heat and the flow of electricity is mediated by the same electrons so that the coupling is unavoidable. In general it will be found that coupling of various fluxes will result from the same physical entities being involved in both.

To study thermoelectric effects consider first two metallic strips which are in contact at two ends which are held at different temperatures (see fig. 21-1). If an ammeter is placed in conductor A a flow of electric current will be

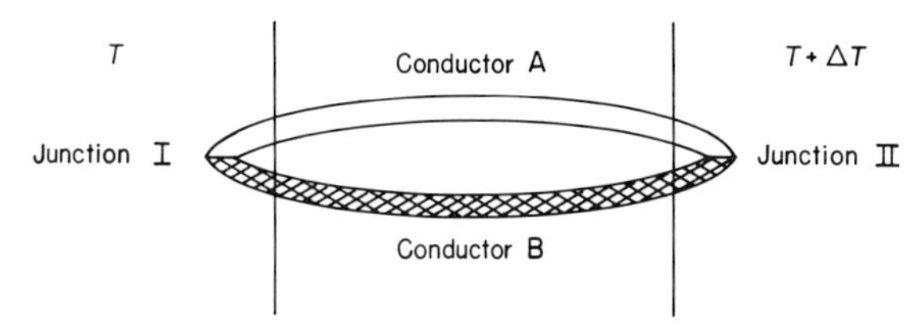

FIGURE 21-1. A simple thermocouple consisting of two different metals with two junctions held at different temperatures.

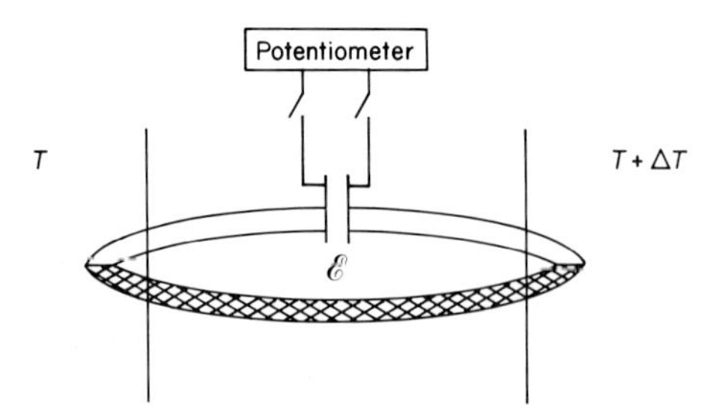

FIGURE 21-2. The thermocouple is now arranged so as to measure the thermoelectric potential on a potentiometer.

observed. An electric potential develops at each junction. Since these potentials are temperature dependent, they will be different at the two junctions and a net potential will exist between the two. To measure this potential we place a capacitor on the circuit and after the system comes to equilibrium measure the potential with a potentiometer (see fig. 21-2). *A series of measurements then shows that the potential across the capacitor is proportional to ΔT, independent of T over a wide range, and dependent on the composition of the two conductors* (empirical generalization). This can be represented as

$$\Delta \mathscr{E} = \mathfrak{s} \, \Delta T \tag{21-1}$$

$$\mathfrak{s} = \frac{\Delta \mathscr{E}}{\Delta T} \tag{21-2}$$

The coefficient $\mathfrak{s}$ is known as the thermoelectric power (definition) and the existence of a thermocouple voltage is called the Seebeck effect.

Note that, in addition to the Seebeck effect, there will be a Fourier heat flow from the hot reservoir to the cold reservoir. Temperature gradients will develop along the wires which will give rise to effects that we will subsequently discuss.

If current is allowed to flow around the circuit as in fig. 21-2 there will be the usual $J_e^2 \mathscr{R}$ electrical heating where J_e is the electric current and $\mathscr{R}$ the resistance. *This dissipation term is known as the Joule heat* (definition).

Next consider a two-metal circuit into which we introduce a voltage source as in fig. 21-3. The flow of current leads to Joule heating along the conductors. In addition *there is a further heat absorption at one junction and heat production at the other. This heat transfer is called the Peltier effect and the rate of heat flow is proportional to the current* (empirical generalization):

$$J_q = \mathfrak{p} J_e \tag{21-3}$$

$$\mathfrak{p} = \frac{J_q}{J_e} \tag{21-4}$$

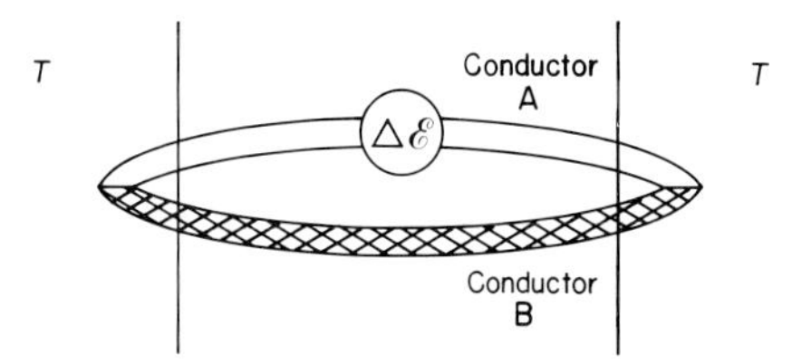

FIGURE 21-3. The Peltier effect is observed when current flows through coupled junctions from metal A to B and then from B to A.

The coefficient $\mathfrak{p}$ *is called the Peltier coefficient* (definition). By convention $\mathfrak{p}_{AB}$ is positive when the flow of current from A to B causes absorption of heat at the junction where the current flow is in that direction. By the symmetry of fig. 21-3 the effect is clearly reversible so that changing the flow of current will reverse the direction of heat transfer. The Peltier coefficient is a function of the materials of the junctions. The Peltier and Seebeck effects are clearly closely linked, being in one case a heat flow due to a current flow and in the other a current flow due to a heat flow.

Yet another thermoelectrical effect can be observed, in a single wire with both a thermal and electrical gradient, as seen in fig. 21-4. Such a system will show a Fourier heat flow due to the temperature gradient and a Joule heating due to the current. If the current carries electrons from a hot region to a cooler region there is an additional heating due to the electrical term. The converse is true if the current is reversed and electrons are carried into a hotter region. *This cross term is known as the Thomson effect* (definition) and takes place uniformally along the entire wire. The infinitesimal rate at which the heat is transferred between regions of temperature difference ΔT is given

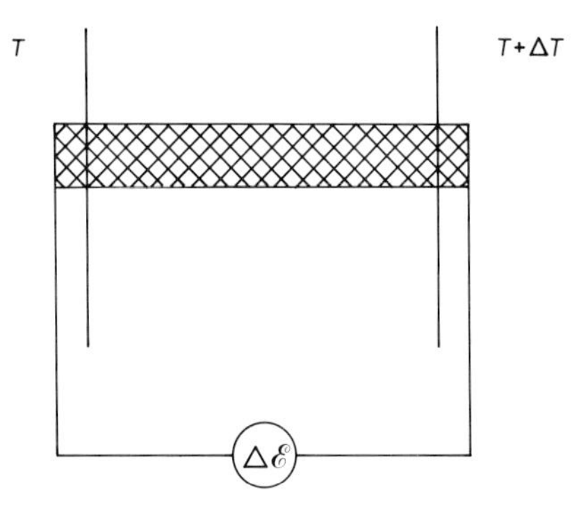

FIGURE 21-4. The Thomson effect is observed when both temperature and potential gradients exist in a metallic wire.

by (empirical generalization)

$$J_q = \mathrm{t} J_e \Delta T \tag{21-5}$$

The Thomson coefficient, t, is dependent on the material of the wire and its average temperature. The convention with regard to the sign of t is that it is positive when a current opposite to the direction of the temperature gradient causes an absorption of heat.

B. ANALYSIS USING THE RECIPROCAL RELATIONS

In our application of the reciprocity relations to thermoelectricity we will concentrate on the metallic junction effects and not consider the Thomson effect further. Consider a wire of length L (see fig. 21-4) with a temperature gradient and an electrical potential gradient, both of which are constant. Using the formalism of eq. (19-57) we can set down the following flux equations:

$$J_q = -L_{qq} \frac{\Delta T}{T^2} - L_{qe} \frac{\Delta \mathscr{E}}{T} \tag{21-6}$$

$$J_e = -L_{eq} \frac{\Delta T}{T^2} - L_{ee} \frac{\Delta \mathscr{E}}{T} \tag{21-7}$$

The term involving L_{qq} is the Fourier heat flow and the L_{ee}-containing term is the ohmic current flow. Next consider the measurement of the Seebeck effect in fig. 21-2. The quantities ΔT and $\Delta \mathscr{E}$ are defined by

$$\Delta T = L \nabla T \tag{21-8}$$

$$\Delta \mathscr{E} = L \nabla \mathscr{E} \tag{21-9}$$

The L_{ij} in eqs. (21-6) and (21-7) are modified from those of (19-60). They are the earlier values divided by the length of the wire L which is a constant for this treatment. Under the balance conditions of measurement the current vanishes and for each wire we can write

$$0 = -L_{eqA} \frac{\Delta T}{T^2} - L_{eeA} \frac{\Delta \mathscr{E}_A}{T} \tag{21-10}$$

$$0 = -L_{eqB} \frac{\Delta T}{T^2} - L_{eeB} \frac{\Delta \mathscr{E}_B}{T} \tag{21-11}$$

In order for the current flow to be zero the potential across the capacitor must be

$$\Delta \mathscr{E} = \Delta \mathscr{E}_B - \Delta \mathscr{E}_A \tag{21-12}$$

Equation (21-12) comes from applying Kirchhoff's voltage law to the circuit under vanishing currents. We then get

$$\Delta \mathscr{E} = \left(\frac{L_{eq_B}}{L_{ee_B}} - \frac{L_{eq_A}}{L_{ee_A}} \right) \frac{\Delta T}{T} \tag{21-13}$$

Comparing (21-3) with (21-2) we see that the thermoelectric power of a thermocouple is related to the various coefficients by

$$\mathfrak{s} = \left(\frac{L_{eq_B}}{L_{ee_B}} - \frac{L_{eq_A}}{L_{ee_A}} \right) \frac{1}{T} \tag{21-14}$$

To analyze the Peltier effect consider fig. 21-3 and the condition that $\Delta T = 0$. We then have

$$J_{q_A} = L_{qe_A} \frac{\Delta \mathscr{E}_A}{T} \tag{21-15}$$

$$J_{e_A} = L_{ee_A} \frac{\Delta \mathscr{E}_A}{T} \tag{21-16}$$

$$J_{q_B} = L_{qe_B} \frac{\Delta \mathscr{E}_B}{T} \tag{21-17}$$

$$J_{e_B} = L_{ee_B} \frac{\Delta \mathscr{E}_B}{T} \tag{21-18}$$

from which we deduce

$$J_{q_A} = \frac{L_{qe_A}}{L_{ee_A}} J_{e_A} \tag{21-19}$$

$$J_{q_B} = \frac{L_{qe_B}}{L_{ee_B}} J_{e_B} \tag{21-20}$$

subject to the condition $J_{e_A} = -J_{e_B} = J_e$ since the same current must flow around the circuit through both conductors. The total heat flow is

$$J_q = J_{q_A} + J_{q_B} = \left(\frac{L_{qe_B}}{L_{ee_B}} - \frac{L_{qe_A}}{L_{ee_A}} \right) J_e \tag{21-21}$$

Comparing (21-21) with (21-4) we see that the Peltier coefficient is given by

$$\mathfrak{p} = \frac{L_{qe_B}}{L_{ee_B}} - \frac{L_{qe_A}}{L_{ee_A}} \tag{21-22}$$

Next note that the Onsager reciprocity relations require $L_{eq} = L_{qe}$ so that (21-22) and (21-14) may be combined to give

$$\mathfrak{p} = \mathfrak{s} T \tag{21-23}$$

Equation (21-23) is subject to experimental check and has been confirmed, thus lending support to the reciprocity among cross coefficients.

More elaborate analysis has been undertaken to relate the Peltier and Thomson coefficients, but it is sufficient for our present purposes to realize that the Onsager relations lead to experimentally testable predictions. The various experiments serve to confirm the validity of the reciprocity relations. These examples occur in electrokinetic effects such as streaming potential and electro-osmosis, in thermomolecular pressure differences as well as in other phenomena. The biologically interesting case of osmotic flow will be considered in the next chapter.

TRANSITION

Applying irreversible theory to a series of thermoelectric effects illustrates the general features of the method and indicates how it may lead to predictions about experimental parameters. In the following pages we utilize the formalism to deepen our understanding of osmotic transport phenomena.

BIBLIOGRAPHY

Katchalsky, A., and Curran, P. F., "Nonequilibrium Thermodynamics in Biophysics." Harvard Univ. Press, Cambridge, Massachusetts, 1965.

A presentation of near-to-equilibrium thermodynamics including a number of biological applications.

Prigogine, I., "Introduction to Thermodynamics of Irreversible Processes." Wiley, New York, 1955.

A short introduction to near-to-equilibrium thermodynamics.

Zemansky, M. W., "Heat and Thermodynamics." McGraw-Hill, New York, 1968.

A detailed discussion of the fundamentals of thermodynamics. This work contains an especially good treatment of experimental background and engineering applications. Includes a wide variety of special topics.

22 Osmotic Flow

A. IRREVERSIBLE THERMODYNAMICS AND OSMOSIS

The problem of the flow of molecules across membranes is clearly central to biology. Organ, cell, and organelle membranes are involved in function at all levels. In this chapter we shall develop the formalism of local equilibrium thermodynamics to study the flow of molecules across barriers. The point of view to be elaborated was developed by A. Katchalsky and co-workers and we will closely follow their treatment in developing the subject. Once we have established the thermodynamic viewpoint we will explore a kinetic approach in order to seek an understanding of the phenomena at a molecular level.

Consider two compartments separated by a membrane as in fig. 22-1. Each compartment is connected to a piston so that pressure may be regulated. Each compartment is well stirred so that we need not consider bulk phase gradients. We will postulate that the membrane is a planar sheet of thickness Δx containing pores of area $\mathscr{A}$ as is shown in fig. 22-2, and confine our interest to steady-state solutions, one- dimensional flows, and nonreacting chemical species. Under these conditions, concentrations in both chambers stay constant in time and we are able to concentrate on the flow process. Stirring cannot take place within the pores. Actually there is an unstirred zone on either side of the membrane, but we will assume that pores are sufficiently far apart so that effects of this layer can be ignored.

The concentration of solute is C_{SL} in the left-hand chamber and C_{SR} in the right. Similarly the solvent concentrations are C_{WL} and C_{WR} on the two sides. Within the membrane we may note the steady-state condition by

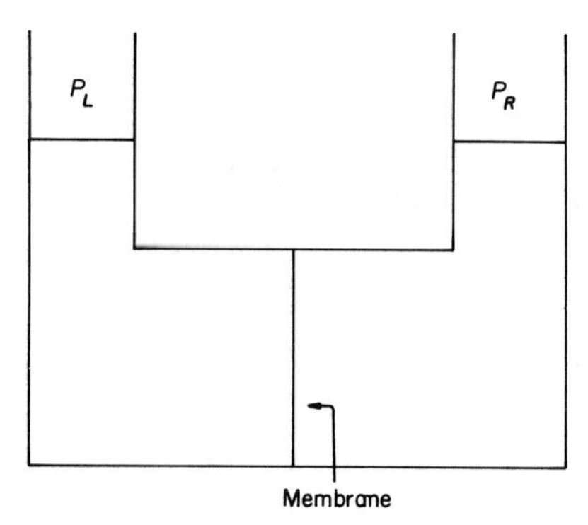

FIGURE 22-1. Two compartments separated by a semipermeable membrane.

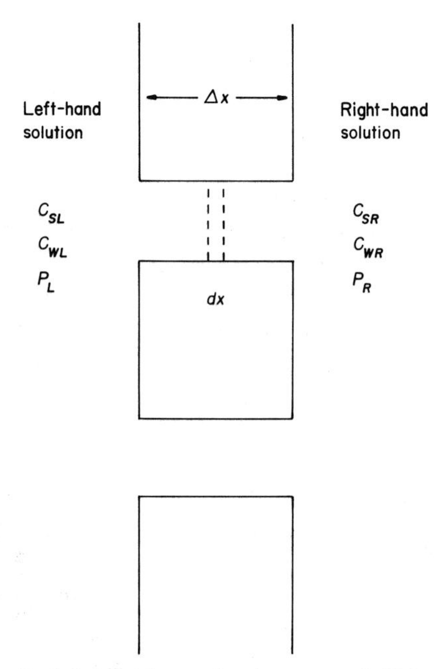

FIGURE 22-2. Detail of the membrane showing pores of thickness Δx and area $\mathscr{A}$. The pore diameters are the order of size of the solute molecules.

noting that at each point

$$-\nabla J_i'' = \frac{\partial C_i}{\partial t} = 0 \tag{22-1}$$

$$J_i'' = \text{constant} \tag{22-2}$$

Equation (22-1) follows from the equation of continuity, (19-37), applied to the steady-state condition of vanishing time derivatives, and eq. (22-2) is the integral of (22-1). J_i'' is the flow per unit area of the ith component within the pores. The entropy production per unit volume in the pore is given from eq. (19-42) by

$$\sigma_s = -\sum J_i'' \cdot \nabla \frac{\mu_i}{T} \tag{22-3}$$

Instead of entropy production we will utilize the dissipation per unit volume which is defined as $T\sigma_s$ and will be written as Φ''. Since T is everywhere constant we can write

$$\Phi'' = \sum_{i=1}^{r} J_i'' \nabla(-\mu_i) \tag{22-4}$$

The total dissipation in each pore is

$$\Phi' = \int_0^{\Delta x} \phi'' \, dV = \int_0^{\Delta x} \sum J_i'' \frac{\partial(-\mu_i)}{\partial x} \mathscr{A} \, dx \tag{22-5}$$

where $\mathscr{A}\, dx$ is the element of volume dV. J_i'' is the flow of the ith component per unit area so that $J_i''\mathscr{A}$ equals J_i', the total flow per pore. Since J_i'' is not a function of x we can directly integrate eq. (22-3) and get

$$\begin{aligned}\Phi' &= \sum J_i' \left(\mu_i(0) - \mu_i(\Delta x)\right) \\ &= \sum J' \Delta\mu_i \end{aligned} \tag{22-6}$$

For the two-component systems under consideration we can write

$$\Phi' = J_S' \Delta\mu_S + J_W' \Delta\mu_W \tag{22-7}$$

The quantity $\Delta\mu$ is the difference in chemical potential across the membrane. If a membrane has N pores per unit area of membrane, the flux per unit area of membrane becomes simply $J_i = NJ_i'$ so that we can write

$$\Phi = N\Phi' = J_S \Delta\mu_S + J_W \Delta\mu_W \tag{22-8}$$

Note that J_S is the flow of the solute, to be distinguished from J_s, the entropy flow, which was introduced in Chapter 19.

Equation (22-8), once formulated, is rather obvious. There are two species that can flow and the dissipation is the sum of two linear terms in the flux and the difference of chemical potential.

In order to work with more convenient quantities we will now transform to a new set of variables with the restrictions that they lead to the same value for entropy production and preserve the bilinear form as in eq. (22-8). The transformation will express the problem in terms of more familiar experimental parameters such as hydrostatic pressure, osmotic pressure, and concentration. We start with the results of (15-67) and (15-68) and rewrite μ_W as

$$\mu_W = \mu_W{}^0 + \bar{V}_W P + \mu_W{}^C \tag{22-9}$$

The last term $\mu_W{}^C$ is the concentration-dependent part of the chemical potential of the solvent, usually written as $RT \ln a_W$. We can go from eq. (22-9) to an evaluation of $\Delta\mu$. Thus

$$\Delta\mu_W = \mu_{WL} - \mu_{WR} = \bar{V}_W \,\Delta P + \Delta\mu_W{}^C \tag{22-10}$$

From our considerations of osmotic pressure [eqs. (15-70) (15-71)] we note that at equilibrium

$$\Delta\mu_W{}^C = -\bar{V}_W \,\Delta\Pi = -\bar{V}_W RT \,\Delta C_S \tag{22-11}$$

The quantity Π is the osmotic pressure that would obtain for this system across a perfect nonleaky osmotic barrier. We also note that

$$\Delta\mu_S = \bar{V}_S \,\Delta P + \Delta\mu_S{}^C \tag{22-12}$$

We next define a new variable $\bar{C}_S$ whose meaning will become clear later. It is here introduced simply as a definition:

$$\bar{C}_S = \frac{\Delta\Pi}{\Delta\mu_S{}^C} \tag{22-13}$$

Equations (22-12) and (22-13) can be combined to give

$$\Delta\mu_S = \bar{V}_S \,\Delta P + \frac{\Delta\Pi}{\bar{C}_S} \tag{22-14}$$

For ideal dilute solutions we also note that

$$\Delta\mu_S{}^C = RT(\ln C_{SL} - \ln C_{SR}) \tag{22-15}$$

$$\Delta\Pi = RT(C_{SL} - C_{SR}) \tag{22-16}$$

Substituting (22-15) and (22-16) with (22-13) we get

$$\bar{C}_S = \frac{(C_{SL} - C_{SR})}{\ln(C_{SL}/C_{SR})} \tag{22-17}$$

If the ratio C_{SL}/C_{SR} is near to unity, that is, if we are near to equilibrium with solute on both sides of the membrane, we can utilize the first term in the expansion of the denominator:

$$\ln x = \frac{2(x-1)}{x+1} + \frac{1}{3}\frac{(x-1)^3}{x+1} + \cdots \tag{22-18}$$

Taking the first term and rearranging,

$$\bar{C}_S \cong \frac{C_{SL} + C_{SR}}{2} \tag{22-19}$$

The new variable under these restricted conditions is approximately equal to the average of the concentrations.

We now utilize eqs. (22-10) and (22-14) and substitute into eq. (22-8). This leads to

$$\Phi = (J_W \bar{V}_W + J_S \bar{V}_S)\,\Delta P + \left(\frac{J_S}{\bar{C}_S} - \bar{V}_W J_W\right)\Delta\Pi \tag{22-20}$$

If we now examine the transformed eq. (22-20) and adopt as new forces ΔP, the hydrostatic pressure across the membrane, and $\Delta\Pi$, the osmotic or concentration force, then we get new flux terms which are linear combinations of the previous fluxes. Conjugate to the hydrostatic pressure is the flux J_P:

$$J_P = J_W \bar{V}_W + J_S \bar{V}_S \tag{22-21}$$

The quantity J_P is the total volume flux across the membrane due to solvent and solute flow. The second flow is J_Π:

$$J_\Pi = \frac{J_S}{\bar{C}_S} - J_W \bar{V}_W \tag{22-22}$$

The flux J_S can be approximated by $N\mathscr{A}v_S\bar{C}_S$ where v_S is the linear velocity of solute and $\bar{C}_S$ is the mean concentration. Similarly J_W is $N\mathscr{A}v_W\bar{C}_W$ and $\bar{V}_W$ is $1/\bar{C}_W$ so that

$$J_\Pi = N\mathscr{A}(v_S - v_W) \tag{22-23}$$

The second flux is proportional to the velocity difference between solvent and solute. It is the type of measure that would be of interest in an ultrafiltration experiment. Having recast the problem in terms of a new set of variables we can now rewrite the phenomenological equations and show that they assume a quite convenient form in terms of experimentally accessible parameters.

We can write

$$J_P = L_{PP}\,\Delta P + L_{P\Pi}\,\Delta\Pi \tag{22-24}$$

$$J_\Pi = L_{\Pi P}\,\Delta P + L_{\Pi\Pi}\,\Delta\Pi \tag{22-25}$$

Thus the volume flow is due to both a pressure difference and a concentration difference as is the selective transport across the membrane. The driving forces are the hydrostatic pressure and the activity differences. The Onsager relations allow us to write

$$L_{P\Pi} = L_{\Pi P} \tag{22-26}$$

and this immediately provides significant extra information.

We consider some experiments relating to eqs. (22-24) and (22-25). First allow the concentration on both sides of the membrane to be the same so that $\Delta\Pi$ goes to zero. The flux equations are then

$$J_P = L_{PP}\,\Delta P \tag{22-27}$$

$$J_\Pi = L_{\Pi P}\,\Delta P \tag{22-28}$$

The first of these is the hydrodynamic flux across the membrane and L_{PP} is the reciprocal of the hydrodynamic resistance and is in principle calculable from fluid flow considerations if the pores are large compared to molecular dimensions. The second flux, J_Π, has been called ultrafiltration. If $L_{\Pi P}$ is positive the solute is moving faster than the solvent and concentrates on the low-pressure side. The more usual case is that $L_{\Pi P}$ is negative and solute molecules are restrained by the membrane and concentrated on the high-pressure side. Such fractionation (dialysis) is frequently used experimentally.

Alternatively if ΔP is zero and a concentration gradient exists we observe

$$J_\Pi = L_{\Pi\Pi}\,\Delta\Pi \tag{22-29}$$

$$J_P = L_{P\Pi}\,\Delta\Pi \tag{22-30}$$

The first expression is for normal diffusional flow with a concentration gradient while the second is that of osmotic flow, the bulk movement in the presence of a concentration gradient. Because of the Onsager conditions we can assert

$$\left(\frac{J_P}{\Delta\Pi}\right)_{\Delta P=0} = L_{P\Pi} = L_{\Pi P} = \left(\frac{J_\Pi}{\Delta P}\right)_{\Delta\Pi=0} \tag{22-31}$$

Equation (22-30) has the additional advantage that it is subject to experimental check by measuring the fluxes across membranes.

Consider a simple laboratory situation to investigate the parameters developed in the preceding analysis. We will employ a capillary osmometer of the type shown in fig. 22-3. If we carried out osmotic flow experiments with both ideal and leaky (solute permeable) membrane, we would observe the results shown in fig. 22-4.

Under these experimental conditions J_P is proportional to $d\,\Delta P/dt$. Thus at equilibrium in the ideal membrane case and at the pressure maximum in the leaky membrane case we have the condition

$$L_{PP}\,\Delta P + L_{P\Pi}\,\Delta\Pi = 0 \tag{22-32}$$

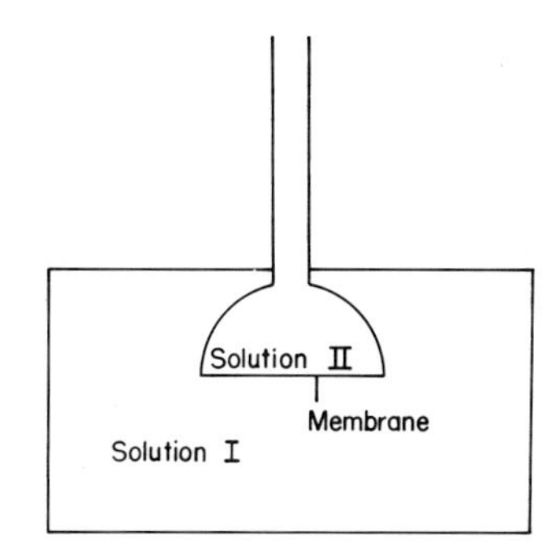

FIGURE 22-3. Capillary osmometer. Solution II has a higher solute concentration than solution I.

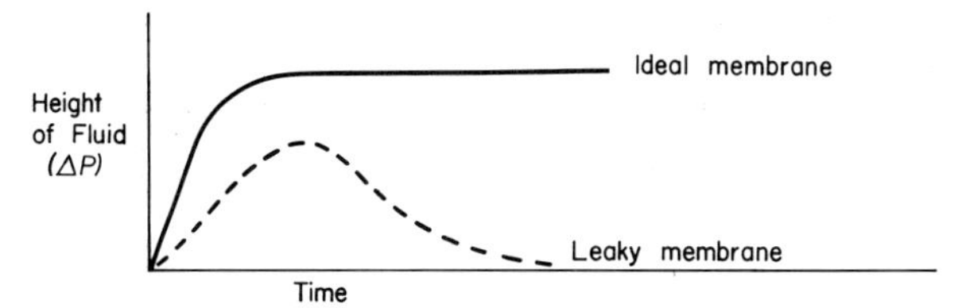

FIGURE 22-4. Pressure–time curves of a capillary osmometer with an ideal membrane and a leaky membrane.

or

$$(\Delta P)_{J_P=0} = \frac{-L_{P\Pi}}{L_{PP}} \Delta\Pi \tag{22-33}$$

For an ideal membrane at equilibrium $\Delta P = \Delta\Pi$ so that

$$-L_{P\Pi(\text{ideal})} = L_{PP(\text{ideal})} \tag{22-34}$$

Ideal membranes are usually regarded as barriers for which J_S, the flow of solute vanishes.

To get an expression for J_S add eqs. (22-21) and (22-22):

$$J_S\left(\bar{V}_S + \frac{1}{\bar{C}_S}\right) = J_P + J_\Pi = (L_{PP} + L_{\Pi P})\Delta P + (L_{P\Pi} + L_{\Pi\Pi})\Delta\Pi \tag{22-35}$$

In order for J_S to vanish for all $\Delta\Pi$ and ΔP both coefficients must vanish:

$$L_{PP} + L_{\Pi P} = 0 \tag{22-36}$$

$$L_{P\Pi} + L_{\Pi\Pi} = 0 \tag{22-37}$$

Equation (22-36) along with the reciprocity condition leads to the same result as (22-34) showing the consistency of the two approaches to the ideal membrane. This result may also be written

$$\frac{-L_{P\Pi}}{L_{PP}} = 1 \tag{22-38}$$

The ratio $-L_{P\Pi}/L_{PP}$ has been defined as the reflection coefficient. When the coefficient is unity we have an ideal osmotic membrane and all solute is reflected. For the reflection coefficient less than one we have a leaky membrane and the value can be measured from eq. (22-32) using ΔP from fig. 22-4 and $\Delta \Pi$ from independent chemical determination.

B. A KINETIC MODEL

To grasp the meaning of the various coefficients we need to go to a kinetic model of osmotic transport. The first case we shall handle is shown in fig. 22-5, a membrane with cylindrical pores having radii large in comparison to the solvent molecules but small in comparison to the solute molecules. In the absence of solute molecules the flow of solvent through the pores is a matter of hydrodynamics and L_{PP} could be calculated directly from the Poiseuille formula

$$J_P = \left(\frac{\pi r^4}{8\eta} \frac{\Delta P}{\Delta x}\right) N \tag{22-39}$$

where r is the pore radius, Δx the membrane thickness, and η the solvent viscosity. Under these conditions

$$L_{PP} = \frac{N \pi r^4}{8\eta \, \Delta x} \tag{22-40}$$

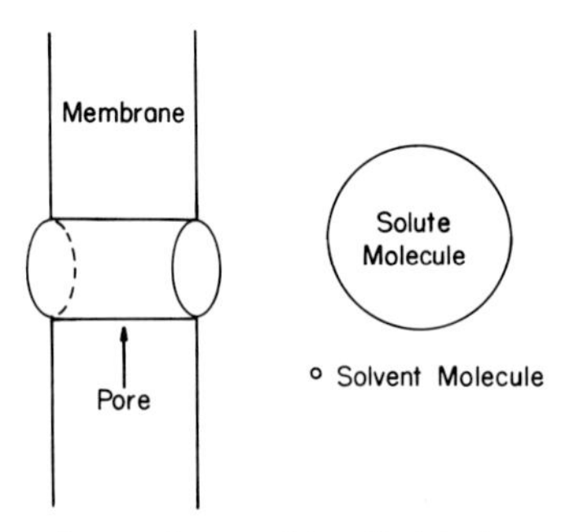

FIGURE 22-5. Ideal osmotic membrane with cylindrical pores small in comparison with the solute molecule and large compared to the solvent molecule.

The hydrodynamic coefficient can be calculated directly from a physical model.

Next set ΔP equal to zero and establish a solute concentration C_{SR} on the right-hand side of the membrane. The kinetic component of the pressure at the membrane face is due to collisions of both solvent and solute particles. At the right-hand face the solute molecules do not transfer momentum to the fluid in the pore but to the membrane which is regarded as rigid and totally reflecting. Hence at the pore face there is an effective pressure deficit which, by the argument given in Chapter 2, eqs. (2-10)–(2-14) will be

$$\Delta P = -CRT \tag{22-41}$$

The solvent in the pore will thus be subject to a real hydrostatic pressure gradient and will flow subject to the Poiseuille relation. Thus

$$J_P = \frac{\pi r^4(-CRT)}{8\eta\,\Delta x}N \tag{22-42}$$

Since $\Delta\Pi$ is equal to CRT we get

$$J_P = L_{P\Pi}\,\Delta\Pi = -\frac{N\pi r^4\,\Delta\Pi}{8\eta\,\Delta x} \tag{22-43}$$

Comparing with (22-40),

$$L_{PP} = -L_{P\Pi} \tag{22-44}$$

The reflection coefficient is thus one.

Next consider the case where the solute molecule is just slightly smaller than the pore. On collision with the pore face there is a probability $\mathscr{P}_{mv}$ that the momentum will be transferred to the wall and a probability $(1 - \mathscr{P}_{mv})$ that the momentum will be transferred to the liquid in the pore. The pressure deficit will now be

$$\Delta P = -C\mathscr{P}_{mv}RT \tag{22-45}$$

Applying the same reasoning as before leads to

$$\text{reflection coefficient} = \mathscr{P}_{mv} < 1 \tag{22-46}$$

and we have a leaky membrane. As we choose solute molecules of progressively smaller sizes, $\mathscr{P}_{mv}$ decreases and leakier membranes result.

When the pore becomes of the same order of magnitude as the solvent molecules the analysis becomes more complex and more sophisticated models are required. Indeed there need be no pore at all as the solute and/or the solvent molecules can be soluble in the membrane and may then diffuse across. We will not treat these cases. A more detailed discussion of osmotic transport will be found in the books listed in the Bibliography.

Equation (22-46) indicates why the reflection coefficient is so named. It is the probability that a solute molecule colliding with the pore area will be reflected. When that probability is unity we have an ideal semipermeable barrier. The kinetic treatment just presented is somewhat idealized but it does give insight into the nature of the phenomenon. For example, the reason that osmotic pressure is a colligative property is that the contribution to pressure depends on the mean kinetic energy which by equipartition is the same for the translational modes of all molecules regardless of their size. Reflection for impermeant molecules must mean interaction with the membrane and hence a pressure deficit on the pore. The reasoning is very general and thus independent of molecular detail.

BIBLIOGRAPHY

House, C. R., "Water Transport in Cells and Tissues." Arnold, London, 1973.

This comprehensive work contains a reasonably detailed theoretical section as well as a wealth of experimental information.

Katchalsky, A., and Curran, P. F., "Nonequilibrium Thermodynamics in Biophysics." Harvard Univ. Press, Cambridge, Massachusetts, 1965.

Contains an extensive section on membrane transport.

APPENDIX I Standard International Units (SI)[†]

A. NAMES OF INTERNATIONAL UNITS

Physical quantity	Name of unit	Symbol	
	BASIC UNITS		
Length	meter	m	
Mass	kilogram	kg	
Time	second	s	
Electric current	ampere	A	
Temperature	kelvin	K	
Luminous intensity	candela	cd	
Amount of substance	mole	mol	
	DERIVED UNITS		
Area	square meter	m^2	
Volume	cubic meter	m^3	
Frequency	hertz	Hz	(s^{-1})
Density	kilogram per cubic meter	kg/m^3	
Velocity	meter per second	m/s	
Angular velocity	radian per second	rad/s	
Acceleration	meter per second squared	m/s^2	
Angular acceleration	radian per second squared	rad/s^2	
Force	newton	N	$(kg \cdot m/s^2)$
Pressure	newton per sq meter	N/m^2	
Kinematic viscosity	sq meter per second	m^2/s	
Dynamic viscosity	newton-second per sq meter	$N \cdot s/m^2$	
Work, energy, quantity of heat	joule	J	$(N \cdot m)$

(continued)

[†] Appendix I is adapted from NASA SP-7012, "The International System of Units", ed. by E. A. Mechtly.

Physical quantity	Name of unit	Symbol	
	DERIVED UNITS		
Power	watt	W	(J/s)
Electric charge	coulomb	C	(A·s)
Voltage, potential difference, electromotive force	volt	V	(W/A)
Electric field strength	volt per meter	V/m	
Electric resistance	ohm	Ω	(V/A)
Electric capacitance	farad	F	(A·s/V)
Magnetic flux	weber	Wb	(V·s)
Inductance	henry	H	(V·s/A)
Magnetic flux density	tesla	T	(Wb/m^2)
Magnetic field strength	ampere per meter	A/m	
Magnetomotive force	ampere	A	
Luminous flux	lumen	lm	(cd·sr)
Luminance	candela per sq meter	cd/m^2	
Illumination	lux	lx	(lm/m^2)
Wave number	1 per meter	m^{-1}	
Entropy	joule per kelvin	J/K	
Specific heat	joule per kilogram kelvin	$J\ kg^{-1}\ K^{-1}$	
Thermal conductivity	watt per meter kelvin	$W\ m^{-1}\ K^{-1}$	
Radiant intensity	watt per steradian	W/sr	
Activity (of a radioactive source)	1 per second	s^{-1}	
	SUPPLEMENTARY UNITS		
Plane angle	radian	rad	
Solid angle	steradian	sr	

B. PREFIXES

The names of multiples and submultiples of SI Units may be formed by application of the prefixes:

Factor by which unit is multiplied	Prefix	Symbol	Factor by which unit is multiplied	Prefix	Symbol
10^{12}	tera	T	10^{-2}	centi	c
10^{9}	giga	G	10^{-3}	milli	m
10^{6}	mega	M	10^{-6}	micro	μ
10^{3}	kilo	k	10^{-9}	nano	n
10^{2}	hecto	h	10^{-12}	pico	p
10	deka	da	10^{-15}	femto	f
10^{-1}	deci	d	10^{-18}	atto	a

C. DEFINITIONS OF INTERNATIONAL UNITS

Definitions of the most important SI Units are given in the following paragraphs. These definitions have been extracted from the records of the International Committee and the General Conferences.

meter (m)

The *meter* is the length equal to 1 650 763.73 wavelengths in vacuum of the radiation corresponding to the transition between the levels 2 p_{10} and 5 d_5 of the krypton-86 atom.

kilogram (kg)

The *kilogram* is the unit of mass; it is equal to the mass of the international prototype of the kilogram. (The international prototype of the kilogram is a particular cylinder of platinum–iridium alloy which is preserved in a vault at Sèvres, France, by the International Bureau of Weights and Measures.)

second (s)

The *second* is the duration of 9 192 631 770 periods of the radiation corresponding to the transition between the two hyperfine levels of the ground state of the cesium-133 atom.

ampere (A)

The *ampere* is that constant current which, if maintained in two straight parallel conductors of infinite length, of negligible circular cross section, and placed 1 meter apart in vacuum, would produce between these conductors a force equal to 2×10^{-7} newton per meter of length.

kelvin (K)

The *kelvin*, unit of thermodynamic temperature, is the fraction 1/273.16 of the thermodynamic temperature of the triple point of water.

candela (cd)

The *candela* is the luminous intensity, in the perpendicular direction, of a surface of 1/600 000 square meter of a blackbody at the temperature of freezing platinum under a pressure of 101 325 newtons per square meter.

mole (mol)

The *mole* is the amount of substance of a system which contains as many elementary entities as there are carbon atoms in 0.012 kg of carbon 12. The elementary entities must be specified and may be atoms, molecules, ions, electrons, other particles, or specified groups of such particles.

newton (N)

The *newton* is that force which gives to a mass of 1 kilogram an acceleration of 1 meter per second per second.

joule (J)

The *joule* is the work done when the point of application of 1 newton is displaced a distance of 1 meter in the direction of the force.

watt (W)

The *watt* is the power which gives rise to the production of energy at the rate of 1 joule per second.

volt (V)

The *volt* is the difference of electric potential between two points of a conducting wire carrying a constant current of 1 ampere, when the power dissipated between these points is equal to 1 watt.

ohm (Ω)

The *ohm* is the electric resistance between two points of a conductor when a constant difference of potential of 1 volt, applied between these two points, produces in this conductor a current of 1 ampere, this conductor not being the source of any electromotive force.

coulomb (C)

The *coulomb* is the quantity of electricity transported in 1 second by a current of 1 ampere.

farad (F)

The *farad* is the capacitance of a capacitor between the plates of which there appears a difference of potential of 1 volt when it is charged by a quantity of electricity equal to 1 coulomb.

henry (H)

The *henry* is the inductance of a closed circuit in which an electromotive force of 1 volt is produced when the electric current in the circuit varies uniformly at a rate of 1 ampere per second.

weber (Wb)

The *weber* is the magnetic flux which, linking a circuit of one turn, produces in it an electromotive force of 1 volt as it is reduced to zero at a uniform rate in 1 second.

lumen (lm)

The *lumen* is the luminous flux emitted in a solid angle of 1 steradian by a uniform point source having an intensity of 1 candela.

D. PHYSICAL CONSTANTS

The following lists of physical constants are from the work of B. N. Taylor, W. H. Parker, and D. N. Langenberg.[†] Their least-squares adjustment of values of the constants depends strongly on a new and highly accurate (2.4 ppm) determination of e/h from the ac Josephson effect in superconductors, and is believed to be more accurate than the 1963 adjustment which appears to suffer from the use of an incorrect value of the fine structure constant as an input datum.

Quantity	Value	Error (ppm)	Prefix	Unit
Speed of light in vacuum	2.997 925 0	0.33	$\times 10^{8}$	$m\ s^{-1}$
Gravitational constant	6.673 2	460	10^{-11}	$N\ m^{2}\ kg^{-2}$
Avogadro constant	6.022 169	6.6	10^{26}	$kmole^{-1}$
Boltzmann constant	1.380 622	43	10^{-23}	$J\ K^{-1}$
Gas constant	8.314 34	42	10^{3}	$J\ kmole^{-1}\ K^{-1}$
Volume of ideal gas, standard conditions	2.241 36	----	10^{1}	$m^{3}\ kmole^{-1}$
Faraday constant	9.648 670	5.5	10^{7}	$C\ kmole^{-1}$
Unified atomic mass unit	1.660 531	6.6	10^{-27}	kg
Planck constant	6.626 196	7.6	10^{-34}	J s
Electron charge	1.602 191 7	4.4	10^{-19}	C
Electron rest mass	9.109 558	6.0	10^{-31}	kg
Proton rest mass	1.672 614	6.6	10^{-27}	kg
Neutron rest mass	1.674 920	6.6	10^{-27}	kg
Electron charge to mass ratio	1.758 802 8	3.1	10^{11}	$C\ kg^{-1}$
Stefan–Boltzmann constant	5.669 61	170	10^{-8}	$W\ m^{-2}\ K^{4}$
First radiation constant	4.992 579	7.6	10^{-24}	J m
Second radiation constant	1.438 833	43	10^{-2}	m K
Rydberg constant	1.097 373 12	.10	10^{7}	m^{-1}
Fine structure constant	7.297 351	1.5	10^{-3}	
Bohr radius	5.291 771 5	1.5	10^{-11}	m
Classical electron radius	2.817 939	4.6	10^{-15}	m
Compton wavelength of electron	2.426 309 6	3.1	10^{-12}	m
Compton wavelength of proton	1.321 440 9	6.8	10^{-15}	m
Compton wavelength of neutron	1.319 621 7	6.8	10^{-15}	m
Electron magnetic moment	9.284 851	7.0	10^{-24}	$J\ T^{-1}$
Proton magnetic moment	1.410 620 3	7.0	10^{-26}	$J\ T^{-1}$
Bohr magneton	9.274 096	7.0	10^{-24}	$J\ T^{-1}$
Nuclear magneton	5.050 951	10	10^{-27}	$J\ T^{-1}$
Gyromagnetic ratio of protons in H_2O	2.675 127 0	3.1	10^{8}	$rad\ s^{-1}\ T^{-1}$
Gyromagnetic ratio of protons in H_2O corrected for diamagnetism of H_2O	2.675 196 5	3.1	10^{8}	$rad\ s^{-1}\ T^{-1}$
Magnetic flux quantum	2.067 853 8	3.3	10^{-15}	Wb
Quantum of circulation	3.636 947	3.1	10^{-4}	$J\ s\ kg^{-1}$

[†] *Rev. Modern Phys.* **41**, 375–496 (1969).

APPENDIX II

Conservation of Mechanical Energy: General Treatment

In order to generalize the mechanical development of the law of conservation of energy, it will first be necessary to present the briefest introduction to vector analysis, to become acquainted with the formal apparatus. In physics vectors are used to represent quantities which possess both magnitude and direction. Examples of such quantities are force and velocity. Quantities which have only magnitude are represented by numbers which are called scalars. Examples of scalar quantities are volume, temperature, and density. To distinguish vectors, we will represent them in boldface type such as the vector **F** for force.

In a Cartesian coordinate system we may represent a vector as a straight line beginning at the origin. Its length represents its magnitude while its position relative to the coordinate system represents its direction. A vector in a two-dimensional coordinate system is shown in fig. A2-1a. Such a vector may also be represented by its projections along the x and y axes, F_x and F_y. Thus the magnitude of **F** is equal to $F_x{}^2 + F_y{}^2$ and the direction $\theta =$ arctan F_y/F_x. *Vector addition is accomplished by placing the tail of one vector adjacent to the arrow of the other, as shown in fig. A2-1b* (definition). Thus $\mathbf{F}_C = \mathbf{F}_A + \mathbf{F}_B$. In terms of the components it can be shown that

$$\mathbf{F}_{Cx} = \mathbf{F}_{Ax} + \mathbf{F}_{Bx}, \qquad \mathbf{F}_{Cy} = \mathbf{F}_{Ay} + \mathbf{F}_{By} \tag{A2-1}$$

The same reasoning can be extended to three dimensions as is indicated in fig. A2-2. We now introduce the concept of a unit vector, a vector which has a magnitude of one in whatever units are used for the coordinate system. Any vector may be written as $F\mathbf{u}$, where $\mathbf{u}$ is a unit vector, which establishes direction, and F is a scalar coefficient, which establishes magnitude. If we

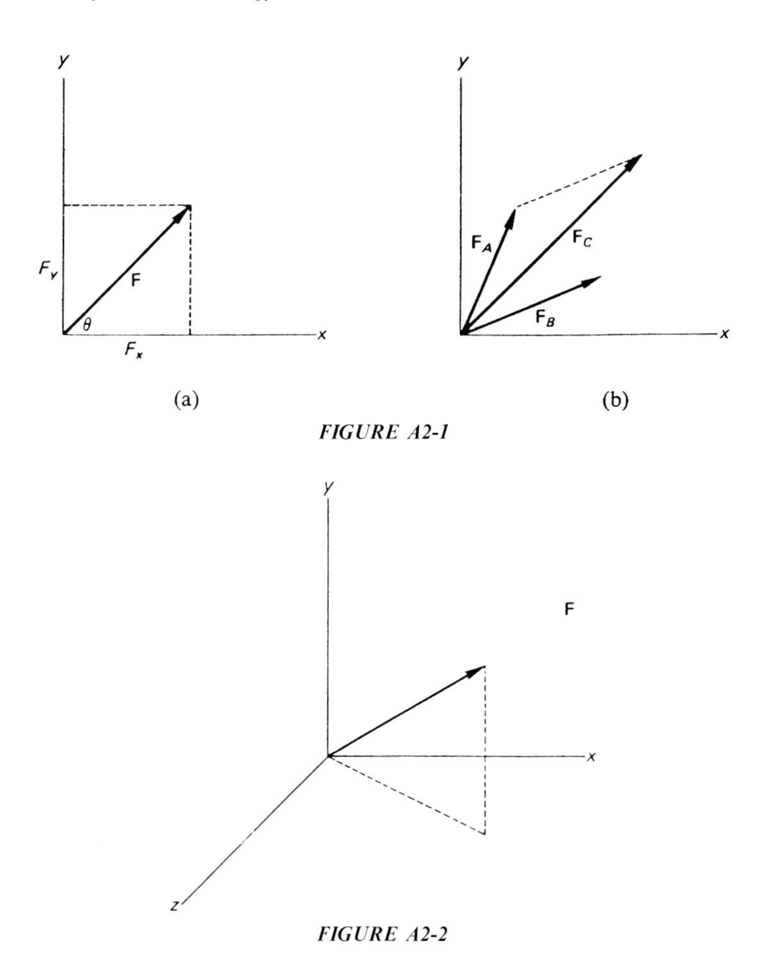

FIGURE A2-1

FIGURE A2-2

represent unit vectors along the x, y, and z axes as **i**, **j**, and **k**, we can then represent any vector as

$$\mathbf{F} = \mathbf{i}F_x + \mathbf{j}F_y + \mathbf{k}F_z \tag{A2-2}$$

Vector addition then becomes

$$\begin{aligned}\mathbf{F}_A + \mathbf{F}_B &= (\mathbf{i}F_{Ax} + \mathbf{j}F_{Ay} + \mathbf{k}F_{Az}) + (\mathbf{i}F_{Bx} + \mathbf{j}F_{By} + \mathbf{k}F_{Bz}) \\ &= \mathbf{i}(F_{Ax} + F_{Bx}) + \mathbf{j}(F_{Ay} + F_{By}) + \mathbf{k}(F_{Az} + F_{Bz})\end{aligned} \tag{A2-3}$$

An important quantity in a number of physical applications is *the scalar product of two vectors, which is a number formed by multiplying the scalar magnitude of two vectors times the cosine of the angle between them* (definition). We represent the operation by a dot (·) between the two vectors.

As a result of this representation, the scalar product is sometimes referred to as the dot product of two vectors; thus,

$$\mathbf{F} \cdot \mathbf{r} = Fr\cos\theta \tag{A2-4}$$

If two vectors are parallel, the angle between them is 0°, the cosine of zero is unity, and the scalar product is simply the product of their magnitudes. If two vectors are perpendicular, the angle is 90°, the cosine of the angle is 0, and the scalar product is zero. Therefore

$$\mathbf{i} \cdot \mathbf{i} = \mathbf{j} \cdot \mathbf{j} = \mathbf{k} \cdot \mathbf{k} = 1 \tag{A2-5}$$

$$\mathbf{i} \cdot \mathbf{j} = \mathbf{i} \cdot \mathbf{k} = \mathbf{j} \cdot \mathbf{k} = 0 \tag{A2-6}$$

and

$$\begin{aligned} \mathbf{F} \cdot \mathbf{r} &= (\mathbf{i}F_x + \mathbf{j}F_y + \mathbf{k}F_z) \cdot (\mathbf{i}r_x + \mathbf{j}r_y + \mathbf{k}r_z) \\ &= F_x r_x + F_y r_y + F_z r_z \end{aligned} \tag{A2-7}$$

The scalar coefficients of vectors are subject to the ordinary rules of differentiation and integration. Consider the position vector $\mathbf{r}$, where

$$r_x = x, \qquad r_y = y, \qquad r_z = z \tag{A2-8}$$

Then

$$\frac{d\mathbf{r}}{dt} = \mathbf{i}\frac{dx}{dt} + \mathbf{j}\frac{dy}{dt} + \mathbf{k}\frac{dz}{dt} \tag{A2-9}$$

The derivative $d\mathbf{r}/dt$ is the velocity vector $\mathbf{v}$ and the second derivative $d^2\mathbf{r}/dt^2$ is the acceleration vector.

We will start our discussion of mechanics by assuming that the reader is familiar with the notions of force, position, velocity, and acceleration as vector quantities and the notion of mass as a scalar quantity. *The work done in moving a mass from one point to another is defined as the distance moved times the force directed along the line of motion* (definition). If the force is not entirely directed along the line of motion, we consider only its component in that direction, which is $F \cos\theta$ if θ is the angle between the force and line of motion. Since the force may change magnitude and direction along a path of motion, we must use differentials. Thus the amount of work in moving a distance dr is

$$dW = F\cos\theta\, dr \tag{A2-10}$$

However, as indicated in the last section, this can be represented in vector notation by

$$dW = \mathbf{F} \cdot d\mathbf{r} \tag{A2-11}$$

APPENDIX **IV**

Stirling's Approximation

We wish to prove

$$\ln x! = x \ln x - x \tag{A4-1}$$

Consider the left-hand side,

$$\ln x! = \sum_{n=1}^{x} \ln n \tag{A4-2}$$

Graph $\ln n$ versus n as in fig. A4-1.

The sum in eq. (A4-2) is the area under the solid line in fig. A4-1. This can be approximated by the area under the dotted curve, which is $\int_1^x \ln n\, dn$. Therefore

$$\ln x! \cong \int_1^x \ln n\, dn = [n \ln n - n]_0^x \tag{A4-3}$$

$$\ln x! \cong x \ln x - x \tag{A4-4}$$

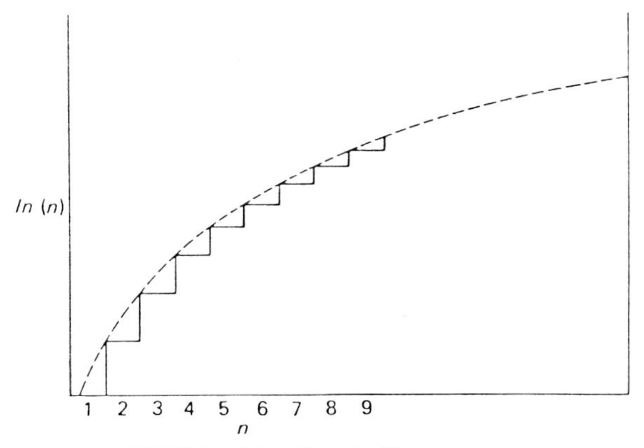

FIGURE A4-1. Graph of $\ln n$ versus n.

APPENDIX **V**

The Geometry and Intuition of Lagrange Multipliers

By John Wyatt and Harold J. Morowitz

In this alternative discussion of Lagrange multipliers, we have emphasized the appealing visual and geometric content of the method while avoiding the more tedious proofs required by mathematical rigor.

A. MOTIVATION AND PICTURES

Let the symbol $\mathbf{x}$ represent a point $(x_1, x_2, \ldots, x_n)$ in n-dimensional Euclidean space. Suppose we wish to find the maximum of a scalar function $f(\mathbf{x})$ subject to the constraint that the point $\mathbf{x}$ must lie in a given smooth curve or surface, C, in the space. If $\mathbf{x}^*$ is the point in C at which the maximum occurs, then it can be seen that ∇f, the gradient of f, must be perpendicular to C at $\mathbf{x}^*$. For if any component of ∇f were parallel to C at $\mathbf{x}^*$, we could move away from $\mathbf{x}^*$ and up the gradient of f while remaining on C, contrary to the assumption that this point was a maximum. Note that $\nabla f \perp C$ is a necessary but not sufficient condition for a maximum.

Since maximizing f is the same as minimizing $-f$, and since $\nabla f \perp C$ is equivalent to $\nabla(-f) \perp C$, it follows that the minimum of f must also occur at a point where $\nabla f \perp C$.

Example 1. Suppose a bead is allowed to slide along a smooth curved wire under the influence of gravity until it comes to rest as is shown in fig. A5-1. The only possible rest points are points where the wire is parallel to the ground, i.e., the interval a and the points b, c, d, and e. These are precisely the points where the wire (i.e., the constraint set C) is perpendicular

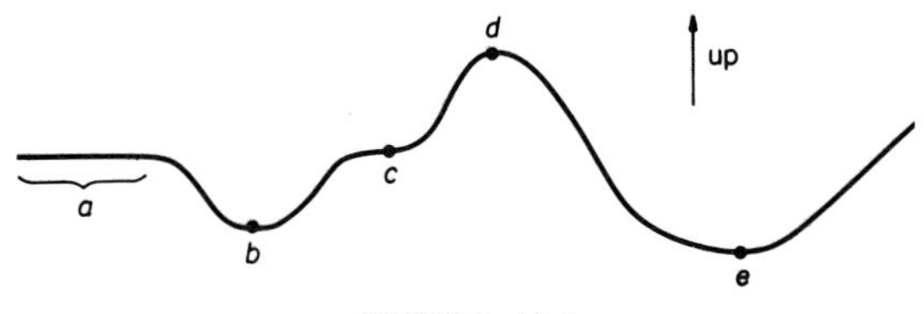

FIGURE A5-1

to ∇f, the gradient of the gravitational potential f. Thus the maximum and minimum of f, points d and e, are both points at which $\nabla f \perp C$. The converse, however, does not hold, since ∇f is also perpendicular to C on the interval a and at the points b and c, which are not global maxima or minima for f.

Example 2. Find the point on the unit circle which maximizes $f(x, y) = x + y$. The result may be seen from inspection of fig. A5-2; $\mathbf{x}^* = (\sqrt{2}/2, \sqrt{2}/2)$. The unit circle is the constraint set C, and the vector field ∇f is sketched in fig. A5-2. Notice that ∇f is perpendicular to C at $\mathbf{x}^*$ and also at the point $-\mathbf{x}^*$, which minimizes f.

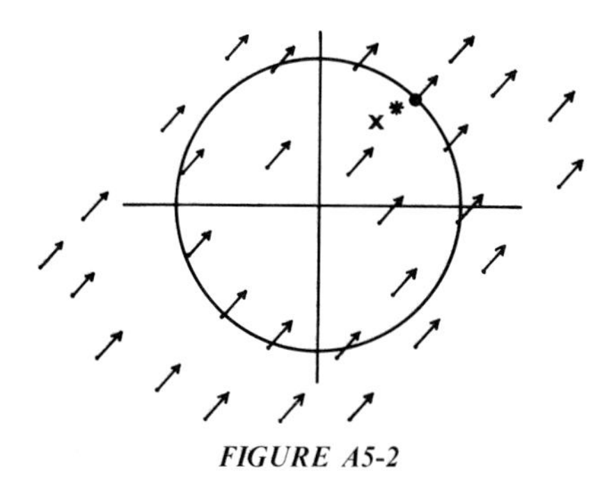

FIGURE A5-2

Example 3. Let C be a surface in 3-dimensional space. We want to find the maximum value attained by f on C. A possible vector field ∇f is sketched in fig. A5-3, and this allows us to look for possible maxima.

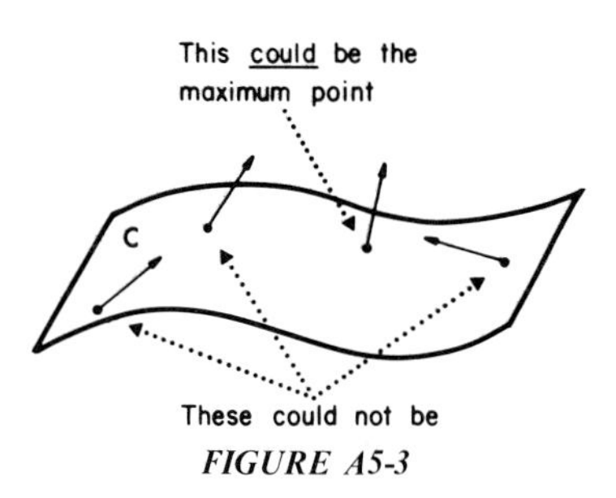

FIGURE A5-3

B. MATHEMATICAL FORMULATION

Suppose the set C in which the solution is constrained to lie is given as the set of points which simultaneously satisfy all the following equations:

$$
\begin{aligned}
&g_1(\mathbf{x}) = 0\\
&g_2(\mathbf{x}) = 0\\
&\quad\vdots\\
&g_k(\mathbf{x}) = 0, \qquad k < n \quad \text{(where } n \text{ is the dimension of the space;}\\
&\qquad\qquad\qquad\qquad\quad \text{i.e., the number of independent variables)}
\end{aligned} \tag{A5-1}
$$

In this case the problem can be formulated as "maximize $f(\mathbf{x})$ subject to the constraints $g_1(\mathbf{x}) = g_2(\mathbf{x}) = \cdots = g_k(\mathbf{x}) = 0$." How do we know whether ∇f is normal (perpendicular) to C when the problem occurs in a space of too high a dimension to visualize ($n \geqslant 4$)? The first clue is that, for any of the constraint functions g_j, $1 \leqslant j \leqslant k$, and any point $\mathbf{x} \in C$, ∇g_j must be normal to C because ∇g_j points in the direction g_j is changing most rapidly, whereas C is a surface on which g_j is constant. (It might be helpful to think of electrostatics—the $\vec{\mathbf{E}}$ field is always normal to surfaces of constant potential.) In order to construct a rigorous proof that $\nabla g_j \perp C$, write an expression for g_j along any curve lying in C and differentiate using the chain rule.

Since each vector $\nabla g_j(\mathbf{x})$ is normal to C at $\mathbf{x}$, any linear combination of the ∇g_j's will also be normal to C. Furthermore, it can be shown that if the ∇g_j's are linearly independent at a point, then every vector normal to C at that point can be written as a linear combination of the ∇g_j's. This should motivate the following definition.

Definition. If $\nabla g_1, \ldots, \nabla g_k$ are all linearly independent at a point $\mathbf{x}$ in C, then the set of all linear combinations of $\nabla g_1, \ldots, \nabla g_k$ at $\mathbf{x}$ is called the normal space to C at $\mathbf{x}$.

Example 4. The unit sphere centered at the origin in 3-space is the set of points satisfying the constraint

$$g_1(x, y, z) = x^2 + y^2 + z^2 - 1 = 0 \tag{A5-2}$$

See fig. A5-4.

The normal space to the sphere at the point $(x, y, z) = (1, 0, 0)$ is just the set of all vectors of the form $\lambda \cdot \nabla g_1(1, 0, 0) = \lambda \cdot (2, 0, 0)$, where the quantity λ is a scalar constant. It is the set of all vectors with tails at $(1, 0, 0)$ which are perpendicular to the sphere; i.e., it is the x axis itself.

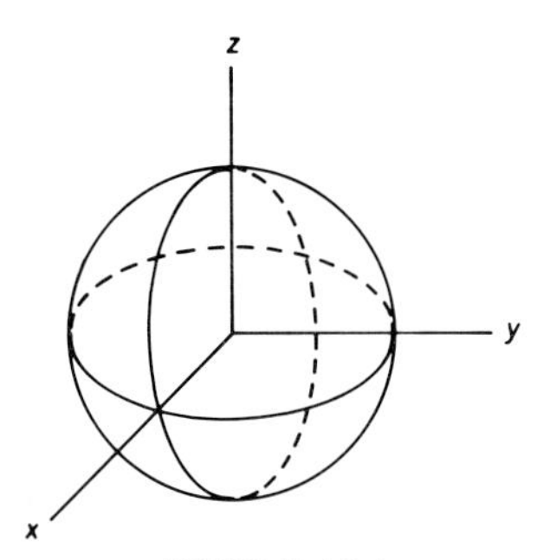

FIGURE A5-4

Example 5. The unit circle lying in the x–y plane in 3-space is the set of points satisfying

$$\begin{aligned} g_1(x, y, z) &= x^2 + y^2 + z^2 - 1 = 0 \\ g_2(x, y, z) &= z = 0 \end{aligned} \tag{A5-3}$$

See fig. A5-5.

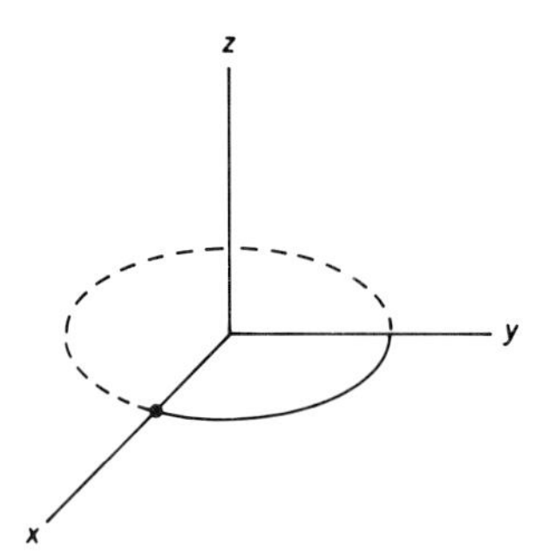

FIGURE A5-5

The normal space to the circle at the point $(1, 0, 0)$ is the set of all linear combinations of $\nabla g_1 = (2, 0, 0)$ and $\nabla g_2 = (0, 0, 1)$, which is just the x–z plane.

The content of the method of Lagrange multipliers is simply that the maximum of f over C must occur at a point where ∇f lies in the normal space to C. We state this fact more precisely in the following theorem.

Theorem.[†] Suppose $f(\mathbf{x})$ and $g_j(\mathbf{x})$, $j = 1, 2, \ldots, k < n$, are defined and continuously differentiable on an open set in n-dimensional Euclidean space.

[†] A rigorous formal proof is given in Chapter 3 of "Calculus of Vector Functions," by Williamson, Crowell, and Trotter, Prentice-Hall, Englewood Cliffs, New Jersey, 1968.

Suppose further that at each point where $g_1 = \cdots = g_k = 0$ we know that $\nabla g_1, \ldots, \nabla g_k$ are linearly independent. Then the maximum value of f subject to the constraints that $g_1 = \cdots = g_k = 0$, if such a value exists, must occur at a point $\mathbf{x}^*$ in C such that

$$\nabla f(\mathbf{x}^*) = \sum_{j=1}^{k} \lambda_j \nabla g_j(\mathbf{x}^*) \tag{A5-4}$$

is satisfied for *some value* of $\lambda_1, \ldots, \lambda_n$.

The λ's are called *Lagrange multipliers.*

C. CAUTIONARY EXAMPLES ON EXISTENCE AND UNIQUENESS OF SOLUTIONS

We have not claimed that a solution of eq. (A5-4) (a) exists, (b) is unique, or (c) is a maximum. The theorem gives a *necessary but not sufficient* condition for a maximum, as the following example will show.

Example 6. Maximize $f(x, y) = y$ subject to the constraint $g(x, y) = y - x = 0$. The problem is just to find the largest value of y on the line $y = x$, and it clearly has no solution. Furthermore, $\nabla f = (0, 1)$, $\nabla g = (-1, 1)$, and there is no point where $\nabla f = \lambda \nabla g$.

Example 7. Find the maximum value of $f(x, y) = xy$ subject to the constraint $g(x, y) = x^2 + 2y^2 - 1 = 0$. Using Lagrange multipliers we write the necessary condition

$$\nabla f = \lambda \nabla g \tag{A5-5}$$

or

$$\begin{pmatrix} y \\ x \end{pmatrix} = \lambda \begin{pmatrix} 2x \\ 4y \end{pmatrix} \tag{A5-6}$$

Equation (A5-6) can be satisfied by some λ if and only if

$$\frac{y}{2x} = \frac{x}{4y} \quad (= \lambda) \tag{A5-7}$$

that is,

$$4y^2 = 2x^2 \tag{A5-8}$$

or

$$x = \pm\sqrt{2}y \tag{A5-9}$$

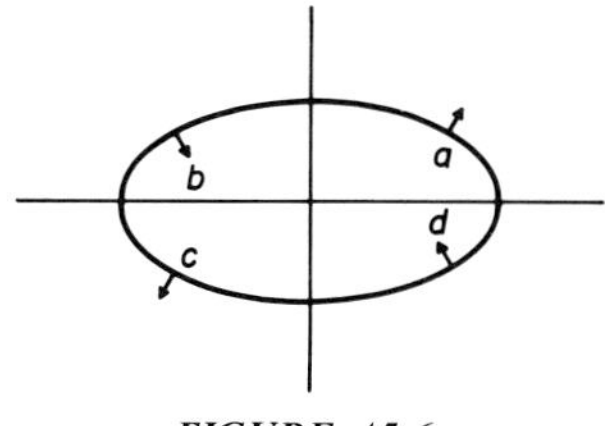

FIGURE A5-6

Thus solutions must satisfy eq. (A5-9) as well as the constraint $x^2 + 2y^2 - 1 = 0$. It is not difficult to verify that there are exactly four points which simultaneously satisfy both equations, and that they are given by $x = \pm\sqrt{2}/2$, $y = \pm\frac{1}{2}$. Figure A5-6 shows the constraint set C (i.e., the ellipse which is the locus of solutions to $x^2 + 2y^2 - 1 = 0$), the four points of C at which eq. (A5-6) is satisfied, and the direction of ∇f at those four points. Which of these points are maxima? The method of Lagrange multipliers per se offers no further help at this point, and we must proceed with a more detailed analysis. One way to find out is to plot f on the ellipse as a function of the angle θ. The calculation goes as follows:

$$\begin{aligned}
&x^2 + 2y^2 = 1\\
&(x^2 + y^2) + y^2 = 1\\
&r^2 + r^2 \sin^2 \theta = 1\\
&r^2 = (1 + \sin^2 \theta)^{-1} \qquad\qquad \text{(A5-10)}\\
&r(\theta) = (1 + \sin^2 \theta)^{-1/2}\\
&f = xy = r\cos\theta\, r\sin\theta = r^2 \sin\theta\cos\theta\\
&f(\theta) = \sin\theta\cos\theta/(1 + \sin^2\theta)
\end{aligned}$$

The result is plotted in fig. A5-7.

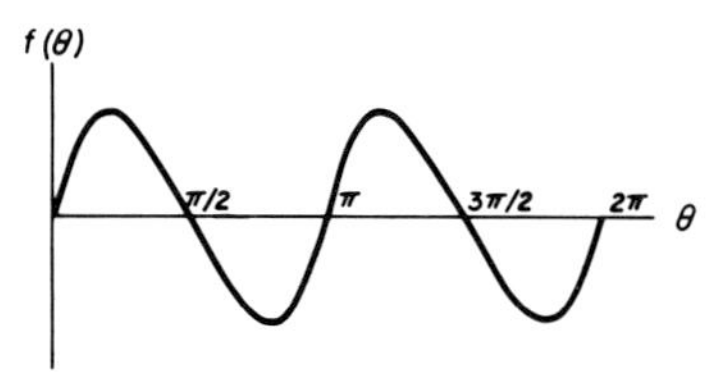

FIGURE A5-7

This outcome should make it clear that the method must be used with caution if one is seeking only maxima.

D. LAGRANGE MULTIPLIERS ARE AN EXTENSION OF THE CONDITION $f'(x) = 0$ FROM CALCULUS

The failure of the method to guarantee existence, uniqueness, or maximality can probably best be understood by comparison with the methods of elementary calculus. To find the maximum of $f(x)$, we look for points where $f'(x) = 0$. In this case it comes as no surprise that there may be no solutions (think of $f(x) = x^3 + x$), many solutions ($f(x) = \sin x$), or misleading solutions which are minima ($f(x) = x^2$) or nonextremal stationary points ($f(x) = x^3$).

But the method of elementary calculus is just a special case of the method of Lagrange multipliers. Consider the problem of maximizing $f(x, y)$ subject to the constraint $g(y) = y = 0$. If we define $h(x) = f(x, 0)$, then this is the same as trying to maximize $h(x)$. The method of Lagrange multipliers tells us to look for points where

$$\nabla f = \lambda \nabla g \tag{A5-11}$$

or

$$\begin{pmatrix} \dfrac{\partial f}{\partial x} \\ \dfrac{\partial f}{\partial y} \end{pmatrix} = \lambda \begin{pmatrix} 0 \\ 1 \end{pmatrix} \tag{A5-12}$$

or

$$\frac{\partial f}{\partial x} = 0 \tag{A5-13}$$

that is, points where $h'(x) = 0$.

The elementary calculus problem of trying to find the maximum value of $f(x)$ is made much easier if $f''(x) < 0$ everywhere. In that case there are only two possibilities:

(i) there is no solution (e.g., $f(x) = -e^{-x}$), or
(ii) there is exactly one solution, and it is a maximum.

There is no possibility of multiple solutions, or of solutions which are not maxima. If $f''(x) < 0$ everywhere, then a straight line drawn between any two points on the graph of f must lie below the graph at every point in between. Consider $f(x) = -x^2$ as an example. A function with this property is called *strictly concave*. In the next section we discuss a generalization that greatly simplifies the use of Lagrange multipliers.

E. A SPECIAL CASE IN WHICH SOLUTIONS ARE UNIQUE AND MAXIMAL

Definition. An $n \times n$ real matrix $\mathbf{A}$ is said to be *negative definite* if, for every nonzero n-dimensional column vector $\mathbf{y}$,

$$\mathbf{y}^{\mathrm{T}}\mathbf{A}\mathbf{y} < 0 \tag{A5-14}$$

where the superscript denotes the transpose operation.

For example, if $\mathbf{A}$ is diagonal and all its diagonal elements are negative, then $\mathbf{A}$ is negative definite.

In one important special case there is at most one point at which eq. (A5-4) is satisfied, and if there is such a point, it is the maximum. That is when the constraint equations are linear (or, more exactly, affine), that is,

$$g_i(x_1, \ldots, x_n) = \sum_{j=1}^{n} g_{ij}x_j + b_i, \qquad i = 1, \ldots, k \tag{A5-15}$$

and in addition the Hessian of f (the matrix of its second partial derivatives) given by

$$\{\mathbf{H}f\}(\mathbf{x}) = \left\{\frac{\partial^2 f}{\partial x_i \, \partial x_j}\right\}(\mathbf{x}) \tag{A5-16}$$

is negative definite at each point $\mathbf{x}$ in C.

In this case f is strictly concave on C. As an example we look at the function $f(x, y) = -x^2 - y^2$ defined on the plane. The Hessian is negative definite at every point and f is strictly concave. Notice that in fact it has a unique maximum. Think about the case of $f(x, y) = -e^{-x} - e^{-y}$.

Example 8. Given an alphabet of n symbols, what choice of probabilities $p_1, \ldots, p_n$ maximizes the average information per symbol? We want to maximize

$$f(p_1, \ldots, p_n) = -\sum_{i=1}^{n} p_i \log_2 p_i = -(0.69) \sum_{i=1}^{n} p_i \ln p_i \tag{A5-17}$$

subject to the constraint

$$g(p_1, \ldots, p_n) = \sum_{i=1}^{n} p_i - 1 = 0 \tag{A5-18}$$

Since

$$\frac{\partial^2 f}{\partial p_i \, \partial p_j} = \frac{-(0.69)}{p_i} \delta_{ij} \tag{A5-19}$$

the Hessian of f is negative definite at every point of interest, and, since g is of the form of eq. (A5-15), we know that a solution of

$$\nabla f = \lambda \nabla g \tag{A5-20}$$

will be the unique maximum. Substituting f and g into eq. (A5-20) yields

$$-(0.69)\begin{pmatrix} 1 + \ln p_1 \\ \vdots \\ 1 + \ln p_n \end{pmatrix} = \lambda \begin{pmatrix} 1 \\ 1 \\ \vdots \\ 1 \end{pmatrix} \tag{A5-21}$$

which is possible only if $p_1 = \cdots = p_n$. Comparison with the constraint equation yields the solution

$$p_i = 1/n, \qquad i = 1, 2, \ldots, n \tag{A5-22}$$

that is, all symbols should be equally probable.

Example 9. The Maxwell–Boltzmann distribution. Let $\mathbf{N}$ represent the vector $(N_1, \ldots, N_n)$, where N_j is the number of systems in state j, characterized by energy E_j. The total number of systems is N_T and the total energy is E_T. The problem is to maximize

$$N_T(\ln N_T) - \sum_{j=1}^{n} N_j(\ln N_j) \tag{A5-23}$$

subject to the constraints

$$\begin{aligned} g_1(\mathbf{N}) &= \sum_{j=1}^{n} N_j - N_T = 0 \\ g_2(\mathbf{N}) &= \sum_{j=1}^{n} E_j N_j - E_T = 0 \end{aligned} \tag{A5-24}$$

Since N_T is not a variable in the optimization problem, we can simply maximize

$$f(\mathbf{N}) = -\sum_{j=1}^{n} N_j(\ln N_j) \tag{A5-25}$$

As in Example 8, the Hessian of f is negative definite and the constraint equations are linear, so we know that a solution of eq. (A5-4) will be the unique maximum. Substituting into eq. (A5-4) yields

$$\nabla f(\mathbf{N}^*) = \lambda_1 \nabla g_1(\mathbf{N}^*) + \lambda_2 \nabla g_2(\mathbf{N}^*) \tag{A5-26}$$

or

$$\begin{pmatrix} -\ln N_1 - 1 \\ -\ln N_2 - 1 \\ \vdots \\ -\ln N_n - 1 \end{pmatrix} = \lambda_1 \begin{pmatrix} 1 \\ 1 \\ \vdots \\ 1 \end{pmatrix} + \lambda_2 \begin{pmatrix} E_1 \\ E_2 \\ \vdots \\ E_n \end{pmatrix} \tag{A5-27}$$

that is,

$$\ln N_j = (-\lambda_1 - \lambda_2 E_j - 1), \qquad j = 1, \ldots, n \tag{A5-28}$$

so

$$N_j = \exp(-\lambda_1 - \lambda_2 E_j - 1) \tag{A5-29}$$

Since

$$\sum_{j=1}^{n} N_j = N_T \tag{A5-30}$$

then

$$\sum_{j=1}^{n} \exp(-\lambda_1 - \lambda_2 E_j - 1) = \exp(-\lambda_1 - 1) \sum_{j=1}^{n} \exp(-\lambda_2 E_j) = N_T \tag{A5-31}$$

Therefore

$$\exp(-\lambda_1 - 1) = \frac{N_T}{\sum_{j=1}^{n} \exp(-\lambda_2 E_j)} \tag{A5-32}$$

and finally

$$N_j = \frac{N_T \exp(-\lambda_2 E_j)}{\sum_{j=1}^{n} \exp(-\lambda_2 E_j)} \tag{A5-33}$$

F. INTERPRETATION OF LAGRANGE MULTIPLIERS AS SENSITIVITY TO CHANGES IN THE CONSTRAINTS

So far in our discussion the actual values of the Lagrange multipliers have been irrelevant; we have only been concerned that there should exist values that will satisfy eq. (A5-4). We will now develop an interpretation of the role of the λ's that assigns a meaning to their numerical values.

Suppose one of our constraints (the lth constraint, say) is altered slightly so that the new constraints are

$$\begin{aligned} g_1(\mathbf{x}) &= 0 \\ &\vdots \\ g_l(\mathbf{x}) &= \varepsilon \\ &\vdots \\ g_k(\mathbf{x}) &= 0 \end{aligned} \tag{A5-34}$$

where ε is a small positive or negative number. We can imagine solving the problem for each value of ε in some small interval around zero. The maximum point $\mathbf{x}^*$ will in general depend on ε, so we write it as $\mathbf{x}^*(\varepsilon)$. If the problem is well posed, then $\mathbf{x}^*(\varepsilon)$ will vary smoothly with ε and the derivative vector $\mathbf{x}^{*\prime}(\varepsilon)$ will be well defined.

How do these small changes in the constraints affect the maximum value f can achieve? In other words, how does $f(\mathbf{x}^*(\varepsilon))$ vary with ε? Using the chain rule we can calculate that

$$\left.\frac{df(\mathbf{x}^*(\varepsilon))}{d\varepsilon}\right|_{\varepsilon=0} = (\nabla f)_{(\mathbf{x}^*(0))} \cdot \mathbf{x}^{*\prime}(0) \tag{A5-35}$$

In order to cast eq. (A5-35) into a more revealing form, we must utilize the following four observations:

(1) Since $\mathbf{x}^*(0)$ is a maximum point when $\varepsilon = 0$,

$$\nabla f_{(\mathbf{x}^*(0))} = \sum_{j=1}^{k} \lambda_j \nabla g_{j(\mathbf{x}^*(0))} \tag{A5-36}$$

(2) By the chain rule,

$$\left.\frac{dg_j(\mathbf{x}^*(\varepsilon))}{d\varepsilon}\right|_{\varepsilon=0} = \nabla g_{j(\mathbf{x}^*(0))} \cdot \mathbf{x}^{*\prime}(0), \qquad j = 1, \ldots, k \tag{A5-37}$$

(3) Since for each value of ε, $\mathbf{x}^*(\varepsilon)$ satisfies the constraints

$$g_j(\mathbf{x}^*(\varepsilon)) = 0, \qquad j \neq l \tag{A5-38}$$

we have

$$\left.\frac{dg_j(\mathbf{x}^*(\varepsilon))}{d\varepsilon}\right|_{\varepsilon=0} = 0, \qquad j \neq l \tag{A5-39}$$

(4) And, since $g_l(\mathbf{x}^*(\varepsilon)) = \varepsilon$, we have

$$\left.\frac{dg_l(\mathbf{x}^*(\varepsilon))}{d\varepsilon}\right|_{\varepsilon=0} = 1, \qquad \text{so} \qquad \left.\frac{dg_j(\mathbf{x}^*(\varepsilon))}{d\varepsilon}\right|_{\varepsilon=0} = \delta_{jl} \tag{A5-40}$$

We now use these four facts to transform eq. (A5-35):

$$\left.\frac{df(\mathbf{x}^*(\varepsilon))}{d\varepsilon}\right|_{\varepsilon=0} = (\nabla f)_{(\mathbf{x}^*(0))} \cdot \mathbf{x}^{*\prime}(0)$$

$$= \sum_{j=1}^{k} \lambda_j \nabla g_{j(\mathbf{x}^*(0))} \cdot \mathbf{x}^{*\prime}(0)$$

$$= \sum_{j=1}^{k} \lambda_j \left.\frac{dg_j(\mathbf{x}^*(\varepsilon))}{d\varepsilon}\right|_{\varepsilon=0}$$

$$= \sum_{j=1}^{k} \lambda_j \delta_{jl} = \lambda_l \qquad \text{(A5-41)}$$

Thus λ_l is the rate of change of the maximum value attained by f with changes in the lth constraint.

G. THE RELATION BETWEEN THE LAGRANGE MULTIPLIER AND THERMODYNAMIC TEMPERATURE IN THE MAXWELL–BOLTZMANN EQUATION

From eq. (A5-23), define

$$F(\mathbf{N}) = N_T(\ln N_T) - \sum_{j=1}^{n} N_j(\ln N_j) \qquad \text{(A5-42)}$$

Then $F(\mathbf{N})$ differs from $f(\mathbf{N})$ in eq. (A5-25) only by an additive constant, and $\mathbf{N}^*$ is the value of $\mathbf{N}$ which maximizes $F(\mathbf{N})$ subject to the constraints in eq. (A5-24). We know from the previous section that λ_2 in eq. (A5-33) is the rate of change of $F(\mathbf{N}^*)$ with respect to changes in the second constraint equation in (A5-24), i.e., altering the constraint to

$$g_2(\mathbf{N}) = \sum_{j=1}^{n} E_j N_j - E_T = \varepsilon \qquad \text{(A5-43)}$$

increases $F(\mathbf{N}^*)$ by an amount

$$\Delta F \approx \left.\frac{\partial F(\mathbf{N}^*(\varepsilon))}{\partial \varepsilon}\right|_{\varepsilon=0} \cdot \varepsilon = \lambda_2 \varepsilon \qquad \text{(A5-44)}$$

to first order in ε. But since eq. (A5-43) can equally well be written

$$g_2(\mathbf{N}) = \sum_{j=1}^{n} E_j N_j - (E_T + \varepsilon) = 0 \qquad \text{(A5-45)}$$

we can regard λ_2 as the rate of change of $F(\mathbf{N}^*)$ with changes in E_T; i.e.,

$$\lambda_2 = \frac{\partial F(\mathbf{N}^*)}{\partial E_T} \qquad \text{(A5-46)}$$

In order to interpret eq. (A5-46), let us write $F(\mathbf{N}^*)$ in terms of the quantities

$$f_j = N_j^*/N_\mathrm{T} \tag{A5-47}$$

the probability that a microsystem chosen at random will be in state j. With this substitution, eq. (A5-42) becomes

$$\begin{aligned} F(\mathbf{N}^*) &= N_\mathrm{T} \ln N_\mathrm{T} - N_\mathrm{T} \sum_{j=1}^{n} (N_j^*/N_\mathrm{T}) \ln[(N_j^*/N_\mathrm{T}) \cdot N_\mathrm{T}] \\ &= N_\mathrm{T} \ln N_\mathrm{T} - N_\mathrm{T} \sum_{j=1}^{n} (N_j^*/N_\mathrm{T})[\ln(N_j^*/N_\mathrm{T}) + \ln(N_\mathrm{T})] \\ &= N_\mathrm{T} \ln N_\mathrm{T} - N_\mathrm{T} \sum_{j=1}^{n} f_j \ln f_j - (\ln N_\mathrm{T}) \sum_{j=1}^{n} N_j^* \\ &= -N_\mathrm{T} \sum_{j=1}^{n} f_j \ln f_j \end{aligned} \tag{A5-48}$$

If we define the average energy of a microsystem by

$$\bar{E} = E_\mathrm{T}/N_\mathrm{T} \tag{A5-49}$$

and substitute eqs. (A5-48) and (A5-49) into eq. (A5-46), we have

$$\lambda_2 = \frac{\partial(-N_\mathrm{T} \sum_{j=1}^{n} f_j \ln f_j)}{\partial E_\mathrm{T}} = \frac{\partial(-\sum_{j=1}^{n} f_j \ln f_j)}{\partial \bar{E}} \tag{A5-50}$$

If we recognize from eq. (8-16) of the text that

$$-\sum_{j=1}^{n} f_j \ln f_j = S/k \tag{A5-51}$$

where S is the entropy of a microsystem and k is Boltzmann's constant, and from classical thermodynamics that

$$T = \frac{\partial \bar{E}}{\partial S} \tag{A5-52}$$

eq. (A5-50) becomes

$$\lambda_2 = \frac{1}{k} \frac{\partial S}{\partial \bar{E}} = \frac{1}{kT} \tag{A5-53}$$

and eq. (A5-33) becomes

$$N_j = \frac{N_\mathrm{T} \exp(-E_j/kT)}{\sum_{j=1}^{n} \exp(-E_j/kT)} \tag{A5-54}$$

APPENDIX **VI**

Evaluation of the Partition Function of an Ideal Gas

We start out with eq. (8-30):

$$Z = \frac{1}{N!h^{3N}} \left[\int \exp\left(-\frac{p_x^2 + p_y^2 + p_z^2}{2mkT} \right) dp_x\, dp_y\, dp_z\, dx\, dy\, dz \right]^N \qquad \text{(A6-1)}$$

The integral over $dx\,dy\,dz$ is simply the volume of the system V. Equation (A6-1) may be simplified to give

$$Z = \frac{V^N}{N!h^{3N}} \left[\int_{-\infty}^{\infty} \exp - \frac{p_x^2}{2mkT}\, dp_x \int_{-\infty}^{\infty} \exp - \frac{p_y^2}{2mkT}\, dp_y \times \int_{-\infty}^{\infty} \exp - \frac{p_z^2}{2mkT}\, dp_z \right]^N \qquad \text{(A6-2)}$$

Since all of the integrals in the bracket are identical, we can further simplify

$$Z = \frac{V^N}{N!h^{3N}} \left[\int_{-\infty}^{\infty} \exp - \frac{p_x^2}{2mkT}\, dp_x \right]^{3N} \qquad \text{(A6-3)}$$

The integral is a standard definite integral and can be shown to have the value

$$\int_{-\infty}^{\infty} \exp - \frac{p_x^2}{2mkT}\, dp_x = \sqrt{2\pi mkT} \qquad \text{(A6-4)}$$

The partition function then becomes

$$Z = \frac{1}{N!} \left[\frac{V}{h^3} (2\pi mkT)^{3/2} \right]^N \qquad \text{(A6-5)}$$

Index

D

E

A 8
B 9
C 0
D 1
E 2
F 3
G 4
H 5
I 6
J 7